Lecture Notes in Mathematics

1690

Editors:
A. Dold, Heidelberg
F. Takens, Groningen
B. Teissier, Paris

T0223241

Springer
Berlin
Heidelberg
New York
Barcelona
Budapest
Hong Kong
London
Milan
Paris
Singapore
Tokyo

M. T. Barlow   D. Nualart

Lectures on Probability Theory and Statistics

Ecole d'Eté de Probabilités
de Saint-Flour XXV – 1995

Editor: P. Bernard

 Springer

Authors

Martin T. Barlow
Department of Mathematics
University of British Columbia
121-1984 Mathematics Road
Vancouver, B.C. Canada V6T 1Z2

David Nualart
Department d'Estadistica
Universitat de Barcelona
Facultat de Matemàtiques
Gran Via de les Corts Catalanes, 585
E-08007 Barcelona, Spain

Editor

Pierre Bernard
Laboratoire de Mathématiques Appliquées
UMR CNRS 6620
Université Blaise Pascal
Clermont-Ferrand
F-63177 Aubière Cedex, France

Cataloging-in-Publication Data applied for

Die Deutsche Bibliothek - CIP-Einheitsaufnahme

Lectures on probability theory and statistics / Ecole d'Eté de
Probabilités de Saint-Flour XXV - 1995. M. T. Barlow ; D. Nualart.
Ed.: P. Bernard. - Berlin ; Heidelberg ; New York ; Barcelona ;
Budapest ; Hong Kong ; London ; Milan ; Paris ; Santa Clara ;
Singapore ; Tokyo : Springer, 1998
    (Lecture notes in mathematics ; Vol. 1690)
    ISBN 3-540-64620-5

Mathematics Subject Classification (1991):
60-01, 60-02, 60-06, 60D05, 60G57, 60H07, 60J15, 60J60, 60J65

ISSN 0075-8434
ISBN 3-540-64620-5 Springer-Verlag Berlin Heidelberg New York

Typesetting: Camera-ready T$_E$X output by the authors
SPIN: 10649898   41/3143-543210 - Printed on acid-free paper

INTRODUCTION

This volume contains lectures given at the Saint-Flour Summer School of Probability Theory during the period 10th - 26th July, 1995.

We thank the authors for all the hard work they accomplished. Their lectures are a work of reference in their domain.

The school brought together 100 participants, 29 of whom gave a lecture concerning their research work.

At the end of this volume you will find the list of participants and their papers.

Finally, to facilitate research concerning previous schools we give here the number of the volume of "Lecture Notes" where they can be found :

Lecture Notes in Mathematics

1971 : n°307 -   1973 : n°390 -   1974 : n°480 -   1975 : n°539 -   1976 : n°598 -
1977 : n°678 -   1978 : n°774 -   1979 : n°876 -   1980 : n°929 -   1981 : n°976 -
1982 : n°1097 -   1983 : n°1117 -   1984 : n°1180 -   1985 - 1986 et   1987 : n°1362 -
1988 : n°1427 -   1989 : n°1464 -   1990 : n°1527 -   1991 : n°1541 -   1992 : n°1581
1993 : n°1608 :   1994 : n°1648 -   1996 : n°1665

Lecture Notes in Statistics

1986 : n°50

TABLE OF CONTENTS

2

1. Introduction.

The notes are based on lectures given in St. Flour in 1995, and cover, in greater detail, most of the course given there.

The word "fractal" was coined by Mandelbrot [Man] in the 1970s, but of course sets of this type have been familiar for a long time – their early history being as a collection of pathological examples in analysis. There is no generally agreed exact definition of the word "fractal", and attempts so far to give a precise definition have been unsatisfactory, leading to classes of sets which are either too large, or too small, or both. This ambiguity is not a problem for this course: a more precise title would be "Diffusions on some classes of regular self-similar sets".

Initial interest in the properties of processes on fractals came from mathematical physicists working in the theory of disordered media. Certain media can be modelled by percolation clusters at criticality, which are expected to exhibit fractal-like properties. Following the initial papers [AO], [RT], [GAM1-GAM3] a very substantial physics literature has developed – see [HBA] for a survey and bibliography.

Let G be an infinite subgraph of $\mathbb{Z}^d$. A simple random walk (SRW) $(X_n, n \geq 0)$ on G is just the Markov chain which moves from $x \in G$ with equal probability to each of the neighbours of x. Write $p_n(x,y) = \mathbb{P}^x(X_n = y)$ for the n-step transition probabilities. If G is the whole of $\mathbb{Z}^d$ then $\mathbb{E}(X_n)^2 = n$ with many familiar consequences – the process moves roughly a distance of order $\sqrt{n}$ in time n, and the probability law $p_n(x, \cdot)$ puts most of its mass on a ball of radius $c_d n$.

If G is not the whole of $\mathbb{Z}^d$ then the movement of the process is on the average restricted by the removal of parts of the space. Probabilistically this is not obvious – but see [DS] for an elegant argument, using electrical resistance, that the removal of part of the state space can only make the process X 'more recurrent'. So it is not unreasonable to expect that for certain graphs G one may find that the process X is sufficiently restricted that for some $\beta > 2$

$$(1.1) \qquad \mathbb{E}^x(X_n - x)^2 \asymp n^{2/\beta}.$$

(Here and elsewhere I use $\asymp$ to mean 'bounded above and below by positive constants', so that (1.1) means that there exist constants c_1, c_2 such that $c_1 n^{2/\beta} \leq \mathbb{E}^x(X_n - x)^2 \leq c_2 n^{2/\beta}$). In [AO] and [RT] it was shown that if G is the Sierpinski gasket (or more precisely an infinite graph based on the Sierpinski gasket – see Fig. 1.1) then (1.1) holds with $\beta = \log 5/ \log 2$.

Figure 1.1: The graphical Sierpinski gasket.

Physicists call behaviour of this kind by a random walk (or a diffusion – they are not very interested in the distinction) *subdiffusive* – the process moves on average slower than a standard random walk on $\mathbb{Z}^d$. Kesten [Ke] proved that the SRW on the 'incipient infinite cluster' C (a percolation cluster at $p = p_c$ but conditioned to be infinite) is subdiffusive. The large scale structure of C is given by taking one infinite path (the 'backbone') together with a collection of 'dangling ends', some of which are very large. Kesten attributes the subdiffusive behaviour of SRW on C to the fact that the process X spends a substantial amount of time in the dangling ends.

However a graph such as the Sierpinski gasket (SG) has no dangling ends, and one is forced to search for a different explanation for the subdiffusivity. This can be found in terms of the existence of 'obstacles at all length scales'. Whilst this holds for the graphical Sierpinski gasket, the notation will be slightly simpler if we consider another example, the graphical Sierpinski carpet (GSC). (Figure 1.2).

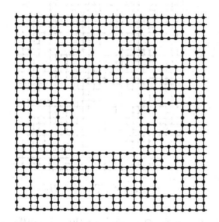

Figure 1.2: The graphical Sierpinski carpet.

This set can be defined precisely in the following fashion. Let $H_0 = \mathbb{Z}^2$. For $x = (n, m) \in H_0$ write n,m in ternary – so $n = \sum_{i=0}^{\infty} n_i 3^i$, where $n_i \in \{0, 1, 2\}$, and $n_i = 0$ for all but finitely many i. Set

$$J_k = \{(m, n) : n_k = 1 \text{ and } m_k = 1\},$$

so that J_k consists of a union of disjoint squares of side 3^k: the square in J_k closest to the origin is $\{3^k, \ldots, 2.3^k - 1\} \times \{3^k, \ldots, 2.3^k - 1\}$. Now set

(1.2) $$H_n = H_0 - \bigcup_{k=1}^{n} J_k, \quad H = \bigcap_{n=0}^{\infty} H_n.$$

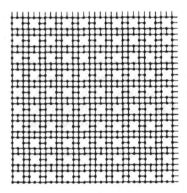

Figure 1.3: The set H_1 .

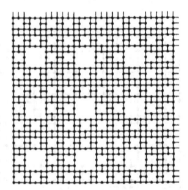

Figure 1.4: The set H_2 .

Note that $H \cap [0, 3^n]^2 = H_n \cap [0, 3^n]^2$, so that the difference between H and H_n will only be detected by a SRW after it has moved a distance of 3^n from the origin. Now let $X^{(n)}$ be a SRW on H_n, started at the origin, and let X be a SRW on H. The process $X^{(0)}$ is just SRW on $\mathbb{Z}_+^2$ and so we have

(1.3)
$$\mathbb{E}(X_n^{(0)})^2 \simeq n.$$

The process $X^{(1)}$ is a random walk on a the intersection of a translation invariant subset of $\mathbb{Z}^2$ with $\mathbb{Z}_+^2$. So we expect 'homogenization': the processes $n^{-1/2} X_{[nt]}^{(1)}$, $t \geq 0$ should converge weakly to a constant multiple of Brownian motion in $\mathbb{R}_+^2$. So, for large n we should have $\mathbb{E}(X_n^{(1)})^2 \sim a_1 n$, and we would expect that $a_1 < 1$, since the obstacles will on average tend to impede the motion of the process.

Similar considerations suggest that, writing $\varphi_n(t) = \mathbb{E}^0 (X_t^{(n)})^2$, we should have

$$\varphi_n(t) \sim a_n t \quad \text{as } t \to \infty.$$

However, for small t we would expect that φ_n and φ_{n+1} should be approximately equal, since the process will not have moved far enough to detect the difference between H_n and H_{n+1}. More precisely, if t_n is such that $\varphi_n(t_n) = (3^n)^2$ then φ_n

and φ_{n+1} should be approximately equal on $[0, t_{n+1}]$. So we may guess that the behaviour of the family of functions $\varphi_n(t)$ should be roughly as follows:

(1.4)
$$\varphi_n(t) = b_n + a_n(t - t_n), \quad t \geq t_n,$$
$$\varphi_{n+1}(s) = \varphi_n(s), \quad 0 \leq s \leq t_{n+1}.$$

If we add the guess that $a_n = 3^{-\alpha}$ for some $\alpha > 0$ then solving the equations above we deduce that
$$t_n \asymp 3^{(2+\alpha)n}, \quad b_n = 3^{2n}.$$

So if $\varphi(t) = \mathbb{E}^0(X_t)^2$ then as $\varphi(t) \simeq \lim_n \varphi_n(t)$ we deduce that φ is close to a piecewise linear function, and that

$$\varphi(t) \asymp t^{2/\beta}$$

where $\beta = 2 + \alpha$. Thus the random walk X on the graph H should satisfy (1.1) for some $\beta > 2$.

The argument given here is not of course rigorous, but (1.1) does actually hold for the set H – see [BB6, BB7]. (See also [Jo] for the case of the graphical Sierpinski gasket. The proofs however run along rather different lines than the heuristic argument sketched above).

Given behaviour of this type it is natural to ask if the random walk X on H has a scaling limit. More precisely, does there exist a sequence of constants τ_n such that the processes

(1.5)
$$(3^{-n} X_{[t/\tau_n]}, t \geq 0)$$

converge weakly to a non-degenerate limit as $n \to \infty$? For the graphical Sierpinski carpet the convergence is not known, though there exist τ_n such that the family (1.5) is tight. However, for the graphical Sierpinski gasket the answer is 'yes'.

Thus, for certain very regular fractal sets $F \subset \mathbb{R}^d$ we are able to define a limiting diffusion process $X = (X_t, t \geq 0, \mathbb{P}^x, x \in F)$ where $\mathbb{P}^x$ is for each $x \in F$ a probability measure on $\Omega = \{\omega \in C([0, \infty), F) : \omega(0) = x\}$. Writing $T_t f(x) = \mathbb{E}^x f(X_t)$ for the semigroup of X we can define a 'differential' operator $\mathcal{L}_F$, defined on a class of functions $\mathcal{D}(\mathcal{L}_F) \subset C(F)$. In many cases it is reasonable to call $\mathcal{L}_F$ the Laplacian on F.

From the process X one is able to obtain information about the solutions to the Laplace and heat equations associated with $\mathcal{L}_F$, the heat equation for example taking the form

(1.6)
$$\frac{\partial u}{\partial t} = \mathcal{L}_F u,$$
$$u(0, x) = u_0(x),$$

where $u = u(t, x)$, $x \in F$, $t \geq 0$. The wave equation is rather harder, since it is not very susceptible to probabilistic analysis. See, however [KZ2] for work on the wave equation on a some manifolds with a 'large scale fractal structure'.

The mathematical literature on diffusions on fractals and their associated infinitesimal generators can be divided into broadly three parts:

1. Diffusions on finitely ramified fractals.
2. Diffusions on generalized Sierpinski carpets, a family of infinitely ramified fractals.
3. Spectral properties of the 'Laplacian' $\mathcal{L}_F$.

These notes only deal with the first of these topics. On the whole, infinitely ramified fractals are significantly harder than finitely ramified ones, and sometimes require a very different approach. See [Bas] for a recent survey.

These notes also contain very little on spectral questions. For finitely ramified fractals a direct approach (see for example [FS1, Sh1-Sh4, KL]), is simpler, and gives more precise information than the heat kernel method based on estimating

$$\int_F p(t, x, x)dx = \sum_i e^{-\lambda_i t}.$$

In this course Section 2 introduces the simplest case, the Sierpinski gasket. In Section 3 I define a class of well-behaved diffusions on metric spaces, "Fractional Diffusions", which is wide enough to include many of the processes discussed in this course. It is possible to develop their properties in a fairly general fashion, without using much of the special structure of the state space. Section 4 contains a brief introduction to the theory of Dirichlet forms, and also its connection with electrical resistances. The remaining chapters, 5 to 8, give the construction and some properties of diffusions on a class of finitely ramified regular fractals. In this I have largely followed the analytic 'Japanese' approach, developed by Kusuoka, Kigami, Fukushima and others. Many things can now be done more simply than in the early probabilistic work – but there is loss as well as gain in added generality, and it is worth pointing out that the early papers on the Sierpinski gasket ([Kus1, Go, BP]) contain a wealth of interesting direct calculations, which are not reproduced in these notes. Any reader who is surprised by the abrupt end of these notes in Section 8 should recall that some, at least, of the properties of these processes have already been obtained in Section 3.

c_i denotes a positive real constant whose value is fixed within each Lemma, Theorem etc. Occasionally it will be necessary to use notation such as $c_{3.5.4}$ – this is simply the constant c_4 in Definition 3.5. c, c', c'' denote positive real constants whose values may change on each appearance. $B(x, r)$ denotes the open ball with centre x and radius r, and if X is a process on a metric space F then

$$T_A = \inf\{t > 0 : X_t \in A\},$$
$$T_y = \inf\{t > 0 : X_t = y\},$$
$$\tau(x, r) = \inf\{t \geq 0 : X_t \notin B(x, r)\}.$$

I have included in the references most of the mathematical papers in this area known to me, and so they contain many papers not mentioned in the text. I am grateful to Gerard Ben Arous for a number of interesting conversations on the physical conditions under which subdiffusive behaviour might arise, to Ben Hambly for checking the final manuscript, and to Ann Artuso and Liz Rowley for their typing.

Acknowledgements. This research is supported by a NSERC (Canada) research grant, by a grant from the Killam Foundation, and by a EPSRC (UK) Visiting Fellowship.

2. The Sierpinski Gasket

This is the simplest non-trivial connected symmetric fractal. The set was first defined by Sierpinski [Sie1], as an example of a pathological curve; the name "Sierpinski gasket" is due to Mandelbrot [Man, p.142].

Let $G_0 = \{(0,0), (1,0), (1/2, \sqrt{3}/2)\} = \{a_0, a_1, a_2\}$ be the vertices of the unit triangle in $\mathbb{R}^2$, and let $\mathcal{H}u(G_0) = H_0$ be the closed convex hull of G_0. The construction of the Sierpinski gasket (SG for short) G is by the following Cantor-type subtraction procedure. Let b_0, b_1, b_2 be the midpoints of the 3 sides of G_0, and let A be the interior of the triangle with vertices $\{b_0, b_1, b_2\}$. Let $H_1 = H_0 - A$, so that H_1 consists of 3 closed upward facing triangles, each of side 2^{-1}. Now repeat the operation on each of these triangles to obtain a set H_2, consisting of 9 upward facing triangles, each of side 2^{-2}.

Figure 2.1: The sets H_1 and H_2.

Continuing in this fashion, we obtain a decreasing sequence of closed non-empty sets $(H_n)_{n=0}^{\infty}$, and set

$$(2.1) \qquad G = \bigcap_{n=0}^{\infty} H_n.$$

Figure 2.2: The set H_4.

It is easy to see that G is connected: just note that $\partial H_n \subset H_m$ for all $m \geq n$, so that no point on the edge of a triangle is ever removed. Since $|H_n| = (3/4)^n |H_0|$, we clearly have that $|G| = 0$.

We begin by exploring some geometrical properties of G. Call an n-*triangle* a set of the form $G \cap B$, where B is one of the 3^n triangles of side 2^{-n} which make up H_n. Let μ_n be Lebesgue measure restricted to H_n, and normalized so that $\mu_n(H_n) = 1$; that is

$$\mu_n(dx) = 2 \cdot (4/3)^n 1_{H_n}(x)\, dx.$$

Let $\mu_G = \text{wlim}\mu_n$; this is the natural "flat" measure on G. Note that μ_G is the unique measure on G which assigns mass 3^{-n} to each n-triangle. Set $d_f = \log 3 / \log 2 \simeq 1.58\ldots$

Lemma 2.1. For $x \in G$, $0 \leq r < 1$

$$(2.2) \qquad\qquad 3^{-1} r^{d_f} \leq \mu_G\big(B(x,r)\big) \leq 18 r^{d_f}.$$

Proof. The result is clear if $r = 0$. If $r > 0$, choose n so that $2^{-(n+1)} < r \leq 2^{-n}$ – we have $n \geq 0$. Since $B(x,r)$ can intersect at most 6 n-triangles, it follows that

$$\mu_G\big(B(x,r)\big) \leq 6.3^{-n} = 18.3^{-(n+1)}$$
$$= 18(2^{-(n+1)})^{d_f} < 18 r^{d_f}.$$

As each $(n+1)$-triangle has diameter $2^{-(n+1)}$, $B(x,r)$ must contain at least one $(n+1)$-triangle and therefore

$$\mu_G\big(B(x,r)\big) \geq 3^{-(n+1)} = 3^{-1}(2^{-n})^{d_f} \geq 3^{-1} r^{d_f}. \qquad\qquad \square$$

Of course the constants 3^{-1}, 18 in (2.2) are not important; what is significant is that the μ_G-mass of balls in G grow as r^{d_f}. Using terminology from the geometry of manifolds, we can say that G has *volume growth* given by r^{d_f}.

Detour on Dimension.

Let (F, ρ) be a metric space. There are a number of different definitions of dimension for F and subsets of F: here I just mention a few. The simplest of these is *box-counting dimension*. For $\varepsilon > 0$, $A \subset F$, let $N(A, \varepsilon)$ be the smallest number of balls $B(x, \varepsilon)$ required to cover A. Then

$$(2.3) \qquad \dim_{BC}(A) = \limsup_{\varepsilon \downarrow 0} \frac{\log N(A, \varepsilon)}{\log \varepsilon^{-1}}.$$

To see how this behaves, consider some examples. We take (F, ρ) to be $\mathbb{R}^d$ with the Euclidean metric.

Examples. 1. Let $A = [0,1]^d \subset \mathbb{R}^d$. Then $N(A, \varepsilon) \asymp \varepsilon^{-d}$, and it is easy to verify that

$$\lim_{\varepsilon \downarrow 0} \frac{\log N([0,1]^d, \varepsilon)}{\log \varepsilon^{-1}} = d.$$

2. The Sierpinski gasket G. Since $G \subset H_n$, and H_n is covered by 3^n triangles of side 2^{-n}, we have, after some calculations similar to those in Lemma 2.1, that $N(G, r) \asymp (1/r)^{\log 3 / \log 2}$. So,

$$\dim_{BC}(G) = \frac{\log 3}{\log 2}.$$

3. Let $A = \mathbb{Q} \cap [0,1]$. Then $N(A, \varepsilon) \asymp \varepsilon^{-1}$, so $\dim_{BC}(A) = 1$. On the other hand $\dim_{BC}(\{p\}) = 0$ for any $p \in A$.

We see that box-counting gives reasonable answers in the first two cases, but a less useful number in the third. A more delicate, but more useful, definition is obtained if we allow the sizes of the covering balls to vary. This gives us *Hausdorff dimension*. I will only sketch some properties of this here – for more detail see for example the books by Falconer [Fa1, Fa2].

Let $h : \mathbb{R}_+ \to \mathbb{R}_+$ be continuous, increasing, with $h(0) = 0$. For $U \subset F$ write $\mathrm{diam}(U) = \sup\{\rho(x, y) : x, y \in U\}$ for the diameter of U. For $\delta > 0$ let

$$\mathcal{H}^h_\delta(A) = \inf\left\{ \sum_i h\big(d(U_i)\big) : A \subset \bigcup_i U_i, \quad \mathrm{diam}(U_i) < \delta \right\}.$$

Clearly $\mathcal{H}^h_\delta(A)$ is decreasing in δ. Now let

$$(2.4) \qquad \mathcal{H}^h(A) = \lim_{\delta \downarrow 0} \mathcal{H}^h_\delta(A);$$

we call $\mathcal{H}^h(\cdot)$ *Hausdorff h-measure* . Let $\mathcal{B}(F)$ be the Borel σ-field of F.

Lemma 2.2. $\mathcal{H}^h$ *is a measure on* $(F, \mathcal{B}(F))$.

For a proof see [Fa1, Chapter 1].

We will be concerned only with the case $h(x) = x^\alpha$: we then write $\mathcal{H}^\alpha$ for $\mathcal{H}^h$. Note that $\alpha \to \mathcal{H}^\alpha(A)$ is decreasing; in fact it is not hard to see that $\mathcal{H}^\alpha(A)$ is either $+\infty$ or 0 for all but at most one α.

Definition 2.3. The Hausdorff dimension of A is defined by

$$\dim_H(A) = \inf\{\alpha : \mathcal{H}^\alpha(A) = 0\} = \sup\{\alpha : \mathcal{H}^\alpha(A) = +\infty\}.$$

Lemma 2.4. $\dim_H(A) \le \dim_{BC}(A)$.

Proof. Let $\alpha > \dim_{BC}(A)$. Then as A can be covered by $N(A, \varepsilon)$ sets of diameter 2ε, we have $\mathcal{H}^\alpha_\delta(A) \le N(A, \varepsilon)(2\varepsilon)^\alpha$ whenever $2\varepsilon < \delta$. Choose θ so that $\dim_{BC}(A) < \alpha - \theta < \alpha$; then (2.3) implies that for all sufficiently small ε, $N(A, \varepsilon) \le \varepsilon^{-(\alpha-\theta)}$. So $\mathcal{H}^\alpha_\delta(A) = 0$, and thus $\mathcal{H}^\alpha(A) = 0$, which implies that $\dim_H(A) \le \alpha$. $\qquad\square$

Consider the set $A = \mathbb{Q} \cap [0, 1]$, and let $A = \{p_1, p_2, \ldots\}$ be an enumeration of A. Let $\delta > 0$, and U_i be an open internal of length $2^{-i} \wedge \delta$ containing p_i. Then (U_i) covers A, so that $\mathcal{H}^\alpha_\delta(A) \le \sum_{i=1}^\infty (\delta \wedge 2^{-i})^\alpha$, and thus $\mathcal{H}^\alpha(A) = 0$. So $\dim_H(A) = 0$. We see therefore that $\dim_H$ can be strictly smaller than $\dim_{BC}$, and that (in this case at least) $\dim_H$ gives a more satisfactory measure of the size of A.

For the other two examples considered above Lemma 2.4 gives the upper bounds $\dim_H([0,1]^d) \le d$, $\dim_H(G) \le \log 3/\log 2$. In both cases equality holds, but a direct proof of this (which is possible) encounters the difficulty that to obtain a lower bound on $\mathcal{H}^\alpha_\delta(A)$ we need to consider all possible covers of A by sets of diameter less than δ. It is much easier to use a kind of dual approach using measures.

Theorem 2.5. *Let* μ *be a measure on* A *such that* $\mu(A) > 0$ *and there exist* $c_1 < \infty$, $r_0 > 0$, *such that*

$$(2.5) \qquad \mu(B(x, r)) \le c_1 r^\alpha, \quad x \in A, \quad r \le r_0.$$

Then $\mathcal{H}^\alpha(A) \ge c_1^{-1} \mu(A)$, *and* $\dim_H(A) \ge \alpha$.

Proof. Let U_i be a covering of A by sets of diameter less than δ, where $2\delta < r_0$. If $x_i \in U_i$, then $U_i \subset B(x_i, \operatorname{diam}(U_i))$, so that $\mu(U_i) \le c_1 \operatorname{diam}(U_i)^\alpha$. So

$$\sum_i \operatorname{diam}(U_i)^\alpha \ge c_1^{-1} \sum_i \mu(U_i) \ge c_1^{-1} \mu(A).$$

Therefore $\mathcal{H}^\alpha_\delta(A) \ge c_1^{-1} \mu(A)$, and it follows immediately that $\mathcal{H}^\alpha(A) > 0$, and $\dim_H(A) \ge \alpha$. $\qquad\square$

Corollary 2.6. $\dim_H(G) = \log 3/\log 2$.

Proof. By Lemma 2.1 μ_G satisfies (2.5) with $\alpha = d_f$. So by Theorem 2.5 $\dim_H(G) \ge d_f$; the other bound has already been proved. $\qquad\square$

Very frequently, when we wish to compute the dimension of a set, it is fairly easy to find directly a near-optimal covering, and so obtain an upper bound on $\dim_H$ directly. We can then use Theorem 2.5 to obtain a lower bound. However, we can also use measures to derive upper bounds on $\dim_H$.

Theorem 2.7. *Let μ be a finite measure on A such that $\mu\big(B(x,r)\big) \geq c_2 r^\alpha$ for all $x \in A$, $r \leq r_0$. Then $\mathcal{H}^\alpha(A) < \infty$, and $\dim_H(A) \leq \alpha$.*

Proof. See [Fa2, p.61]. ∎

In particular we may note:

Corollary 2.8. *If μ is a measure on A with $\mu(A) \in (0,\infty)$ and*

$$(2.6) \qquad c_1 r^\alpha \leq \mu\big(B(x,r)\big) \leq c_2 r^\alpha, \qquad x \in A, \quad r \leq r_0$$

then $\mathcal{H}^\alpha(A) \in (0,\infty)$ and $\dim_H(A) = \alpha$.

Remarks. 1. If A is a k-dimensional subspace of $\mathbb{R}^d$ then $\dim_H(A) = \dim_{BC}(A) = k$.

2. Unlike $\dim_{BC}$ $\dim_H$ is stable under countable unions: thus

$$\dim_H\left(\bigcup_{i=1}^\infty A_i\right) = \sup_i \dim_H(A_i).$$

3. In [Tri] Tricot defined "packing dimension" $\dim_P(\cdot)$, which is the largest reasonable definition of "dimension" for a set. One has $\dim_P(A) \geq \dim_H(A)$; strict inequality can hold. The hypotheses of Corollary 2.8 also imply that $\dim_P(A) = \alpha$. See [Fa2, p.48].

4. The sets we consider in these notes will be quite regular, and will very often satisfy (2.6): that is they will be "α-dimensional" in every reasonable sense.

5. Questions concerning Hausdorff measure are frequently much more delicate than those relating just to dimension. However, the fractals considered in this notes will all be sufficiently regular so that there is a direct construction of the Hausdorff measure. For example, the measure μ_G on the Sierpinski gasket is a constant multiple of the Hausdorff x^{d_f}-measure on G.

We note here how $\dim_H$ changes under a change of metric.

Theorem 2.9. *Let ρ_1, ρ_2 be metrics on F, and write $\mathcal{H}^{\alpha,i}$, $\dim_{H,i}$ for the Hausdorff measure and dimension with respect to ρ_i, $i = 1, 2$.*

(a) *If $\rho_1(x,y) \leq \rho_2(x,y)$ for all $x,y \in A$ with $\rho_2(x,y) \leq \delta_0$, then $\dim_{H,1}(A) \geq \dim_{H,2}(A)$.*

(b) *If $1 \wedge \rho_1(x,y) \asymp (1 \wedge \rho_2(x,y))^\theta$ for some $\theta > 0$, then*

$$\dim_{H,2}(A) = \theta \dim_{H,1}(A).$$

Proof. Write $d_j(U)$ for the ρ_j-diameter of U. If (U_i) is a cover of A by sets with $\rho_2(U_i) < \delta < \delta_0$, then

$$\sum_i d_1(U_i)^\alpha \leq \sum_i d_2(U_i)^\alpha$$

so that $\mathcal{H}_\delta^{\alpha,1}(A) \leq \mathcal{H}_\delta^{\alpha,2}(A)$. Then $\mathcal{H}^{\alpha,1}(A) \leq \mathcal{H}^{\alpha,2}(A)$ and $\dim_{H,1}(A) \geq \dim_{H,2}(A)$, proving (a).

(b) If U_i is any cover of A by sets of small diameter, we have

$$\sum_i d_1(U_i)^\alpha \asymp \sum_i d_2(U_i)^{\theta\alpha}.$$

Hence $\mathcal{H}^{\alpha,1}(A) = 0$ if and only if $\mathcal{H}^{\theta\alpha,2}(A) = 0$, and the conclusion follows. □

Metrics on the Sierpinski gasket.

Since we will be studying continuous processes on G, it is natural to consider the metric on G given by the shortest path in G between two points. We begin with a general definition.

Definition 2.10. Let $A \subset \mathbb{R}^d$. For $x, y \in A$ set

$$d_A(x,y) = \inf\{|\gamma| : \gamma \text{ is a path between } x \text{ and } y \text{ and } \gamma \subset A\}.$$

If $d_A(x,y) < \infty$ for all $x, y \in A$ we call d_A the *geodesic metric* on A.

Lemma 2.11. *Suppose A is closed, and that $d_A(x,y) < \infty$ for all $x, y \in A$. Then d_A is a metric on A and (A, d_A) has the geodesic property:*

For each $x, y \in A$ there exists a map $\Phi(t) : [0,1] \to A$ such that
$$d_A(x, \Phi(t)) = t d_A(x,y), \quad d_A(\Phi(t), y) = (1-t) d_A(x,y).$$

Proof. It is clear that d_A is a metric on A. To prove the geodesic property, let $x, y \in A$, and $D = d_A(x,y)$. Then for each $n \geq 1$ there exists a path $\gamma_n(t)$, $0 \leq t \leq 1 + D$ such that $\gamma_n \subset A$, $|d\gamma_n(t)| = dt$, $\gamma_n(0) = x$ and $\gamma_n(t_n) = y$ for some $D \leq t_n \leq D + n^{-1}$. If $p \in [0, D] \cap \mathbb{Q}$ then since $|x - \gamma_n(p)| \leq p$ the sequence $(\gamma_n(p))$ has a convergent subsequence. By a diagonalization argument there exists a subsequence n_k such that $\gamma_{n_k}(p)$ converges for each $p \in [0, D] \cap \mathbb{Q}$; we can take $\Phi = \lim \gamma_{n_k}$. □

Lemma 2.12. *For $x, y \in G$,*

$$|x - y| \leq d_G(x,y) \leq c_1|x - y|.$$

Proof. The left hand inequality is evident.

It is clear from the structure of H_n that if A, B are n-triangles and $A \cap B = \emptyset$, then

$$|a - b| \geq (\sqrt{3}/2)2^{-n} \quad \text{for } a \in A, \, b \in B.$$

Let $x, y \in G$ and choose n so that

$$(\sqrt{3}/2)2^{-(n+1)} \leq |x - y| < (\sqrt{3}/2)2^{-n}.$$

So x, y are either in the same n-triangle, or in adjacent n-triangles. In either case choose $z \in G_n$ so it is in the same n-triangle as both x and y.

Let $z_n = z$, and for $k > n$ choose $z_k \in G_k$ such that x, z_k are in the same k-triangle. Then since z_k and z_{k+1} are in the same k-triangle, and both are contained in H_{k+1}, we have $d_G(z_k, z_{k+1}) = d_{H_{k+1}}(z_k, z_{k+1}) \leq 2^{-k}$. So,

$$d_G(z, x) \leq \sum_{k=n}^{\infty} d_G(z_k, z_{k+1}) \leq 2^{1-n} \leq 4|x - y|.$$

Hence $d_G(x, y) \leq d_G(x, z) + d_G(z, y) \leq 8|x - y|.$ □

Construction of a diffusion on the Sierpinski gasket.

Let G_n be the set of vertices of n-triangles. We can make G_n into a graph in a natural way, by taking $\{x, y\}$ to be an edge in G_n if x, y belong to the same n-triangle. (See Fig. 2.3). Write E_n for the set of edges.

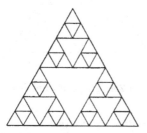

Figure 2.3: The graph G_3.

Let $Y_k^{(n)}$, $k = 0, 1, \ldots$ be a simple random walk on G_n. Thus from $x \in G_n$, the process $Y^{(n)}$ jumps to each of the neighbours of x with equal probability. (Apart from the 3 points in G_0, all the points in G_n have 4 neighbours). The obvious way to construct a diffusion process $(X_t, t \geq 0)$ on G is to use the graphs G_n, which provide a natural approximation to G, and to try to define X as a weak limit of the processes $Y^{(n)}$. More precisely, we wish to find constants $(\alpha_n, n \geq 0)$ such that

(2.7) $$\left(Y_{[\alpha_n t]}^{(n)}, t \geq 0\right) \Rightarrow (X_t, t \geq 0).$$

We have two problems:
(1) How do we find the right (α_n)?
(2) How do we prove convergence?

We need some more notation.

Definition 2.13. Let S_n be the collection of sets of the form $G \cap A$, where A is an n-triangle. We call the elements of S_n n-*complexes*. For $x \in G_n$ let $D_n(x) = \bigcup\{S \in S_n : x \in S\}$.

The key properties of the SG which we use are, first that it is very symmetric, and secondly, that it is finitely ramified. (In general, a set A in a metric space F is finitely ramified if there exists a finite set B such that $A - B$ is not connected). For the SG, we see that each n-complex A is disconnected from the rest of the set if we remove the set of its corners, that is $A \cap G_n$.

The following is the key observation. Suppose $Y_0^{(n)} = y \in G_{n-1}$ (take $y \notin G_0$ for simplicity), and let $T = \inf\{k > 0 : Y_k^{(n)} \in G_{n-1} - \{y\}\}$. Then $Y^{(n)}$ can only escape from $D_{n-1}(y)$ at one of the 4 points, $\{x_1, \ldots, x_4\}$ say, which are neighbours of y in the graph (G_{n-1}, E_{n-1}). Therefore $Y_T^{(n)} \in \{x_1, \ldots, x_4\}$. Further the symmetry of the set $G_n \cap D_n(y)$ means that each of the events $\{Y_T^{(n)} = x_i\}$ is equally likely.

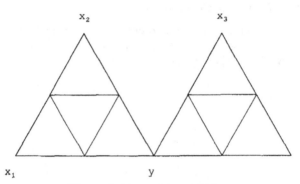

Figure 2.4: y and its neighbours.

Thus

$$\mathbb{P}\left(Y_T^{(n)} = x_i \,\middle|\, Y_0^{(n)} = y\right) = \tfrac{1}{4},$$

and this is also equal to $\mathbb{P}(Y_1^{(n-1)} = x_i | Y_0^{(n-1)} = y)$. (Exactly the same argument applies if $y \in G_0$, except that we then have only 2 neighbours instead of 4). It follows that $Y^{(n)}$ looked at at its visits to G_{n-1} behaves exactly like $Y^{(n-1)}$. To state this precisely, we first make a general definition.

Definition 2.14. Let $\mathbb{T} = \mathbb{R}_+$ or $\mathbb{Z}_+$, let $(Z_t, t \in \mathbb{T})$ be a cadlag process on a metric space F, and let $A \subset F$ be a discrete set. Then *successive disjoint hits* by Z on A are the stopping times $T_0, T_1, \ldots$ defined by

(2.8)
$$T_0 = \inf\{t \geq 0 : Z_t \in A\},$$
$$T_{n+1} = \inf\left\{t > T_n : Z_t \in A - \{Z_{T_n}\}\right\}, \qquad n \geq 0.$$

With this notation, we can summarize the observations above.

Lemma 2.15. *Let $(T_i)_{i \geq 0}$ be successive disjoint hits by $Y^{(n)}$ on G_{n-1}. Then $(Y_{T_i}^{(n)}, i \geq 0)$ is a simple random walk on G_{n-1} and is therefore equal in law to $(Y_i^{(n-1)}, i \geq 0)$.*

Using this, it is clear that we can build a sequence of "nested" random walks on G_n. Let $N \geq 0$, and let $Y_k^{(N)}$, $k \geq 0$ be a SRW on G_N with $Y_0^{(N)} = 0$. Let $0 \leq m \leq N - 1$ and $(T_i^{N,m})_{i \geq 0}$ be successive disjoint hits by $Y^{(N)}$ on G_m, and set

$$Y_i^{(m)} = Y^{(N)}(T_i^{N,m}) = Y_{T_i^{N,m}}^{(N)}, \qquad i \geq 0.$$

It follows from Lemma 2.15 that $Y^{(m)}$ is a SRW on G_m, and for each $0 \leq n \leq m \leq N$ we have that $Y^{(m)}$, sampled at its successive disjoint hits on G_n, equals $Y^{(n)}$.

We now wish to construct a sequence of SRWs with this property holding for $0 \leq n \leq m < \infty$. This can be done, either by using the Kolmogorov extension theorem, or directly, by building $Y^{(N+1)}$ from $Y^{(N)}$ with a sequence of independent "excursions". The argument in either case is not hard, and I omit it.

Thus we can construct a probability space $(\Omega, \mathcal{F}, \mathbb{P})$, carrying random variables $(Y_k^{(n)}, n \geq 0, k \geq 0)$ such that

(a) For each n, $(Y_k^{(n)}, k \geq 0)$ is a SRW on G_n starting at 0.
(b) Let $T_i^{n,m}$ be successive disjoint hits by $Y^{(n)}$ on G_m. (Here $m \leq n$). Then

$$(2.9) \qquad Y^{(n)}(T_i^{n,m}) = Y_i^{(m)}, \quad i \geq 0, \quad m \leq n.$$

If we just consider the paths of the processes $Y^{(n)}$ in G, we see that we are viewing successive discrete approximations to a continuous path. However, to define a limiting process we need to rescale time, as was suggested by (2.7).

Write $\tau = T_1^{1,0} = \min\{k \geq 0 : |Y_k^{(1)}| = 1\}$, and set $f(s) = \mathbb{E}s^\tau$, for $s \in [0,1]$.

Lemma 2.16. $f(s) = s^2/(4 - 3s)$, $\mathbb{E}\tau = f'(1) = 5$, and $\mathbb{E}\tau^k < \infty$ for all k.

Proof. This is a simple exercise in finite state Markov chains. Let a_1, a_2 be the two non-zero elements of G_0, let $b = \frac{1}{2}(a_1 + a_2)$, and $c_i = \frac{1}{2}a_i$. Writing $f_c(s) = \mathbb{E}^{c_i}s^\tau$, and defining f_b, f_a similarly, we have $f_a(s) = 1$,

$$
\begin{aligned}
f(s) &= s f_c(s), \\
f_c(s) &= \tfrac{1}{4}s\big(f(s) + f_c(s) + f_b(s) + f_a(s)\big), \\
f_b(s) &= \tfrac{1}{2}s\big(f_c(s) + f_a(s)\big),
\end{aligned}
$$

and solving these equations we obtain $f(s)$.

The remaining assertions follow easily from this. □

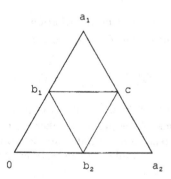

Figure 2.5: The graph G_1.

Now let $Z_n = T_1^{n,0}$, $n \geq 0$. The nesting property of the random walks $Y^{(n)}$ implies that Z_n is a simple branching process, with offspring distribution (p_n), where

$$(2.10) \qquad f(s) = \sum_{k=2}^{\infty} s^k p_k.$$

To see this, note that $Y_k^{(n+1)}$, for $T_i^{n+1,n} \leq k \leq T_{i+1}^{n+1,n}$ is a SRW on $G_{n+1} \cap D_n(Y_i^{(n)})$, and that therefore $T_{i+1}^{n+1,n} - T_i^{n+1,n} \overset{(d)}{=} \tau$. Also, by the Markov property, the r.v. $\xi_i = T_{i+1}^{n+1,n} - T_i^{n+1,n}$, $i \geq 0$, are independent. Since

$$Z_{n+1} = \sum_{i=0}^{Z_n - 1} \xi_i,$$

(Z_n) is a branching process.

As $E\tau^2 < \infty$, and $E\tau = 5$, the convergence theorem for simple branching processes implies that

$$5^{-n} Z_n \xrightarrow{a.s.} W$$

for some strictly positive r.v. W. (See [Har, p. 13]). The convergence is easy using a martingale argument: proving that $W > 0$ a.s. takes a little more work. (See [Har, p. 15]). In addition, if

$$\varphi(u) = E e^{-uW}$$

then φ satisfies the functional equation

$$(2.11) \qquad \varphi(5u) = f(\varphi(u)), \quad \varphi'(0) = -1.$$

We have a similar result in general.

Proposition 2.17. *Fix $m \geq 0$. The processes*

$$Z_n^{(i)} = T_i^{n,m} - T_{i-1}^{n,m}, \quad n \geq m$$

are branching processes with offspring distribution τ, and $Z^{(i)}$ are independent. Thus there exist $W_i^{(m)}$ such that for each m $(W_i^{(m)}, i \geq 0)$ are independent, $W_i^{(m)} \overset{(d)}{=} 5^{-m} W$, and

$$5^{-n} \left(T_i^{n,m} - T_{i-1}^{n,m} \right) \to W_i^{(m)} \quad a.s.$$

Note in particular that $\mathbb{E}(T_1^{n,0}) = 5^n$, that is that the mean time taken by $Y^{(n)}$ to cross G_n is 5^n. In terms of the graph distance on G_n we have therefore that $Y^{(n)}$ requires roughly 5^n steps to move a distance 2^n; this may be compared with the corresponding result for a simple random walk on $\mathbb{Z}^d$, which requires roughly 4^n steps to move a distance 2^n.

The slower movement of $Y^{(n)}$ is not surprising — to leave $G_n \cap B(0, 1/2)$, for example, it has to find one of the two 'gateways' $(1/2, 0)$ or $(1/4, \sqrt{3}/4)$. Thus the movement of $Y^{(n)}$ is impeded by a succession of obstacles of different sizes, which act to slow down its diffusion.

Given the space-time scaling of $Y^{(n)}$ it is no surprise that we should take $\alpha_n = 5^n$ in (2.7). Define

$$X_t^n = Y_{[5^n t]}^{(n)}, \quad t \geq 0.$$

In view of the fact that we have built the $Y^{(n)}$ with the nesting property, we can replace the weak convergence of (2.7) with a.s. convergence.

Theorem 2.18. *The processes X^n converge a.s., and uniformly on compact intervals, to a process X_t, $t \geq 0$. X is continuous, and $X_t \in G$ for all $t \geq 0$.*

Proof. For simplicity we will use the fact that W has a non-atomic distribution function. Fix for now $m > 0$. Let $t > 0$. Then, a.s., there exists $i = i(\omega)$ such that

$$\sum_{j=1}^{i} W_j^{(m)} < t < \sum_{j=1}^{i+1} W_j^{(m)}.$$

As $W_j^{(m)} = \lim_{n \to \infty} 5^{-n} \left(T_j^{n,m} - T_{j-1}^{n,m} \right)$ it follows that for $n \geq n_0(\omega)$,

(2.12) $$T_i^{n,m} < 5^n t < T_{i+1}^{n,m}.$$

Now $Y^{(n)}(T_i^{n,m}) = Y_i^{(m)}$ by (2.9). Since $Y_k^{(n)} \in D_m(Y_i^{(m)})$ for $T_i^{n,m} \leq k \leq T_{i+1}^{n,m}$, we have

$$|Y_{[5^n t]}^{(n)} - Y_i^{(m)}| \leq 2^{-m} \quad \text{for all } n \geq n_0.$$

This implies that $|X_t^n - X_t^{n'}| \leq 2^{-m+1}$ for $n, n' \geq n_0$, so that X_t^n is Cauchy, and converges to a r.v. X_t. Since $X_t^n \in G_n$, we have $X_t \in G$.

With a little extra work, one can prove that the convergence is uniform in t, on compact time intervals. I give here a sketch of the argument. Let $a \in \mathbb{N}$, and let

$$\xi_m = \min_{1 \leq i \leq a5^m} W_i^{(m)}.$$

Then $\xi_m > 0$ a.s. Choose n_0 such that for $n \geq n_0$

$$\left| 5^{-n} T_i^{n,m} - \sum_{j=1}^{i} W_j^{(m)} \right| < \tfrac{1}{3}\xi_n, \quad 1 \leq i \leq a5^m.$$

Then if $i = i(t, \omega)$ is such that $W_i^m \leq t < W_{i+1}^m$, and $i \leq a5^m$ we have $5^{-n} T_{i-1}^{n,m} < t < 5^{-n} T_{i+2}^{n,m}$ for all $n \geq n_0$. So, $|X_t^n - Y_i^m| \leq 2^{-m+1}$ for all $n \geq n_0$. This implies that if $T_m = \sum_{j=1}^{a5^m} W_i^{(m)}$, and $S < T_m$, then

$$\sup_{0 \leq t \leq S} |X_t^n - X_t^{n'}| \leq 2^{-m+2}$$

for all $n, n' \geq n_0$. If $S < \liminf_m T_m$ then the uniform a.s. convergence on the (random) interval $[0, S]$ follows. If $s, t < T_m$ and $|t - s| < \xi_m$, then we also have $|X_t^n - X_s^n| \leq 2^{-m+2}$ for $n \geq n_0$. Thus X is uniformly continuous on $[0, S]$. Varying a we also obtain uniform a.s. convergence on fixed intervals $[0, t_0]$. $\qquad\square$

Although the notation is a little cumbersome, the ideas underlying the construction of X given here are quite simple. The argument above is given in [BP], but Kusuoka [Kus1], and Goldstein [Go], who were the first to construct a diffusion on G, used a similar approach. It is also worth noting that Knight [Kn] uses similar methods in his construction of 1-dimensional Brownian motion.

The natural next step is to ask about properties of the process X. But unfortunately the construction given above is not quite strong enough on its own to give us much. To see this, consider the questions

(1) Is $W = \lim_{n\to\infty} 5^{-n} T_1^{n,0} = \inf\{t \geq 0 : X_t \in G - \{0\}\}$?

(2) Is X Markov or strong Markov?

For (1), we certainly have $X_W \in G - \{0\}$. However, consider the possibility that each of the random walks Y_n moves from 0 to a_2 on a path which does not include a_1, but includes an approach to a distance 2^{-n}. In this case we have $a_1 \notin \{X_t^n, 0 \leq t \leq W\}$, but $X_T = a_1$ for some $T < W$. Plainly, some estimation of hitting probabilities is needed to exclude possibilities like this.

(2). The construction above does give a Markov property for X at stopping times of the form $\sum_{j=1}^i W_j^{(m)}$. But to obtain a good Markov process $X = (X_t, t \geq 0, \mathbb{P}^x, x \in G)$ we need to construct X at arbitrary starting points $x \in G$, and to show that (in some appropriate sense) the processes started at close together points x and y are close.

This can be done using the construction given above — see [BP, Section 2]. However, the argument, although not really hard, is also not that simple.

In the remainder of this section, I will describe some basic properties of the process X, for the most part without giving detailed proofs. Most of these theorems will follow from more general results given later in these notes.

Although G is highly symmetric, the group of global isometries of G is quite small. We need to consider maps restricted to subsets.

Definition 2.19. Let (F, ρ) be a metric space. A *local isometry* of F is a triple (A, B, φ) where A, B are subsets of F and φ is an isometry (i.e. bijective and distance preserving) between A and B, and between ∂A and ∂B.

Let $(X_t, t \geq 0, \mathbb{P}^x, x \in F)$ be a Markov process on F. For $H \subset F$, set $T_H = \inf\{t \geq 0 : X_t \in H\}$. X is *invariant with respect to* a local isometry (A, B, φ) if

$$\mathbb{P}^x \left(\varphi(X_{t \wedge T_{\partial A}}) \in \cdot, t \geq 0 \right) = \mathbb{P}^{\varphi(x)} \left(X_{t \wedge T_{\partial B}} \in \cdot, t \geq 0 \right).$$

X is *locally isotropic* if X is invariant with respect to the local isometries of F.

Theorem 2.20. (a) *There exists a continuous strong Markov process* $X = (X_t, t \geq 0, \mathbb{P}^x, x \in G)$ *on* G.
(b) *The semigroup on* $C(G)$ *defined by*

$$P_t f(x) = \mathbb{E}^x f(X_t)$$

is Feller, and is μ_G-*symmetric:*

$$\int_G f(x) P_t g(x) \mu_G(dx) = \int g(x) P_t f(x) \mu_G(dx).$$

(c) X is locally isotropic on the spaces $(G, |\cdot - \cdot|)$ and (G, d_G).

(d) For $n \geq 0$ let $T_{n,i}$, $i \geq 0$ be successive disjoint hits by X on G_n. Then $\widehat{Y}_i^{(n)} = X_{T_{n,i}}$, $i \geq 0$ defines a SRW on G_n, and $\widehat{Y}_{[5^n t]}^{(n)} \to X_t$ uniformly on compacts, a.s. So, in particular $(X_t, t \geq 0, \mathbb{P}^0)$ is the process constructed in Theorem 2.18.

This theorem will follow from our general results in Sections 6 and 7; a direct proof may be found in [BP, Sect. 2]. The main labour is in proving (a); given this (b), (c), (d) all follow in a relatively straightforward fashion from the corresponding properties of the approximating random walks $\widehat{Y}^{(n)}$.

The property of local isotropy on (G, d_G) characterizes X:

Theorem 2.21. *(Uniqueness).* Let $(Z_t, t \geq 0, \mathbb{Q}^x, x \in \mathcal{G})$ be a non-constant locally isotropic diffusion on (G, d_G). Then there exists $a > 0$ such that

$$\mathbb{Q}^x(Z_t \in \cdot, t \geq 0) = \mathbb{P}^x(X_{at} \in \cdot, t \geq 0).$$

(So Z is equal in law to a deterministic time change of X).

The beginning of the proof of Theorem 2.21 runs roughly along the lines one would expect: for $n \geq 0$ let $(\widetilde{Y}_i^{(n)}, i \geq 0)$ be $\widetilde{Z}$ sampled at its successive disjoint hits on G_n. The local isotropy of $\widetilde{Z}$ implies that $\widetilde{Y}^{(n)}$ is a SRW on G_n. However some work (see [BP, Sect. 8]) is required to prove that the process Y does not have traps, i.e. points x such that $\mathbb{Q}^x(Y_t = x \text{ for all } t) = 1$.

Remark 2.22. The definition of invariance with respect to local isometries needs some care. Note the following examples.

1. Let $x, y \in G_n$ be such that $D_n(x) \cap G_0 = a_0$, $D_n(y) \cap G_0 = \emptyset$. Then while there exists an isometry φ from $D_n(x) \cap G$ to $D_n(y) \cap G$, φ does not map $\partial_R D_n(x) \cap G$ to $\partial_R D_n(y) \cap G$. (∂_R denotes here the relative boundary in the set G).

2. Recall the definition of H_n, the n-th stage in the construction of G, and let $B_n = \partial H_n$. We have $G = \text{cl}(\cup B_n)$. Consider the process Z_t on G, whose local motion is as follows. If $Z_t \in H_n - H_{n-1}$, then Z_t runs like a standard 1-dimensional Brownian motion on H_n, until it hits H_{n-1}. After this it repeats the same procedure on H_{n-1} (or H_{n-k} if it has also hit H_{n-k} at that time). This process is also invariant with respect to local isometries (A, B, φ) of the metric space $(G, |\cdot - \cdot|)$. See [He] for more on this and similar processes.

To discuss scale invariant properties of the process X it is useful to extend G to an unbounded set $\widetilde{G}$ with the same structure. Set

$$\widetilde{G} = \bigcup_{n=0}^{\infty} 2^n G,$$

and let $\widetilde{G}_n$ be the set of vertices of n-triangles in $\widetilde{G}_n$, for $n \geq 0$. We have

$$\widetilde{G}_n = \bigcup_{k=0}^{\infty} 2^k G_{n+k},$$

and if we define $G_m = \{0\}$ for $m < 0$, this definition also makes sense for $n < 0$. We can, almost exactly as above, define a limiting diffusion $\widetilde{X} = (\widetilde{X}_t, t \geq 0, \widetilde{\mathbb{P}}^x, x \in \widetilde{G})$ on $\widetilde{G}$:

$$\widetilde{X}_t = \lim_{n \to \infty} \widetilde{Y}^{(n)}_{[5^n t]}, \qquad t \geq 0, \text{ a.s.}$$

where $(\widetilde{Y}^{(n)}_k, n \geq 0, k \geq 0)$ are a sequence of nested simple random walks on $\widetilde{G}_n$, and the convergence is uniform on compact time intervals.

The process $\widetilde{X}$ satisfies an analogous result to Theorem 2.20, and in addition satisfies the scaling relation

$$(2.13) \qquad \mathbb{P}^x(2\widetilde{X}_t \in \cdot, t \geq 0) = \mathbb{P}^{2x}(\widetilde{X}_{5t} \in \cdot, t \geq 0).$$

Note that (2.13) implies that $\widetilde{X}$ moves a distance of roughly $t^{\log 2/\log 5}$ in time t. Set

$$d_w = d_w(G) = \log 5/\log 2.$$

We now turn to the question: "What does this process look like?"

The construction of X, and Theorem 2.20(d), tells us that the 'crossing time' of a 0-triangle is equal in law to the limiting random variable W of a branching process with offspring p.g.f. given by $f(s) = s^2/(4-3s)$. From the functional equation (2.11) we can extract information about the behaviour of $\varphi(u) = E \exp(-uW)$ as $u \to \infty$, and from this (by a suitable Tauberian theorem) we obtain bounds on $\mathbb{P}(W \leq t)$ for small t. These translate into bounds on $\mathbb{P}^x(|X_t - x| > \lambda)$ for large λ. (One uses scaling and the fact that to move a distance in $\widetilde{G}$ greater than 2, X has to cross at least one 0-triangle). These bounds give us many properties of X. However, rather than following the development in [BP], it seems clearer to first present the more delicate bounds on the transition densities of $\widetilde{X}$ and X obtained there, and derive all the properties of the process from them. Write $\widetilde{\mu}_G$ for the analogue of μ_G for $\widetilde{G}$, and $\widetilde{P}_t$ for the semigroup of $\widetilde{X}$. Let $\widetilde{\mathcal{L}}$ be the infinitesimal generator of $\widetilde{P}_t$.

Theorem 2.23. $\widetilde{P}_t$ and P_t have densities $\widetilde{p}(t, x, y)$ and $p(t, x, y)$ respectively.
(a) $\widetilde{p}(t, x, y)$ is continuous on $(0, \infty) \times \widetilde{G} \times \widetilde{G}$.
(b) $\widetilde{p}(t, x, y) = \widetilde{p}(t, y, x)$ for all t, x, y.
(c) $t \to \widetilde{p}(t, x, y)$ is C^∞ on $(0, \infty)$ for each (x, y).
(d) For each t, y

$$|\widetilde{p}(t, x, y) - \widetilde{p}(t, x', y)| \leq c_1 t^{-1} |x - x'|^{d_w - d_f}, \qquad x, x' \in \widetilde{G}.$$

(e) For $t \in (0, \infty)$, $x, y \in \widetilde{G}$

$$(2.14) \qquad c_2 t^{-d_f/d_w} \exp\left(-c_3 \left(\frac{|x-y|^{d_w}}{t}\right)^{1/(d_w-1)}\right) \leq \widetilde{p}(t, x, y)$$

$$\leq c_4 t^{-d_f/d_w} \exp\left(-c_5 \left(\frac{|x-y|^{d_w}}{t}\right)^{1/(d_w-1)}\right).$$

(f) For each $y_0 \in \tilde{G}$, $\tilde{p}(t, x, y_0)$ is the fundamental solution of the heat equation on $\tilde{G}$ with pole at y_0:

$$\frac{\partial}{\partial t}\tilde{p}(t, x, y_0) = \tilde{\mathcal{L}}\tilde{p}(t, x, y_0), \qquad \tilde{p}(0, \cdot, y_0) = \delta_{y_0}(\cdot).$$

(g) $p(t, x, y)$ satisfies (a)–(f) above (with $\tilde{G}$ replaced by G and $t \in (0, \infty]$ replaced by $t \in (0, 1]$).

Remarks. 1. The proof of this in [BP] is now largely obsolete — simpler methods are now available, though these are to some extent still based on the ideas in [BP]. 2. If $d_f = d$ and $d_w = 2$ we have in (2.14) the form of the transition density of Brownian motion in $\mathbb{R}^d$. Since $d_w = \log 5 / \log 2 > 2$, the tail of the distribution of $|X_t - x|$ under $\mathbb{P}^x$ decays more rapidly than an exponential, but more slowly than a Gaussian.

It is fairly straightforward to integrate the bounds (2.14) to obtain information about X. At this point we just present a few simple calculations; we will give some further properties of this process in Section 3.

Definition 2.24. For $x \in \tilde{G}$, $n \in \mathbb{Z}$, let x_n be the point in $\tilde{G}_n$ closest to x in Euclidean distance. (Use some procedure to break ties). Let $D_n(x) = D_n(x_n)$.

Note that $\tilde{\mu}_G\big(D_n(x_n)\big)$ is either 3^{-n} or 2.3^{-n}, that

$$(2.15) \qquad |x - y| \le 2.2^{-n} \quad \text{if } y \in D_n(x),$$

and that

$$(2.16) \qquad |x - y| \ge \tfrac{\sqrt{3}}{4}2^{-(n+1)} \quad \text{if } y \in \tilde{G}\bigcap D_n(x)^c.$$

The sets $D_n(x)$ form a convenient collection of neighbourhoods of points in $\tilde{G}$. Note that $\bigcup_{n\in\mathbb{Z}}D_n(x) = \tilde{G}$.

Corollary 2.25. For $x \in \tilde{G}$,

$$c_1 t^{2/d_w} \le \mathbb{E}^x|X_t - x|^2 \le c_2 t^{2/d_w}, \qquad t \ge 0.$$

Proof. We have

$$\mathbb{E}^x|X_t - x|^2 = \int_{\tilde{G}}(y - x)^2\tilde{p}(t, x, y)\tilde{\mu}_G(dy).$$

Set $A_m = D_m(x) - D_{m+1}(x)$. Then

$$(2.17) \quad \int_{A_m}(y - x)^2\tilde{p}(t, x, y)\tilde{\mu}_G(dy)$$

$$\le c(2^{-m})^2 t^{-d_f/d_w}\exp\left(-c'\left((2^{-m})^{d_w}/t\right)^{1/(d_w-1)}\right)3^{-m}$$

$$= c(2^{-m})^{2+d_f}t^{-d_f/d_w}\exp\left(-c'(5^{-m}/t)^{1/(d_w-1)}\right).$$

Choose n such that $5^{-n} \leq t < 5^{-n+1}$, and write $a_m(t)$ for the final term in (2.17). Then

$$\mathbb{E}^x(X_t - x)^2 \leq \sum_{m=-\infty}^{n-1} a_m(t) + \sum_{m=n}^{\infty} a_m(t).$$

For $m < n$, $5^{-m}/t > 1$ and the exponential term in (2.17) is dominant. After a few calculations we obtain

$$\sum_{m=-\infty}^{n-1} a_m(t) \leq c(2^{-n})^{2+d_f} t^{-d_f/d_w}$$

$$\leq ct^{(2+d_f)/d_w - d_f/d_w} \leq ct^{(2+d_f)/d_w - d_f/d_w} \leq ct^{2/d_w},$$

where we used the fact that $(2^{-n})^{d_w} \asymp t$. For $m \geq n$ we neglect the exponential term, and have

$$\sum_{m=n}^{\infty} a_m(t) \leq c\, t^{-d_f/d_w} \sum_{m=n}^{\infty} (2^{-m})^{2+d_f}$$

$$\leq ct^{-d_f/d_w}(2^{-n})^{2+d_f} \leq c't^{2/d_w}.$$

Similar calculations give the lower bound. $\square$

Remarks 2.26. 1. Since $2/d_w = \log 4/\log 5 < 1$ this implies that X is subdiffusive.

2. Since $\tilde{\mu}_G(B(x,r)) \asymp r^{d_f}$, for $x \in \tilde{G}$, it is tempting to try and prove Corollary 2.25 by the following calculation:

$$(2.18) \quad \mathbb{E}^x|\tilde{X}_t - x|^2 = \int_0^\infty r^2\, dr \int_{\partial B(x,r)} \tilde{p}(t,x,y)\tilde{\mu}_G(dy)$$

$$\asymp \int_0^\infty dr\, r^2\, r^{d_f-1} t^{-d_f/d_w} \exp\left(-c(r^{d_w}/t)^{1/d_w-1}\right)$$

$$= t^{2/d_w} \int_0^\infty s^{1+d_f} \exp\left(-c(s^{d_w})^{1/d_w-1}\right) ds = ct^{2/d_w}.$$

Of course this calculation, as it stands, is not valid: the estimate

$$\tilde{\mu}_B(B(x,r+dr) - B(x,r)) \asymp r^{d_f-1} dr$$

is certainly not valid for all r. But it does hold on average over length scales of $2^n < r < 2^{n+1}$, and so splitting $\tilde{G}$ into suitable shells, a rigorous version of this calculation may be obtained – and this is what we did in the proof of Corollary 2.25.

The λ-potential kernel density of $\tilde{X}$ is defined by

$$u_\lambda(x,y) = \int_0^\infty e^{-\lambda t} \tilde{p}(t,x,y)\, dt.$$

From (2.14) it follows that u_λ is continuous, that $u_\lambda(x,x) \leq c\lambda^{d_f/d_w-1}$, and that $u_\lambda \to \infty$ as $\lambda \to 0$. Thus the process $\tilde{X}$ (and also X) "hits points" – that is if

$T_y = \inf\{t > 0 : \widetilde{X}_t = y\}$ then

$$(2.19) \qquad\qquad \mathbb{P}^x(T_y < \infty) > 0.$$

It is of course clear that X must be able to hit points in G_n – otherwise it could not move, but (2.19) shows that the remaining points in G have a similar status. The continuity of $u_\lambda(x, y)$ in a neighbourhood of x implies that

$$\mathbb{P}^x(T_x = 0) = 1,$$

that is that x is regular for $\{x\}$ for all $x \in \widetilde{G}$.

The following estimate on the distribution of $|\widetilde{X}_t - x|$ can be obtained easily from (2.14) by integration, but since this bound is actually one of the ingredients in the proof, such an argument would be circular.

Proposition 2.27. *For $x \in \widetilde{G}$, $\lambda > 0$, $t > 0$,*

$$c_1 \exp\left(-c_2(\lambda^{d_w}/t)^{1/d_w - 1}\right) \le \mathbb{P}^x(|\widetilde{X}_t - x| > \lambda)$$
$$\le c_3 \exp\left(-c_4(\lambda^{d_w}/t)^{(1/d_w - 1)}\right).$$

From this, it follows that the paths of $\widetilde{X}$ are Hölder continuous of order $1/d_w - \varepsilon$ for each $\varepsilon > 0$. In fact we can (up to constants) obtain the precise modulus of continuity of $\widetilde{X}$. Set

$$h(t) = t^{1/d_w}(\log t^{-1})^{(d_w - 1)/d_w}.$$

Theorem 2.28. *(a) For $x \in G$*

$$c_1 \le \lim_{\delta \downarrow 0} \sup_{\substack{0 \le s \le t \le 1 \\ |t-s| < \delta}} \frac{|\widetilde{X}_s - \widetilde{X}_t|}{h(s-t)} \le c_2, \qquad \mathbb{P}^x - \text{a.s.}$$

(b) The paths of $\widetilde{X}$ are of infinite quadratic variation, a.s., and so in particular $\widetilde{X}$ is not a semimartingale.

The proof of (a) is very similar to that of the equivalent result for Brownian motion in $\mathbb{R}^d$.

For (b), Proposition 2.23 implies that $|X_{t+h} - X_t|$ is of order h^{1/d_w}; as $d_w > 2$ this suggests that X should have infinite quadratic variation. For a proof which fills in the details, see [BP, Theorem 4.5]. $\qquad\square$

So far in this section we have looked at the Sierpinski gasket, and the construction and properties of a symmetric diffusion X on G (or $\widetilde{G}$). The following three questions, or avenues for further research, arise naturally at this point.

1. Are there other natural diffusions on the SG?
2. Can we do a similar construction on other fractals?
3. What finer properties does the process X on G have? (More precisely: what about properties which the bounds in (2.17) are not strong enough to give information on?)

The bulk of research effort in the years since [Kus1, Go, BP] has been devoted to (2). Only a few papers have looked at (1), and (apart from a number of works on spectral properties), the same holds for (3).

Before discussing (1) or (2) in greater detail, it is worth extracting one property of the SRW $Y^{(1)}$ which was used in the construction.

Let $V = (V_n, n \geq 0, \mathbb{P}^a, a \in G_0)$ be a Markov chain on G_0: clearly V is specified by the transition probabilities

$$p(a_i, a_j) = \mathbb{P}^{a_i}(V_1 = a_j), \quad 0 \leq i, j \leq 2.$$

We take $p(a, a) = 0$ for $a \in G_0$, so V is determined by the three probabilities $p(a_i, a_j)$, where $j = i + 1 \pmod 3$.

Given V we can define a Markov Chain V' on G_1 by a process we call *replication*. Let $\{b_{01}, b_{02}, b_{12}\}$ be the 3 points in $G_1 - G_0$, where $b_{ij} = \frac{1}{2}(a_i + a_j)$. We consider G_1 to consist of three 1-cells $\{a_i, b_{ij}, j \neq i\}$, $0 \leq i \leq 2$, which intersect at the points $\{b_{ij}\}$. The law of V' may be described as follows: V' moves inside each 1-cell in the way same as V does; if V_0' lies in two 1-cells then it first chooses a 1-cell to move in, and chooses each 1-cell with equal probability. More precisely, writing $V' = (V_n', n \geq 0, \overline{\mathbb{P}}^a, a \in G_1)$, and

$$p'(a, b) = \overline{\mathbb{P}}^a(V_1' = b),$$

we have

(2.20)
$$p'(a_i, b_{ij}) = p(a_i, a_j),$$
$$p'(b_{ij}, b_{ik}) = \tfrac{1}{2}p(a_j, a_k), \quad p'(b_{ij}, a_i) = \tfrac{1}{2}p(a_j, a_i).$$

Now let T_k, $k \geq 0$ be successive disjoint hits by V' on G_0, and let $U_k = V_{T_k}'$, $k \geq 0$. Then U is a Markov Chain on G_0; we say that V is *decimation invariant* if U is equal in law to V.

We saw above that the SRW $Y^{(0)}$ on G_0 was decimation invariant. A natural question is:

What other decimation invariant Markov chains are there on G_0?

Two classes have been found:

1. (See [Go]). Let $p(a_0, a_1) = p(a_1, a_0) = 1$, $p(a_2, a_0) = \frac{1}{2}$.
2. "p-stream random walks" ([Kum1]). Let $p \in (0, 1)$ and

$$p(a_0, a_1) = p(a_1, a_2) = p(a_2, a_0) = p.$$

From each of these processes we can construct a limiting diffusion in the same way as in Theorem 2.18. The first process is reasonably easy to understand: essentially its paths consist of a downward drift (when this is possible), and a behaviour like 1-dimensional Brownian motion on the portions on G which consist of line segments parallel to the x-axis.

For $p > \frac{1}{2}$ Kumagai's p-stream diffusions tend to rotate in an anti-clockwise direction, so are quite non-symmetric. Apart from the results in [Kum1] nothing is known about this process.

Two other classes of diffusions on G, which are not decimation invariant, have also been studied. The first are the "asymptotically 1-dimensional diffusions" of [HHW4], the second the diffusions, similar to that described in Remark 2.22, which are $(G, |\cdot - \cdot|)$-isotropic but not (G, d_G)- isotropic – see [He]. See also [HH1, HK1, HHK] for work on the self-avoiding random walk on the SG.

Diffusions on other fractal sets.

Of the three questions above, the one which has received most attention is that of making similar constructions on other fractals. To see the kind of difficulties which can arise, consider the following two fractals, both of which are constructed by a Cantor type procedure, based on squares rather than triangles. For each curve the figure gives the construction after two stages.

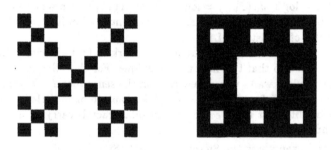

Figure 2.6: The Vicsek set and the Sierpinski carpet.

The first of these we will call the "Vicsek set" (VS for short). We use similar notation as for the SG, and write G_0, $G_1, \ldots$ for the succession of sets of vertices of corners of squares. We denote the limiting set by $F = F_{VS}$. One difficulty arises immediately. Let Y_r be the SRW on G_0 which moves from any point $x \in G_0$ to each of its neighbours with equal probability. (The neighbours of x are the 2 points y in G_0 with $|x - y| = 1$). Then $Y^{(0)}$ is not decimation invariant. This is easy to see: $Y^{(0)}$ cannot move in one step from $(0,0)$ to $(1,1)$, but $Y^{(1)}$ can move from $(0,0)$ to $(1,1)$ without hitting any other point in G_0.

However it is not hard to find a decimation invariant random walk on G_0. Let $p \in [0,1]$, and consider the random walk $(Y_r, r \geq 0, \mathbb{E}_p^x, x \in G_0)$ on G_0 which moves diagonally with probability p, and horizontally or vertically with probability $\frac{1}{2}(1-p)$. Let $(Y_r', r \geq 0, \mathbb{E}_p^x, x \in G_1)$ be the Markov chain on G_1 obtained by replication, and let T_k, $k \geq 0$ be successive disjoint hits by Y' on G_0.

Then writing $f(p) = \mathbb{P}_p^0(Y_{T_1}' = (1,1))$ we have (after several minutes calculation)

$$f(p) = \frac{1}{4 - 3p}.$$

The equation $f(p) = p$ therefore has two solutions: $p = \frac{1}{3}$ and $p = 1$, each of which corresponds to a decimation invariant walk on G_0. (The number $\frac{1}{3}$ here has no general significance: if we had looked at the fractal similar to the Vicsek set, but based on a 5×5 square rather than a 3×3 square, then we would have obtained a different number).

One may now carry through, in each of these cases, the construction of a diffusion on the Vicsek set F, very much as for the Sierpinski gasket. For $p = 1$ one gets a rather uninteresting process, which, if started from $(0,0)$, is (up to a constant time change) 1-dimensional Brownian motion on the diagonal $\{(t,t), 0 \le t \le 1\}$. It is worth remarking that this process is not strong Markov: for each $x \in F$ one can take $\mathbb{P}^x$ to be the law of a Brownian motion moving on a diagonal line including x, but the strong Markov property will fail at points where two diagonals intersect, such as the point $(\frac{1}{2}, \frac{1}{2})$.

For $p = \frac{1}{3}$ one obtains a process $(X_t, t \ge 0)$ with much the same behaviour as the Brownian motion on the SG. We have for the Vicsek set (with $p = \frac{1}{3}$) $d_f(F_{VS}) = \log 5 / \log 3$, $d_w(F_{VS}) = \log 15 / \log 3$. This process was studied in some detail by Krebs [Kr1, Kr2]. The Vicsek set was mentioned in [Go], and is one of the "nested fractals" of Lindstrøm [L1].

This example shows that one may have to work to find a decimation invariant random walk, and also that this may not be unique. For the VS, one of the decimation invariant random walks was degenerate, in the sense that $P^x(Y \text{ hits } y) = 0$ for some $x, y \in G_0$, and we found the associated diffusion to be of little interest. But it raises the possibility that there could exist regular fractals carrying more than one "natural" diffusion.

The second example is the Sierpinski carpet (SC). For this set a more serious difficulty arises. The VS was finitely ramified, so that if Y_t is a diffusion on F_{VS}, and $(T_k, k \ge 0)$ are successive disjoint hits on G_n, for some $n \ge 0$, then $(Y_{T_k}, k \ge 0)$ is a Markov chain on G_n. However the SC is not finitely ramified: if $(Z_t, t \ge 0)$ is a diffusion on F_{SC}, then the first exit of Z from $[0, \frac{1}{3}]^2$ could occur anywhere on the line segments $\{(\frac{1}{3}, y), 0 \le y \le \frac{1}{3}\}$, $\{(x, \frac{1}{3}), 0 \le x \le \frac{1}{3}\}$. It is not even clear that a diffusion on F_{SC} will hit points in G_n. Thus to construct a diffusion on F_{SC} one will need very different methods from those outlined above. It is possible, and has been done: see [BB1-BB6], and [Bas] for a survey.

On the third question mentioned above, disappointingly little has been done: most known results on the processes on the Sierpinski gasket, or other fractals, are of roughly the same depth as the bounds in Theorem 2.23. Note however the results on the spectrum of $\mathcal{L}$ in [FS1, FS2, Sh1–Sh4], and the large deviation results in [Kum5]. Also, Kusuoka [Kus2] has very interesting results on the behaviour of harmonic functions, which imply that the measure defined formally on G by

$$\nu(dx) = |\nabla f|^2(x)\mu(dx)$$

is singular with respect to μ. There are many open problems here.

3. Fractional Diffusions.

In this section I will introduce a class of processes, defined on metric spaces, which will include many of the processes on fractals mentioned in these lectures. I have chosen an axiomatic approach, as it seems easier, and enables us to neglect (for the time being!) much of fine detail in the geometry of the space.

A metric space (F, ρ) has the *midpoint property* if for each $x, y \in F$ there exists $z \in F$ such that $\rho(x, z) = \rho(z, y) = \frac{1}{2}\rho(x, y)$. Recall that the geodesic metric d_G in Section 2 had this property. The following result is a straightforward exercise:

Lemma 3.1. *(See [Blu]). Let (F, ρ) be a complete metric space with the midpoint property. Then for each $x, y \in F$ there exists a geodesic path $(\gamma(t), 0 \leq t \leq 1)$ such that $\gamma(0) = x$, $\gamma(1) = y$ and $\rho(\gamma(s), \gamma(t)) = |t - s|d(x, y)$, $0 \leq s \leq t \leq 1$.*

For this reason we will frequently refer to a metric ρ with the midpoint property as a *geodesic metric*. See [Stu1] for additional remarks and references on spaces of this type.

Definition 3.2. Let (F, ρ) be a complete metric space, and μ be a Borel measure on $(F, \mathcal{B}(F))$. We call (F, ρ, μ) a *fractional metric space* (FMS for short) if

$$(3.1a) \qquad (F, \rho) \text{ has the midpoint property},$$

and there exist $d_f > 0$, and constants c_1, c_2 such that if $r_0 = \sup\{\rho(x, y) : x, y \in F\} \in (0, \infty]$ is the diameter of F then

$$(3.1b) \qquad c_1 r^{d_f} \leq \mu(B(x, r)) \leq c_2 r^{d_f} \quad \text{for} \quad x \in F,\ 0 < r \leq r_0.$$

Here $B(x, r) = \{y \in F : \rho(x, y) < r\}$.

Remarks 3.3. 1. $\mathbb{R}^d$, with Euclidean distance and Lebesgue measure, is a FMS, with $d_f = d$ and $r_0 = \infty$.
2. If G is the Sierpinski gasket, d_G is the geodesic metric on G, and $\mu = \mu_G$ is the measure constructed in Section 2, then Lemma 2.1 shows that (G, d_G, μ) is a FMS, with $d_f = d_f(G) = \log 3 / \log 2$ and $r_0 = 1$. Similarly $(\widetilde{G}, d_{\widetilde{G}}, \widetilde{\mu})$ is a FMS with $r_0 = \infty$.
3. If (F_k, d_k, μ_k), $k = 1, 2$ are FMS with the same diameter r_0 and $p \in [1, \infty]$, then setting $F = F_1 \times F_2$, $d((x_1, x_2), (y_1, y_2)) = (d_1(x_1, y_1)^p + d_2(x_2, y_2)^p)^{1/p}$, $\mu = \mu_1 \times \mu_2$, it is easily verified that (F, d, μ) is also a FMS with $d_f(F) = d_f(F_1) + d_f(F_2)$.
4. For simplicity we will from now on take either $r_0 = \infty$ or $r_0 = 1$. We will write $r \in (0, r_0]$ to mean $r \in (0, r_0] \cap (0, \infty)$, and define $r_0^\alpha = \infty$ if $\alpha > 0$ and $r_0 = \infty$.

A number of properties of (F, ρ, μ) follow easily from the definition.

Lemma 3.4. (a) $\dim_H(F) = \dim_P(F) = d_f$.
(b) F is locally compact.
(c) $d_f \geq 1$.

Proof. (a) is immediate from Corollary 2.8.
(b) Let $x \in F$, $A = \overline{B}(x, 1)$, and consider a maximal packing of disjoint balls $B(x_i, \varepsilon)$, $x_i \in A$, $1 \leq i \leq m$. As $\mu(A) \leq c_2$, and $\mu(B(x_i, \varepsilon)) \geq c_1 \varepsilon^{d_f}$, we have $m \leq c_2(c_1 \varepsilon^{d_f})^{-1} < \infty$. Also $A = \cup_{i=1}^m B(x_i, 2\varepsilon)$. Thus any bounded set in F can be

covered by a finite number of balls radius ε; this, with completeness, implies that F is locally compact.

(c) Take $x, y \in F$ with $\rho(x, y) = D > 0$. Applying the midpoint property repeatedly we obtain, for $m = 2^k$, $k \geq 1$, a sequence $x = z_0, z_1, \ldots, z_m = y$ with $\rho(z_i, z_{i+1}) = D/m$. Set $r = D/2m$: the balls $B(z_i, r)$ must be disjoint, or, using the triangle inequality, we would have $\rho(x, y) < D$. But then

$$\bigcup_{i=0}^{m-1} B(z_i, r) \subset B(x, D),$$

so that

$$c_2 D^{d_f} \geq \mu\big(B(x, D)\big) \geq \sum_{i=0}^{m-1} \mu\big(B(z_i, r)\big)$$
$$\geq m c_1 D^{d_f} (2m)^{-d_f} = c m^{1-d_f}.$$

If $d_f < 1$ a contradiction arises on letting $m \to \infty$. $\qquad\square$

Definition 3.5. Let (F, ρ, μ) be a fractional metric space. A Markov process $X = (\mathbb{P}^x, x \in F, X_t, t \geq 0)$ is a *fractional diffusion* on F if

(3.2a) X is a conservative Feller diffusion with state space F.

(3.2b) X is μ-symmetric.

(3.2c) X has a symmetric transition density $p(t, x, y) = p(t, y, x)$, $t > 0$, $x, y \in F$, which satisfies, the Chapman-Kolmogorov equations and is, for each $t > 0$, jointly continuous.

(3.2d) There exist constants $\alpha, \beta, \gamma, c_1 - c_4$, $t_0 = r_0^\beta$, such that

$$(3.3) \qquad \begin{aligned} c_1 t^{-\alpha} \exp\left(-c_2 \rho(x, y)^{\beta\gamma} t^{-\gamma}\right) &\leq p(t, x, y) \\ &\leq c_3 t^{-\alpha} \exp\left(-c_4 \rho(x, y)^{\beta\gamma} t^{-\gamma}\right), \quad x, y \in F, \ 0 < t \leq t_0. \end{aligned}$$

Examples 3.6. 1. If F is $\mathbb{R}^d$, and $a(x) = a_{ij}(x)$, $1 \leq i, j \leq d$, $x \in \mathbb{R}^d$ is bounded, symmetric, measurable and uniformly elliptic, let $\mathcal{L}$ be the divergence form operator

$$\mathcal{L} = \sum_{ij} \frac{\partial}{\partial x_i} a_{ij}(x) \frac{\partial}{\partial x_j}.$$

Then Aronsen's bounds [Ar] imply that the diffusion with infinitesimal generator $\mathcal{L}$ is a FD, with $\alpha = d/2$, $\beta = 2$, $\gamma = 1$.

2. By Theorem 2.23, the Brownian motion on the Sierpinski gasket described in Section 2 is a FD, with $\alpha = d_f(SG)/d_w(SG)$, $\beta = d_w(SG)$ and $\gamma = 1/(\beta - 1)$.

The hypotheses in Definition 3.5 are quite strong ones, and (as the examples suggest) the assertion that a particular process is an FD will usually be a substantial theorem. One could of course consider more general bounds than those in (3.3) (with a correspondingly larger class of processes), but the form (3.3) is reasonably natural, and already contains some interesting examples.

In an interesting recent series of papers Sturm [Stu1-Stu4] has studied diffusions on general metric spaces. However, the processes considered there turn out to have an essentially Gaussian long range behaviour, and so do not include any FDs with $\beta \neq 2$.

In the rest of this section we will study the general properties of FDs. In the course of our work we will find some slightly easier sufficient conditions for a process to be a FD than the bounds (3.3), and this will be useful in Section 8 when we prove that certain diffusions on fractals are FDs. We begin by obtaining two relations between the indices d_f, α, β, γ, so reducing the parameter space of FDs to a two-dimensional one.

We will say that F is a $FMS(d_f)$ if F is a FMS and satisfies (3.1b) with parameter d_f (and constants c_1, c_2). Similarly, we say X is a $FD'(d_f, \alpha, \beta, \gamma)$ if X is a FD on a $FMS(d_f)$, and X satisfies (3.3) with constants α, β, γ. (This is temporary notation — hence the $'$).

It what follows we fix a FMS (F, ρ, μ), with parameters r_0 and d_f.

Lemma 3.7. Let $\alpha, \gamma, x > 0$ and set

$$I(\gamma, x) = \int_1^\infty e^{-xt^\gamma} dt,$$

$$S(\alpha, \gamma, x) = \sum_{n=0}^\infty \alpha^n e^{-x\alpha^{n\gamma}}.$$

Then

(3.4) $$(\alpha - 1)S(\alpha, \gamma, \alpha^\gamma x) \leq I(\gamma, x) \leq (\alpha - 1)S(\alpha, \gamma, x),$$

and

(3.5) $$I(\gamma, x) \asymp x^{-1/\gamma} \quad \text{for } x \leq 1,$$
(3.6) $$I(\gamma, x) \asymp x^{-1} e^{-x} \quad \text{for } x \geq 1,$$

Proof. We have

$$I(\gamma, x) = \sum_{n=0}^\infty \int_{\alpha^n}^{\alpha^{n+1}} e^{-xt^\gamma} dt,$$

and estimating each term in the sum (3.4) is evident.

If $0 < x \leq 1$ then since

$$x^{1/\gamma} I(\gamma, x) = \int_{x^{1/\gamma}}^\infty e^{-s^\gamma} ds \rightarrow c(\gamma) \quad \text{as } x \rightarrow 0,$$

(3.5) follows.

If $x \geq 1$ then (3.6) follows from the fact that

$$xe^x I(\gamma, x) = \gamma^{-1} \int_0^\infty e^{-u}((x + u)/x)^{-1+1/\gamma} du \rightarrow \gamma^{-1} \quad \text{as } x \rightarrow \infty. \qquad \square$$

Lemma 3.8. ("Scaling relation"). Let X be a $FD'(d_f, \alpha, \beta, \gamma)$ on F. Then $\alpha = d_f/\beta$.

Proof. From (3.1) we have

$$p(t, x, y) \geq c_1 t^{-\alpha} e^{-c_2} = c_3 t^{-\alpha} \quad \text{for } \rho(x, y) \leq t^{1/\beta}.$$

Set $t_0 = r_0^\beta$. So if $A = B(x, t^{1/\beta})$, and $t \le t_0$

$$1 \ge \mathbb{P}^x(\rho(x, X_t) \le t^{1/\beta}) = \int_A p(t, x, y)\mu(dy) \ge c_3 t^{-\alpha}\mu(A) \ge ct^{-\alpha + d_f/\beta}.$$

If $r_0 = \infty$ then since this holds for all $t > 0$ we must have $\alpha = d_f/\beta$. If $r_0 = 1$ then we only deduce that $\alpha \le d_f/\beta$.

Let now $r_0 = 1$, let $\lambda > 0$, $t < 1$, and $A = B(x, \lambda t^{1/\beta})$. We have $\mu(F) \le c_{3.1.2}$, and therefore

$$\begin{aligned}
1 &= \mathbb{P}^x(X_t \in A) + \mathbb{P}^x(X_t \in A^c) \\
&\le \mu(A) \sup_{y \in A} p(t, x, y) + \mu(F - A) \sup_{y \in A^c} p(t, x, y) \\
&\le c_4 t^{-\alpha + d_f/\beta} \lambda^{d_f/\beta} + c_5 t^{-\alpha} e^{-c_6 \lambda^{\beta\gamma}}.
\end{aligned}$$

Let $\lambda = \left((d_f/\beta)c_6^{-1}\log(1/t)\right)^{1/\beta\gamma}$; then we have for all $t < 1$ that

$$1 \le ct^{-\alpha + d_f/\beta}(1 + (\log(1/t))^{1/\beta\gamma},$$

which gives a contradiction unless $\alpha \ge d_f/\beta$. $\qquad\square$

The next relation is somewhat deeper: essentially it will follow from the fact that the long–range behaviour of $p(t, x, y)$ is fixed by the exponents d_f and β governing its short–range behaviour. Since γ only plays a role in (3.3) when $\rho(x, y)^\beta \gg t$, we will be able to obtain γ in terms of d_f and β (in fact, it turns out, of β only).

We begin by deriving some consequences of the bounds (3.3).

Lemma 3.9. Let X be a $FD'(d_f, d_f/\beta, \beta, \gamma)$. Then
(a) For $t \in (0, t_0]$, $r > 0$

$$\mathbb{P}^x\left(\rho(x, X_t) > r\right) \le c_1 \exp\left(-c_2 r^{\beta\gamma} t^{-\gamma}\right).$$

(b) There exists $c_3 > 0$ such that

$$c_4 \exp\left(-c_5 r^{\beta\gamma} t^{-\gamma}\right) \le \mathbb{P}^x\left(\rho(x, X_t) > r\right) \quad \text{for } r < c_3 r_0, \, t < r^\beta.$$

(c) For $x \in F$, $0 < r < c_3 r_0$, if $\tau(x, r) = \inf\{s > 0 : X_s \notin B(x, r)\}$ then

(3.7)
$$c_6 r^\beta \le \mathbb{E}^x \tau(x, r) \le c_7 r^\beta.$$

Proof. Fix $x \in F$, and set $D(a, b) = \{y \in F : a \le \rho(x, y) \le b\}$. Then by (3.1b)

$$c_{3.1.2} b^{d_f} \ge \mu\left(D(a, b)\right) \ge c_{3.1.1} b^{d_f} - c_{3.1.2} a^{d_f}.$$

Choose $\theta \ge 2$ so that $c_{3.1.1}\theta^{d_f} \ge 2c_{3.1.2}$: then we have

(3.8)
$$c_8 a^{d_f} \le \mu\left(D(a, \theta a)\right) \le c_9 a^{d_f}.$$

Therefore, writing $D_n = D(\theta^n r, \theta^{n+1} r)$, we have $\mu(D_n) \asymp \theta^{nd_f}$ provided $r\theta^{n+1} \le r_0$. Now

$$(3.9) \qquad \mathbb{P}^x\big(\rho(x, X_t) > r\big) = \int_{B(x,r)^c} p(t, x, y)\mu(dy)$$

$$= \sum_{n=0}^{\infty} \int_{D_n} p(t, x, y)\mu(dy)$$

$$\leq \sum_{n=0}^{\infty} c(r\theta^i)^{d_f} t^{-d_f/\beta} \exp\big(-c_{10}t^{-\gamma}(r\theta^n)^{\beta\gamma}\big)$$

$$= c(r^\beta/t)^{d_f/\beta} S(\theta, \beta\gamma, c_{10}(r^\beta/t)^\gamma).$$

If $c_{10}r^\beta > t$ then using (3.6) we deduce that this sum is bounded by

$$c_{11} \exp\left(-c_{12}(r^\beta/t)^\gamma\right),$$

while if $c_{10}r^\beta \leq t$ then (as $\mathbb{P}^x\big(\rho(x, X_t) > r\big) \leq 1$) we obtain the same bound, on adjusting the constant c_{11}.

For the lower bound (b), choose $c_3 > 0$ so that $c_3\theta < 1$. Then $\mu(D_0) \geq cr^{d_f}$, and taking only the first term in (3.9) we deduce that, since $r^\beta > t$,

$$\mathbb{P}^x\big(\rho(x, X_t) > r\big) \geq c(r^\beta/t)^{d_f/\beta} \exp(-c_{13}(r^\beta/t)^\gamma)$$

$$\geq c \exp(-c_{13}(r^\beta/t)^\gamma).$$

(c) Note first that

$$(3.10) \qquad \mathbb{P}^y(\tau(x, r) > t) \leq \mathbb{P}^y(X_t \in B(x, r))$$

$$= \int_{B(x,r)} p(t, y, z)\mu(dz)$$

$$\leq ct^{-d_f/\beta}r^{d_f}.$$

So, for a suitable c_{14}

$$\mathbb{P}^y(\tau(x, r) > c_{14}r^\beta) \leq \tfrac{1}{2}, \quad y \in F.$$

Applying the Markov property of X we have for each $k \geq 1$

$$\mathbb{P}^y(\tau(x, r) > kc_{14}r^\beta) \leq 2^{-k}, \quad y \in F,$$

which proves the upper bound in (3.7).

For the lower bound, note first that

$$\mathbb{P}^x(\tau(x, 2r) < t) = \mathbb{P}^x\left(\sup_{0 \leq s \leq t} \rho(x, X_t) \geq 2r\right)$$

$$\leq \mathbb{P}^x\big(\rho(x, X_t) > r\big) + \mathbb{P}^x\big(\tau(x, 2r) < t, \rho(x, X_t) < r\big)$$

Writing $S = \tau(x, 2r)$, the second term above equals

$$\mathbb{E}^x 1_{(S<t)}\mathbb{P}^{X_S}\big(\rho(x, X_{t-S}) < r\big) \leq \sup_{y \in \partial B(x, 2r)} \sup_{s \leq t} \mathbb{P}^y\big(\rho(y, X_{t-s}) > r\big),$$

so that, using (a),

$$(3.11) \qquad \mathbb{P}^x\big(\tau(x,2r) < t\big) \leq 2\sup_{s\leq t}\sup_{y\in F}\mathbb{P}^y\big(\rho(y,X_s) > r\big)$$

$$\leq 2c_1\exp\big(-c_2(r^\beta/t)^\gamma\big).$$

So if $4c_1 e^{-c_2 a^\gamma} = 1$ then $\mathbb{P}^x(\tau(x,2r) < ar^\beta) \leq \frac{1}{2}$, which proves the left hand side of (3.7). $\qquad\square$

Remark 3.10. Note that the bounds in (c) only used the upper bound on $p(t,x,y)$.

The following result gives sufficient conditions for a diffusion on F to be a fractional diffusion: these conditions are a little easier to verify than (3.3).

Theorem 3.11. Let (F,ρ,μ) be a $FMS(d_f)$. Let $(Y_t, t \geq 0, \mathbb{P}^x, x \in F)$ be a μ-symmetric diffusion on F which has a transition density $q(t,x,y)$ with respect to μ which is jointly continuous in x,y for each $t > 0$. Suppose that there exists a constant $\beta > 0$, such that

$$(3.12) \qquad q(t,x,y) \leq c_1 t^{-d_f/\beta} \quad \text{for all } x,y \in F,\ t \in (0,t_0],$$

$$(3.13) \qquad q(t,x,y) \geq c_2 t^{-d_f/\beta} \quad \text{if } \rho(x,y) \leq c_3 t^{1/\beta},\ t \in (0,t_0],$$

$$(3.14) \qquad c_4 r^\beta \leq \mathbb{E}^x \tau(x,r) \leq c_5 r^\beta, \quad \text{for } x \in F,\ 0 < r < c_6 r_0,$$

where $\tau(x,r) = \inf\{t \geq 0 : Y_t \notin B(x,r)\}$. Then $\beta > 1$ and Y is a FD with parameters d_f, d_f/β, β and $1/(\beta-1)$.

Corollary 3.12. Let X be a $FD'(d_f, d_f/\beta, \beta, \gamma)$ on a $FMS(d_f)$ F. Then $\beta > 1$ and $\gamma = 1/(\beta-1)$.

Proof. By Lemma 3.8, and the bounds (3.3), the transition density $p(t,x,y)$ of X satisfies (3.12) and (3.13). By Lemma 3.9(c) X satisfies (3.14). So, by Theorem 3.11 $\beta > 1$, and X is a $FD'(d_f, d_f/\beta, \beta, (\beta-1)^{-1})$. Since $p(t,x,y)$ cannot satisfy (3.3) for two distinct values of γ, we must have $\gamma = (\beta-1)^{-1}$. $\qquad\square$

Remark 3.13. Since two of the four parameters are now seen to be redundant, we will shorten our notation and say that X is a $FD(d_f, \beta)$ if X is a $FD'(d_f, d_f/\beta, \beta, \gamma)$.

The proof of Theorem 3.11 is based on the derivation of transition density bounds for diffusions on the Sierpinski carpet in [BB4]: most of the techniques there generalize easily to fractional metric spaces. The essential idea is "chaining": in its classical form (see e.g. [FaS]) for the lower bound, and in a slightly different more probabilistic form for the upper bound. We begin with a some lemmas.

Lemma 3.14. [BB1, Lemma 1.1] Let $\xi_1, \xi_2, \ldots, \xi_n, V$ be non-negative r.v. such that $V \geq \sum_1^n \xi_i$. Suppose that for some $p \in (0,1)$, $a > 0$,

$$(3.15) \qquad P\big(\xi_i \leq t | \sigma(\xi_1, \ldots, \xi_{i-1})\big) \leq p + at, \qquad t > 0.$$

Then

$$(3.16) \qquad \log P(V \leq t) \leq 2\left(\frac{ant}{p}\right)^{1/2} - n\log\frac{1}{p}.$$

Proof. If η is a r.v. with distribution function $P(\eta \leq t) = (p + at) \wedge 1$, then

$$E\left(e^{-\lambda \xi_i}|\sigma(\xi_1,\ldots,\xi_{i-1})\right) \leq Ee^{-\lambda \eta}$$

$$= p + \int_0^{(1-p)/a} e^{-\lambda t} a \, dt$$

$$\leq p + a\lambda^{-1}.$$

So

$$P(V \leq t) = P\left(e^{-\lambda V} \geq e^{-\lambda t}\right) \leq e^{\lambda t} Ee^{-\lambda V}$$

$$\leq e^{\lambda t} E \exp \lambda \sum_1^n \xi_i \leq e^{\lambda t}(p + a\lambda^{-1})^n$$

$$\leq p^n \exp\left(\lambda t + \frac{an}{\lambda p}\right).$$

The result follows on setting $\lambda = (an/pt)^{1/2}$. $\qquad\square$

Remark 3.15. The estimate (3.16) appears slightly odd, since it tends to $+\infty$ as $p \downarrow 0$. However if $p = 0$ then from the last but one line of the proof above we obtain $\log P(V \leq t) \leq \lambda t + n \log \frac{a}{\lambda}$, and setting $\lambda = n/t$ we deduce that

$$(3.17) \qquad\qquad \log P(V \leq t) \leq n \log(\frac{ate}{n}).$$

Lemma 3.16. Let $(Y_t, t \geq 0)$ be a diffusion on a metric space (F, ρ) such that, for $x \in F, r > 0$,

$$c_1 r^\beta \leq \mathbb{E}^x \tau(x, r) \leq c_2 r^\beta.$$

Then for $x \in F, t > 0$,

$$\mathbb{P}^x\left(\tau(x, r) \leq t\right) \leq (1 - c_1/(2^\beta c_2)) + c_3 r^{-\beta} t.$$

Proof. Let $x \in F$, and $A = B(x, r)$, $\tau = \tau(x, r)$. Since $\tau \leq t + (\tau - t)1_{(\tau > t)}$ we have

$$\mathbb{E}^x \tau \leq t + \mathbb{E}^x 1_{(\tau > t)} \mathbb{E}^{Y_t}(\tau - t)$$

$$\leq t + \mathbb{P}^x(\tau > t) \sup_y \mathbb{E}^y \tau.$$

As $\tau \leq \tau(y, 2r)$ $\mathbb{P}^y$-a.s. for any $y \in F$, we deduce

$$c_1 r^\beta \leq \mathbb{E}^x \tau \leq t + \mathbb{P}^x(\tau > t)c_2(2r)^\beta,$$

so that

$$c_2 2^\beta \mathbb{P}^x(\tau \leq t) \leq (2^\beta c_2 - c_1) + tr^{-\beta}. \qquad\square$$

The next couple of results are needed to show that the diffusion Y in Theorem 3.11 can reach distant parts of the space F in an arbitrarily short time.

Lemma 3.17. *Let Y_t be a μ-symmetric diffusion with semigroup T_t on a complete metric space (F, ρ). If $f, g \geq 0$ and there exist $a < b$ such that*

$$(3.18) \qquad \int f(x) \mathbb{E}^x g(Y_t) \mu(dx) = 0 \ \text{for } t \in (a, b),$$

then $\int f(x) \mathbb{E}^x g(Y_t) \mu(dx) = 0$ for all $t > 0$.

Proof. Let $(E_\lambda, \lambda \geq 0)$ be the spectral family associated with T_t. Thus (see [FOT, p. 17]) $T_t = \int_0^\infty e^{-\lambda t} dE_\lambda$, and

$$(f, T_t g) = \int_0^\infty e^{-\lambda t} d(f, E_\lambda g) = \int_0^\infty e^{-\lambda t} \nu(d\lambda),$$

where ν is of finite variation. (3.18) and the uniqueness of the Laplace transform imply that $\nu = 0$, and so $(f, T_t g) = 0$ for all t. $\qquad \square$

Lemma 3.18. *Let F and Y satisfy the hypotheses of Theorem 3.11. If $\rho(x, y) < c_3 r_0$ then $\mathbb{P}^x(Y_t \in B(y, r)) > 0$ for all $r > 0$ and $t > 0$.*

Remark. The restriction $\rho(x, y) < c_3 r_0$ is of course unnecessary, but it is all we need now. The conclusion of Theorem 3.11 implies that $\mathbb{P}^x(Y_t \in B(y, r)) > 0$ for all $r > 0$ and $t > 0$, for all $x, y \in F$.

Proof. Suppose the conclusion of the Lemma fails for x, y, r, t. Choose $g \in C(F, \mathbb{R}_+)$ such that $\int_F g d\mu = 1$ and $g = 0$ outside $B(y, r)$. Let $t_1 = t/2$, $r_1 = c_3(t_1)^\beta$, and choose $f \in C(F, \mathbb{R}_+)$ so that $\int_F f d\mu = 1$, $f(x) > 0$ and $f = 0$ outside $A = B(x, r_1)$. If $0 < s < t$ then the construction of g implies that

$$0 = \mathbb{E}^x g(Y_t) = \int_F q(s, x, x') \mathbb{E}^{x'} g(Y_{t-s}) \mu(dx').$$

Since by (3.13) $q(s, x, x') > 0$ for $t/2 < s < t$, $x' \in B(x, r_1)$, we deduce that $\mathbb{E}^{x'} g(Y_u) = 0$ for $x' \in B(x, r_1)$, $u \in (0, t/2)$. Thus as $\mathrm{supp}(f) \subset B(x, r_1)$

$$\int_F f(x') \mathbb{E}^{x'} g(Y_u) d\mu = 0$$

for all $u \in (1, t/2)$, and hence, by Lemma 3.17, for all $u > 0$. But by (3.13) if $u = (\rho(x, y)/c_3)^\beta$ then $q(u, x, y) > 0$, and by the continuity of f, g and q it follows that $\int f \mathbb{E}^x g(Y_u) d\mu > 0$, a contradiction. $\qquad \square$

Proof of Theorem 3.11. For simplicity we give full details of the proof only in the case $r_0 = \infty$; the argument in the case of bounded F is essentially the same. We begin by obtaining a bound on

$$\mathbb{P}^x(\tau(x, r) \leq t).$$

Let $n \geq 1$, $b = r/n$, and define stopping times S_i, $i \geq 0$, by

$$S_0 = 0, \quad S_{i+1} = \inf\{t \geq S_i : \rho(Y_{S_i}, Y_t) \geq b\}.$$

Let $\xi_i = S_i - S_{i-1}$, $i \geq 1$. Let $(\mathcal{F}_t)$ be the filtration of Y_t, and let $\mathcal{G}_i = \mathcal{F}_{S_i}$. We have by Lemma 3.16

$$\mathbb{P}^x(\xi_{i+1} \leq t | \mathcal{G}_i) = \mathbb{P}^{Y_{S_i}}\left(\tau(Y_{S_i}, b) \leq t\right) \leq p + c_6 b^{-\beta} t,$$

where $p \in (0,1)$. As $\rho(Y_{S_i}, Y_{S_{i+1}}) = b$, we have $\rho(Y_0, Y_{S_n}) \leq r$, so that $S_n = \sum_1^n \xi_i \leq \tau(Y_0, r)$. So, by Lemma 3.14, with $a = c_6(r/n)^{-\beta}$,

$$\log \mathbb{P}^x\left(\tau(x, r) \leq t\right) \leq 2p^{-\frac{1}{2}}\left(c_6 r^{-\beta} n^{1+\beta} t\right)^{\frac{1}{2}} - n \log \frac{1}{p}$$

(3.19)

$$= c_7(r^{-\beta} n^{1+\beta} t)^{\frac{1}{2}} - c_8 n.$$

If $\beta \leq 1$ then taking t small enough the right hand side of (3.17) is negative, and letting $n \to \infty$ we deduce $\mathbb{P}^x\left(\tau(x,r) \leq t\right) = 0$, which contradicts the fact that $\mathbb{P}^x\left(Y_t \in B(y,r)\right) > 0$ for all t. So we have $\beta > 1$. (If $r_0 = 1$ then we take r small enough so that $r < c_3$).

If we neglect for the moment the fact that $n \in \mathbb{N}$, and take $n = n_0$ in (3.19) so that

$$\tfrac{1}{2} c_8 n_0 = c_7 \left(n_0^{1+\beta} t r^{-\beta}\right)^{1/2},$$

then

(3.20)

$$n_0^{\beta-1} = (c_8^2/4c_7^2) r^\beta t^{-1},$$

and

$$\log \mathbb{P}^x\left(\tau(x,r) \leq t\right) \leq -\tfrac{1}{2} c_8 n_0.$$

So if $r^\beta t^{-1} \geq 1$, we can choose $n \in \mathbb{N}$ so that $1 \leq n \leq n_0 \vee 1$, and we obtain

(3.21)

$$\mathbb{P}^x\left(\tau(x,r) \leq t\right) \leq c_9 \exp\left(-c_{10}\left(\frac{r^\beta}{t}\right)^{1/(\beta-1)}\right).$$

Adjusting the constant c_9 if necessary, this bound also clearly holds if $r^\beta t^{-1} < 1$.

Now let $x, y \in F$, write $r = \rho(x, y)$, choose $\varepsilon < r/4$, and set $C_z = B(z, \varepsilon)$, $z = x, y$. Set $A_x = \{z \in F : \rho(z, x) \leq \rho(z, y)\}$, $A_y = \{z : \rho(z, x) \geq \rho(z, y)\}$. Let ν_x, ν_y be the restriction of μ to C_x, C_y respectively.

We now derive the upper bound on $q(t, x, y)$ by combining the bounds (3.12) and (3.21): the idea is to split the journey of Y from C_x to C_y into two pieces, and use one of the bounds on each piece. We have

$$(3.22) \quad \mathbb{P}^{\nu_x}(Y_t \in C_y) = \int_{C_y}\int_{C_x} q(t, x', y')\mu(dx')\mu(dy')$$

$$\leq \mathbb{P}^{\nu_x}\left(Y_t \in C_y, Y_{t/2} \in A_x\right) + \mathbb{P}^{\nu_x}(Y_t \in C_y, Y_{t/2} \in A_y).$$

We begin with second term in (3.22):

$$(3.23) \quad \mathbb{P}^{\nu_x}(Y_t \in C_y, Y_{t/2} \in A_y) = \mathbb{P}^{\nu_x}\left(\tau(Y_0, r/4) \leq t/2, Y_{t/2} \in A_y, Y_t \in C_y\right)$$

$$\leq \mathbb{P}^{\nu_x}\left(\tau(Y_0, r/4) \leq t/2\right) \sup_{y' \in A_y} \mathbb{P}^{y'}\left(Y_{t/2} \in C_y\right)$$

$$\leq \nu_x(C_x)c_9 \exp\left(-c_{10}\left(\frac{(r/4)^\beta}{t/2}\right)^{1/(\beta-1)}\right) c_1\nu_y(C_y)t^{-d_f/\beta}$$

$$= \mu(C_x)\mu(C_y)c_{11}t^{-d_f/\beta} \exp\left(-c_{12}(r^\beta/t)^{1/(\beta-1)}\right),$$

where we used (3.21) and (3.12) in the last but one line.

To handle the first term in (3.22) we use symmetry:

$$\mathbb{P}^{\nu_x}(Y_t \in C_y, Y_{t/2} \in A_x) = \mathbb{P}^{\nu_y}(Y_t \in C_x, Y_{t/2} \in A_x),$$

and this can now be bounded in exactly the same way. We therefore have

$$\int\limits_{C_y}\int\limits_{C_x} q(t,x',y')\mu(dx')\mu(dy')$$

$$\leq \mu(C_x)\mu(C_y)2c_{11}t^{-d_f/\beta} \exp\left(-c_{12}(r^\beta/t)^{1/(\beta-1)}\right),$$

so that as $q(t,\cdot,\cdot)$ is continuous

(3.24) $$q(t,x,y) \leq 2c_{11}t^{-d_f/\beta} \exp\left(-c_{12}(r^\beta/t)^{1/(\beta-1)}\right).$$

The proof of the lower bound on q uses the technique of "chaining" the Chapman-Kolmogorov equations. This is quite classical, except for the different scaling.

Fix x, y, t, and write $r = \rho(x,y)$. If $r \leq c_3 t^{1/\beta}$ then by (3.13)

$$q(t,x,y) \geq c_2 t^{-d_f/\beta},$$

and as $\exp(-(r^\beta/t)^{1/(\beta-1)}) \geq \exp(-c_3^{1/(\beta-1)})$, we have a lower bound of the form (3.3). So now let $r > c_3 t^{1/\beta}$. Let $n \geq 1$. By the mid-point hypothesis on the metric ρ, we can find a chain $x = x_0, x_1, \ldots, x_n = y$ in F such that $\rho(x_{i-1}, x_i) = r/n$, $1 \leq i \leq n$. Let $B_i = B(x_i, r/2n)$; note that if $y_i \in B_i$ then $\rho(y_{i-1}, y_i) \leq 2r/n$. We have by the Chapman-Kolmogorov equation, writing $y_0 = x_0$, $y_n = y$,

(3.25) $$q(t,x,y) \geq \int\limits_{B_1} \mu(dy_1) \ldots \int\limits_{B_{n-1}} \mu(dy_{n-1}) \prod_{i=1}^{n} q(t/n, y_{i-1}, y_i).$$

We wish to choose n so that we can use the bound (3.13) to estimate the terms $q(t/n, y_{i-1}, y_i)$ from below. We therefore need:

(3.26) $$\frac{2r}{n} \leq c_3\left(\frac{t}{n}\right)^{1/\beta}$$

which holds provided

(3.27) $$n^{\beta-1} \geq 2^\beta c_3^{-\beta}\frac{r^\beta}{t}.$$

As $\beta > 1$ it is certainly possible to choose n satisfying (3.27). By (3.25) we then obtain, since $\mu(B_i) \geq c(r/2n)^{d_f}$,

$$(3.28) \qquad q(t,x,y) \geq c(r/2n)^{d_f(n-1)} \left(c_2(t/n)^{-d_f/\beta} \right)^n$$

$$= c(r/2n)^{-d_f} \left(c_2(t/n)^{-1/\beta}(r/2n)^{d_f} \right)^n$$

$$= c'(r/n)^{-d_f} \left((t/n)^{-1/\beta}(r/n) \right)^n.$$

Recall that n satisfies (3.27): as $r > c_3 t^{1/\beta}$ we can also ensure that for some $c_{13} > 0$

$$(3.29) \qquad \frac{r}{n} \geq c_{13}(t/n)^{1/\beta},$$

so that $n^{\beta-1} \leq 2^\beta c_{13}^{-\beta} r^\beta/t$. So, by (3.28)

$$q(t,x,y) \geq c(t/n)^{-d_f/\beta} c_{14}^n$$

$$\geq c_{15} t^{-d_f/\beta} \exp\left(n \log c_{14} \right)$$

$$\geq c_{15} t^{-d_f/\beta} \exp\left(-c_{16}(r^\beta/t)^{1/(\beta-1)} \right). \qquad \square$$

Remarks 3.19.

1. Note that the only point at which we used the "midpoint" property of ρ is in the derivation of the lower bound for q.

2. The essential idea of the proof of Theorem 3.11 is that we can obtain bounds on the long range behaviour of Y provided we have good enough information about the behaviour of Y over distances of order $t^{1/\beta}$. Note that in each case, if $r = \rho(x,y)$, the estimate of $q(t,x,y)$ involves splitting the journey from x to y into n steps, where $n \asymp (r^\beta/t)^{1/(\beta-1)}$.

3. Both the arguments for the upper and lower bounds appear quite crude: the fact that they yield the same bounds (except for constants) indicates that less is thrown away than might appear at first sight. The explanation, very loosely, is given by "large deviations". The off-diagonal bounds are relevant only when $r^\beta \gg t$ – otherwise the term in the exponential is of order 1. If $r^\beta \gg t$ then it is difficult for Y to move from x to y by time t and it is likely to do so along more or less the shortest path. The proof of the lower bound suggests that the process moves in a 'sausage' of radius $r/n \asymp t/r^{\beta-1}$.

The following two theorems give additional bounds and restrictions on the parameters d_f and β. Unlike the proofs above the results use the symmetry of the process very strongly. The proofs should appear in a forthcoming paper.

Theorem 3.20. *Let F be a $FMS(d_f)$, and X be a $FD(d_f,\beta)$ on F. Then*

$$(3.30) \qquad 2 \leq \beta \leq 1 + d_f.$$

Theorem 3.21. *Let F be a $FMS(d_f)$. Suppose X^i are $FD(d_f,\beta_i)$ on F, for $i = 1,2$. Then $\beta_1 = \beta_2$.*

Remarks 3.22. 1. Theorem 3.21 implies that the constant β is a property of the metric space F, and not just of the FD X. In particular any FD on $\mathbb{R}^d$, with the usual metric and Lebesgue measure, will have $\beta = 2$. It is very unlikely that every FMS F carries a FD.

2. I expect that (3.30) is the only general relation between β and d_f. More precisely, set

$$A = \{(d_f, \beta) : \text{ there exists a } FD(d_f, \beta)\},$$

and $\Gamma = \{(d_f, \beta) : 2 \le \beta \le 1 + d_f\}$. Theorem 3.20 implies that $A \subset \Gamma$, and I conjecture that int $\Gamma \subset A$. Since $BM(\mathbb{R}^d)$ is a $FD(d, 2)$, the points $(d, 2) \in A$ for $d \ge 1$. I also suspect that

$$\{d_f : (d_f, 2) \in A\} = \mathbb{N},$$

that is that if F is an FMS of dimension d_f, and d_f is not an integer, then any FD on F will not have Brownian scaling.

Properties of Fractional Diffusions.

In the remainder of this section I will give some basic analytic and probabilistic properties of FDs. I will not give detailed proofs, since for the most part these are essentially the same as for standard Brownian motion. In some cases a more detailed argument is given in [BP] for the Sierpinski gasket.

Let F be a $FMS(d_f)$, and X be a $FD(d_f, \beta)$ on F. Write $T_t = \mathbb{E}^x f(X_t)$ for the semigroup of X, and $\mathcal{L}$ for the infinitesimal generator of T_t.

Definition 3.23. Set

$$d_w = \beta, \qquad d_s = \frac{2d_f}{d_w}.$$

This notation follows the physics literature where (for reasons we will see below) d_w is called the "walk dimension" and d_s the "spectral dimension". Note that (3.3) implies that

$$p(t, x, x) \asymp t^{-d_s/2}, \qquad 0 < t \le t_0,$$

so that the on-diagonal bounds on p can be expressed purely in terms of d_s. Since many important properties of a process relate solely to the on-diagonal behaviour of its density, d_s is the most significant single parameter of a FD.

Integrating (3.3), as in Corollary 2.25, we obtain:

Lemma 3.24. $\mathbb{E}^x \rho(X_t, x)^p \asymp t^{p/d_w}$, $x \in F$, $t \ge 0$, $p > 0$.

Since by Theorem 3.20 $d_w \ge 2$ this shows that FDs are diffusive or subdiffusive.

Lemma 3.25. *(Modulus of continuity).* Let $\varphi(t) = t^{1/d_w} (\log(1/t))^{(d_w-1)/d_w}$. Then

$$(3.31) \qquad c_1 \le \lim_{\delta \downarrow 0} \sup_{\substack{0 \le s < t \le 1 \\ |t-s| < \delta}} \frac{\rho(X_s, X_t)}{\varphi(t-s)} \le c_2.$$

So, in the metric ρ, the paths of X just fail to be Hölder $(1/d_w)$. The example of divergence form diffusions in $\mathbb{R}^d$ shows that one cannot hope to have $c_1 = c_2$ in general.

Lemma 3.26. (*Law of the iterated logarithm – see [BP, Thm. 4.7]*). *Let* $\psi(t) = t^{1/d_w} (\log\log(1/t))^{(d_w-1)/d_w}$. *There exist* c_1, c_2 *and constants* $c(x) \in [c_1, c_2]$ *such that*

$$\limsup_{t\downarrow 0} \frac{\rho(X_t, X_0)}{\psi(t)} = c(x) \quad \mathbb{P}^x\text{-a.s.}$$

Of course, the 01 law implies that the limit above is non-random.

Lemma 3.27. (*Dimension of range*).

(3.32) $$\dim_H (\{X_t : 0 \le t \le 1\}) = d_f \wedge d_w.$$

This result helps to explain the terminology "walk dimension" for d_w. Provided the space the diffusion X moves in is large enough, the dimension of range of the process (called the "dimension of the walk" by physicists) is d_w.

Potential Theory of Fractional Diffusions.

Let $\lambda \ge 0$ and set

$$u_\lambda(x, y) = \int_0^\infty e^{-\lambda s} p(s, x, y)\, ds.$$

Then if

$$U_\lambda f(x) = \mathbb{E}^x \int_0^\infty e^{-\lambda s} f(X_s)\, d_s$$

is the λ-resolvent of X, u_λ is the density of U_λ:

$$U_\lambda f(x) = \int_F u_\lambda(x, y)\mu(dy).$$

Write u for u_0.

Proposition 3.28. *Let* $\lambda_0 = 1/r_0$. (*If* $r_0 = \infty$ *take* $\lambda_0 = 0$).
(a) *If* $d_s < 2$ *then* $u_\lambda(x, y)$ *is jointly continuous on* $F \times F$ *and for* $\lambda > \lambda_0$

(3.33) $$c_1 \lambda^{d_s/2-1} \exp\left(-c_2 \lambda^{1/d_w} \rho(x, y)\right) \le u_\lambda(x, y)$$

$$\le c_3 \lambda^{d_s/2-1} \exp\left(-c_4 \lambda^{1/d_w} \rho(x, y)\right).$$

(b) *If* $d_s = 2$ *and* $\lambda > \lambda_0$ *then writing* $R = \rho(x, y)\lambda^{1/d_w}$

(3.34) $$c_5 \left(\log_+(1/R) + e^{-c_6 R}\right) \le u_\lambda(x, y) \le c_7 \left(\log_+(1/R) + e^{-c_8 R}\right).$$

(c) *If* $d_s > 2$ *then*

(3.35) $$c_9 \rho(x, y)^{d_w-d_f} \le u_{\lambda_0}(x, y) \le c_{10}\rho(x, y)^{d_w-d_f}.$$

These bounds are obtained by integrating (3.3): for (a) and (b) one uses Laplace's method. (The continuity in (b) follows from the continuity of p and the uniform bounds on p in (3.3)). Note in particular that:

(i) if $d_s < 2$ then $u_\lambda(x,x) < +\infty$ and $\lim_{\lambda \to 0} u_\lambda(x,y) = +\infty$.

(ii) if $d_s > 2$ then $u(x,x) = +\infty$, while $u(x,y) < \infty$ for $x \neq y$

Since the polarity or non-polarity of points relates to the on-diagonal behaviour of u, we deduce from Proposition 3.28

Corollary 3.29. (a) If $d_s < 2$ then for each $x, y \in F$

$$\mathbb{P}^x(X \text{ hits } y) = 1.$$

(b) If $d_s \geq 2$ then points are polar for X.

(c) If $d_s \leq 2$ then X is set-recurrent: for $\varepsilon > 0$

$$\mathbb{P}^y(\{t : X_t \in B(y,\varepsilon)\} \text{ is non-empty and unbounded}) = 1.$$

(d) If $d_s > 2$ and $r_0 = \infty$ then X is transient.

In short, X behaves like a Brownian motion of dimension d_s; but in this context a continuous parameter range is possible.

Lemma 3.30. (Polar and non-polar sets). Let A be a Borel set in F.

(a) $\mathbb{P}^x(T_A < \infty) > 0$ if $\dim_H(A) > d_f - d_w$,

(b) A is polar for X if $\dim_H(A) < d_f - d_w$.

Since X is symmetric any semipolar set is polar. As in the Brownian case, a more precise condition in terms of capacity is true, and is needed to resolve the critical case $\dim_H(A) = d_f - d_w$.

If X, X' are independent $FD(d_f, \beta)$ on F, and $Z_t = (X_t, X'_t)$, then it follows easily from the definition that Z is a FD on $F \times F$, with parameters $2d_f$ and β. If $D = \{(x,x) : x \in F\} \subset F \times F$ is the diagonal in $F \times F$, then $\dim_H(D) = d_f$, and so Z hits D (with positive probability) if

$$d_f > 2d_f - d_w,$$

that is if $d_s < 2$. So

(3.36) $$\mathbb{P}^x(X_t = X'_t \text{ for some } t > 0) > 0 \quad \text{if } d_s < 2,$$

and

(3.37) $$\mathbb{P}^x(X_t = X'_t \text{ for some } t > 0) = 0 \quad \text{if } d_s > 2.$$

No doubt, as in the Brownian case, X and X' do not collide if $d_s = 2$.

Lemma 3.31. X has k-multiple points if and only if $d_s < 2k/(k-1)$.

Proof. By [Rog] X has k-multiple points if and only if

$$\int_{B(x,1)} u_1(x,y)^k \mu(dy) < \infty;$$

the integral above converges or diverges with

$$\int_0^1 r^{kd_w - (k-1)d_f} r^{-1} \, dr,$$

by a calculation similar to that in Corollary 2.25. □

The bounds on the potential kernel density $u_\lambda(x,y)$ lead immediately to the existence of local times for X – see [Sha, p. 325].

Theorem 3.32. *If $d_s < 2$ then X has jointly measurable local times $(L_t^x, x \in F, t \geq 0)$ which satisfy the density of occupation formula with respect to μ:*

$$(3.38) \qquad \int_0^t f(X_s)ds = \int_F f(a)L_t^a \mu(da), \quad f \text{ bounded and measurable.}$$

In the low-dimensional case (that is when $d_s < 2$, or equivalently $d_f < d_w$) we can obtain more precise estimates on the Hölder continuity of $u_\lambda(x,y)$, and hence on the local times L_t^x. The main lines of the argument follow that of [BB4, Section 4], but on the whole the arguments here are easier, as we begin with stronger hypotheses. We work only in the case $r_0 = \infty$: the same results hold in the case $r_0 = 1$, with essentially the same proofs.

For the next few results we fix F, a $FMS(d_f)$ with $r_0 = \infty$, and X, a $FD(d_f, d_w)$ on F. For $A \subset F$ write

$$\tau_A = T_{A^c} = \inf\{t \geq 0 : X_t \in A^c\}.$$

Let R_λ be an independent exponential time with mean λ^{-1}. Set for $\lambda \geq 0$

$$u_\lambda^A(x,y) = \mathbb{E}^x \int_0^{\tau_A} e^{-\lambda s} dL_s^y = \mathbb{E}^x L_{\tau_A \wedge R_\lambda}^y,$$

$$U_\lambda^A f(x) = \int_F u_\lambda^A(x,y)\mu(dy).$$

Let

$$p_\lambda^A(x,y) = \mathbb{P}^x(T_y \leq \tau_A \wedge R_\lambda);$$

note that

$$(3.39) \qquad u_\lambda^A(x,y) = p_\lambda^A(x,y)u_\lambda^A(y,y) \leq u_\lambda^A(y,y).$$

Write $u^A(x,y) = u_0^A(x,y)$, $U^A = U_0^A$, and note that $u_\lambda(x,y) = u_\lambda^F(x,y)$, $U_\lambda = U_\lambda^F$. As in the case of u we write p^A, p_λ for p_0^A, p_λ^F. As $(\mathbb{P}^x, X_t)$ is μ-symmetric we have $u_\lambda^A(x,y) = u_\lambda^A(y,x)$ for all x, $y \in F$.

The following Lemma enables us to pass between bounds on u_λ and u^A.

Lemma 3.33. *Suppose $A \subset F$, A is bounded, For x, $y \in F$ we have*

$$u^A(x,y) = u_\lambda^B(x,y) + \mathbb{E}^x \left(1_{(R_\lambda \leq \tau_A)} u^A(X_{R_\lambda}, y)\right) - \mathbb{E}^x \left(1_{(R_\lambda > \tau_A)} u_\lambda^B(X_{\tau_A}, y)\right).$$

Proof. From the definition of u^A,

$$
\begin{aligned}
u^A(x,y) &= \mathbb{E}^x\left(L^y_{\tau_A}\;;R_\lambda \le \tau_A\right) + \mathbb{E}^x\left(L^y_{\tau_A}\;;R_\lambda > \tau_A\right)\\
&= \mathbb{E}^x\left(L^y_{R_\lambda}\;;R_\lambda \le \tau_A\right) + \mathbb{E}^x\left(1_{(R_\lambda \le \tau_A)}\mathbb{E}^{X_{R_\lambda}}L^y_{\tau_A}\right)\\
&\quad + \mathbb{E}^x\left(L^y_{R_\lambda}\;;R_\lambda > \tau_A\right) - \mathbb{E}^x\left(L^y_{R_\lambda \wedge \tau_B} - L^y_{\tau_A}\;;R_\lambda > \tau_A\right)\\
&= u_\lambda(x,y) + \mathbb{E}^x\left(1_{(R_\lambda \le \tau_A)}u^A(X_{R_\lambda},y)\right) - \mathbb{E}^x\left(1_{(R_\lambda > \tau_A)}u_\lambda(X_{\tau_A},y)\right). \qquad \square
\end{aligned}
$$

Corollary 3.34. *Let $x \in F$, and $r > 0$. Then*

$$
c_1 r^{d_w - d_f} \le u^{B(x,r)}(x,x) \le c_2 r^{d_w - d_f}.
$$

Proof. Write $A = B(x,r)$, and let $\lambda = \theta r^{-d_w}$, where θ is to be chosen. We have from Lemma 3.33, writing $\tau = \tau(x,r)$,

$$
u^A(x,y) \le u_\lambda(x,y) + \mathbb{E}^x 1_{(R_\lambda < \tau)}u^A(X_{R_\lambda},y).
$$

So if $v = \sup_x u^A(x,y)$ then using (3.33)

$$
(3.40) \qquad v \le c_3 \lambda^{d_s/2-1} + \mathbb{P}^x(R_\lambda < \tau)v.
$$

Let $t_0 > 0$. Then by (3.10)

$$
\begin{aligned}
\mathbb{P}^x(R_\lambda < \tau) &= \mathbb{P}^x(R_\lambda < \tau, \tau \le t_0) + \mathbb{P}^x(R_\lambda < \tau, \tau > t_0)\\
&\le \mathbb{P}^x(R_\lambda < t_0) + \mathbb{P}^x(\tau > t_0)\\
&\le (1 - e^{-\lambda t_0}) + c t_0^{-d_f/d_w}r^{d_f}.
\end{aligned}
$$

Choose first t_0 so that the second term is less than $\frac{1}{4}$, and then λ so that the first term is also less than $\frac{1}{4}$. We have $t_0 \asymp r^{d_w} \asymp \lambda^{-1}$, and the upper bound now follows from (3.40).

The lower bound is proved in the same way, using the bounds on the lower tail of τ given in (3.11). $\qquad \square$

Lemma 3.35. *There exist constants $c_1 > 1$, c_2 such that if $x,y \in F$, $r = \rho(x,y)$, $t_0 = r^{d_w}$ then*

$$
\mathbb{P}^x\left(T_y < t_0 < \tau(x,c_1 r)\right) \ge c_2.
$$

Proof. Set $\lambda = (\theta/r)^{d_w}$; we have $p_\lambda(x,y) \ge c_3 \exp(-c_4 \theta)$ by (3.33). So since

$$
p_\lambda(x,y) = \mathbb{E}^x e^{-\lambda T_v} \le \mathbb{P}^x(T_y < t) + e^{-\lambda t},
$$

we deduce that

$$
\mathbb{P}^x(T_y < t) \ge c_3 \exp(-c_4 \theta) - \exp(-\theta^{d_w}).
$$

As $d_w > 1$ we can choose θ (depending only on c_3, c_4 and d_w) such that $\mathbb{P}^x(T_y < t) \ge \frac{1}{2}c_3 \exp(-c_4 \theta) = c_5$. By (3.11) for $a > 0$

$$
\mathbb{P}^x(\tau(x,aR) < R^{d_w}) \le c_6 \exp(-c_7 a^{d_w/(d_w-1)}),
$$

so there exists $c_1 > 1$ such that $\mathbb{P}^x(\tau(x,c_1 r) < t_0) \le \frac{1}{2}c_5$. So

$$
\mathbb{P}^x\left(T_y < t_0 < \tau(x,c_1 r)\right) \ge \mathbb{P}^x(T_y < t_0) - \mathbb{P}^x(\tau(x,c_1 r) < t_0) \ge \frac{1}{2}c_5. \qquad \square
$$

Definition 3.36. We call a function h *harmonic* (with respect to X) in an open subset $A \subset F$ if $\mathcal{L}h = 0$ on A, or equivalently, $h(X_{t \wedge T_{A^c}})$ is a local martingale.

Proposition 3.37. (*Harnack inequality*). *There exist constants $c_1 > 1$, $c_2 > 0$, such that if $x_0 \in F$, and $h \geq 0$ is harmonic in $B(x_0, c_1 r)$, then*

$$h(x) \geq c_2 h(y), \quad x, y \in B(x_0, r).$$

Proof. Let $c_1 = 1 + c_{3.35.1}$, so that $B(x, c_{3.35.1} r) \subset B(x_0, c_1 r)$ if $\rho(x, x_0) \leq r$. Fix x, y, write $r = \rho(x, y)$, and set $S = T_y \wedge \tau(x, c_{3.35.1} r)$. As $h(X_{\cdot \wedge S})$ is a supermartingale, we have by Lemma 3.35,

$$h(x) \geq \mathbb{E}^x h(X_S) \geq h(y) \mathbb{P}^x(T_y < \tau(x, c_{3.35.1} r)) \geq c_{3.35.2} h(y). \qquad \square$$

Corollary 3.38. *There exists $c_1 > 0$ such that if $x_0 \in F$, and $h \geq 0$ is harmonic in $B(x_0, r)$, then*

$$h(x) \geq c_1 h(y), \quad x, y \in B(x_0, \tfrac{3}{4} r).$$

Proof. This follows by covering $B(x_0, \tfrac{3}{4} r)$ by balls of the form $B(y, c_2 r)$, where c_2 is small enough so that Proposition 3.37 can be applied in each ball. (Note we use the geodesic property of the metric ρ here, since we need to connect each ball to a fixed reference point by a chain of overlapping balls). $\qquad \square$

Lemma 3.39. *Let $x, y \in F$, $r = \rho(x, y)$. If $R > r$ and $B(y, R) \subset A$ then*

$$u^A(y, y) - u^A(x, y) \leq c_1 r^{d_w - d_f}.$$

Proof. We have, writing $\tau = \tau(y, r)$, $T = T_{A^c}$,

$$u^A(y, y) = \mathbb{E}^y L_\tau^y + \mathbb{E}^y \mathbb{E}^{X_\tau} L_T^y = u^B(y, y) + \mathbb{E}^y u^A(X_\tau, y),$$

so by Corollary 3.34

$$(3.41) \qquad \mathbb{E}^y \left(u^A(y, y) - u^A(X_\tau, y) \right) = u^B(y, y) \leq c_2 r^{d_w - d_f}.$$

Set $\varphi(x') = u^A(y, y) - u^A(x', y)$; φ is harmonic on $A - \{y\}$. As $\rho(x, y) = r$ and ρ has the geodesic property there exists z with $\rho(y, z) = \tfrac{1}{4} r$, $\rho(x, z) = \tfrac{3}{4} r$. By Corollary 3.38, since φ is harmonic in $B(x, r)$,

$$\varphi(z) \geq c_{3.38.1} \varphi(x).$$

Now set $\psi(x') = \mathbb{E}^{x'} \varphi(X_\tau)$ for $x' \in B$. Then ψ is harmonic in B and $\varphi \leq \psi$ on B. Applying Corollary 3.38 to ψ in B we deduce

$$\psi(y) \geq c_{3.38.1} \psi(z) \geq c_{3.38.1} \varphi(z) \geq (c_{3.38.1})^2 \varphi(x).$$

Since $\psi(y) = \mathbb{E}^y(u^A(y, y) - u^A(X_\tau, y))$ the conclusion follows from (3.41). $\qquad \square$

Theorem 3.40. (a) *Let $\lambda > 0$. Then for x, x', $y \in F$, and $f \in L^1(F)$, $g \in L^\infty(F)$,*

$$(3.42) \qquad |u_\lambda(x, y) - u_\lambda(x', y)| \leq c_1 \rho(x, x')^{d_w - d_f},$$

$$(3.43) \qquad |U_\lambda f(x) - U_\lambda f(x')| \leq c_1 \rho(x, x')^{d_w - d_f} \|f\|_1.$$

$$(3.44) \qquad |U_\lambda g(x) - U_\lambda g(x')| \leq c_2 \lambda^{-d_s/2} \rho(x, x')^{d_w - d_f} \|g\|_\infty.$$

Proof. Let $x, x' \in F$, write $r = \rho(x, x')$ and let $R > r$, $A = B(x, R)$. Since $u_\lambda^A(y, x') \geq p_\lambda^A(y, x) u_\lambda^A(x, x')$, we have using the symmetry of X that

(3.45)
$$u_\lambda^A(x, y) - u_\lambda^A(x', y) \leq u_\lambda^A(y, x) - p_\lambda^A(y, x) u_\lambda^A(x, x')$$
$$= p_\lambda^A(y, x)\left(u_\lambda^A(x, x) - u_\lambda^A(x, x')\right).$$

Thus
$$|u_\lambda^A(x, y) - u_\lambda^A(x', y)| \leq |u_\lambda^A(x, x) - u_\lambda^A(x, x')|.$$

Setting $\lambda = 0$ and using Lemma 3.39 we deduce

(3.46)
$$|u^A(x, y) - u^A(x', y)| \leq c_3 r^{d_w - d_f}.$$

So
$$|U^A f(x) - U^A f(x')| \leq \int_A |u^A(x, y) - u^A(x', y)| \, |f(y)| \mu(dy)$$
$$\leq c_3 r^{d_w - d_f} \|f 1_A\|_1.$$

To obtain estimates for $\lambda > 0$ we apply the resolvent equation in the form
$$u_\lambda^A(x, y) = u^A(x, y) - \lambda U^A v(x),$$

where $v(x) = u_\lambda^A(x, y)$. (Note that $\|v\|_1 = \lambda^{-1}$). Thus
$$|u_\lambda^A(x, y) - u_\lambda^A(x', y)| \leq |u^A(x, y) - u^A(x', y)| + \lambda |U^A v(x) - U^A v(x')|$$
$$\leq c_3 r^{d_w - d_f} + \lambda c_1 r^{d_w - d_f} \|v\|_1$$
$$= 2 c_3 r^{d_w - d_f}.$$

Letting $R \to \infty$ we deduce (3.42), and (3.43) then follows, exactly as above, by integration.

To prove (3.46) note first that $p_\lambda(y, x) = u_\lambda(y, x)/u_\lambda(x, x)$. So by (3.33)

(3.47)
$$\int_A p_\lambda^A(y, x) |f(y)| \mu(dy) \leq \|f\|_\infty u_\lambda(x, x)^{-1} \int_A u_\lambda(y, x) \mu(dy)$$
$$= \|f\|_\infty u_\lambda(x, x)^{-1} \lambda^{-1}$$
$$\leq c_4 \|f\|_\infty \lambda^{-d_s/2}.$$

From (3.45) and (3.46) we have
$$|u_\lambda^A(x, y) - u_\lambda^A(x', y)| \leq c_2 \left(p_\lambda^A(y, x) + p_\lambda^A(y, x')\right) r^{d_w - d_f},$$

and (3.44) then follows by intergation, using (3.47). $\qquad\square$

The following modulus of continuity for the local times of X then follows from the results in [MR].

Theorem 3.41. *If $d_s < 2$ then X has jointly continuous local times $(L_t^x, x \in F, t \geq 0)$. Let $\varphi(u) = u^{(d_w - d_f)/2}(\log(1/u))^{1/2}$. The modulus of continuity in space of L is given by:*

$$\lim_{\delta \downarrow 0} \sup_{0 \leq s \leq t} \sup_{\substack{0 \leq s \leq t \\ |x-y| < \delta}} \frac{|L_s^x - L_s^y|}{\varphi(\rho(x,y))} \leq c(\sup_{x \in F} L_t^x)^{1/2}.$$

It follows that X is space-filling: for each $x, y \in F$ there exists a r.v. T such that $\mathbb{P}^x(T < \infty) = 1$ and

$$B(y,1) \subset \{X_t, 0 \leq t \leq T\}.$$

The following Proposition helps to explain why in early work mathematical physicists found that for simple examples of fractal sets one has $d_s < 2$. (See also [HHW]).

Proposition 3.42. *Let F be a FMS, and suppose F is finitely ramified. Then if X is a $FD(d_f, d_w)$ on F, $d_s(X) < 2$.*

Proof. Let F_1, F_2 be two connected components of F, such that $D = F_1 \cap F_2$ is finite. If $D = \{y_1, \ldots, y_n\}$, fix $\lambda > 0$ and set

$$M_t = e^{-\lambda t} \sum_{i=1}^n u_\lambda(X_t, y_i).$$

Then M is a supermartingale. Let $T_D = \inf\{t \geq 0 : X_t \in D\}$, and let $x_0 \in F_1 - D$. Since $\mathbb{P}^{x_0}(X_1 \in F_2) > 0$, we have $\mathbb{P}^{x_0}(T_D \leq 1) > 0$. So

$$\infty > \mathbb{E}^{x_0} M_0 \geq \mathbb{E}^{x_0} M_{T_D},$$

and thus $M_{T_D} < \infty$ a.s. So $u_\lambda(X_{T_D}, y_i) < \infty$ for each $y_i \in D$, and thus we must have $u_\lambda(y_i, y_i) < \infty$ for some $y_i \in D$. So, by Proposition 3.25, $d_s < 2$. $\square$

Remark 3.43. For $k = 1, 2$ let (F_k, d_k, μ_k) be FMS with dimension $d_f(k)$, and common diameter r_0. Let $F = F_1 \times F_2$, let $p \geq 1$ and set $d((x_1, x_2), (y_1, y_2)) = (d_1(x_1, y_1)^p + d_2(x_2, y_2)^p)^{1/p}$, $\mu = \mu_1 \times \mu_2$. Then (F, d, μ) is a FMS with dimension $d_f = d_f(1) + d_f(2)$. Suppose that for $k = 1, 2$ X^k is a $FD(d_f(k), d_w(k))$ on F_k. Then if $X = (X^1, X^2)$ it is clear from the definition of FDs that if $d_w(1) = d_w(2) = \beta$ then X is a $FD(d_f, \beta)$ on F. However, if $d_w(1) \neq d_w(2)$ then X is not a FD on F. (Note from (3.3) that the metric ρ can, up to constants, be extracted from the transition density $p(t, x, y)$ by looking at limits as $t \downarrow 0$.) So the class of FDs is not stable under products.

This suggests that it might be desirable to consider a wider class of diffusions with densities of the form:

$$(3.48) \qquad p(t, x, y) \simeq t^{-\alpha} \exp\left(-\sum_1^n \rho_i(x, y)^{\beta_i \gamma_i} t^{-\gamma_i}\right),$$

where ρ_i are appropriate non-negative functions on $F \times F$. Such processes would have different space-time scalings in the different 'directions' in the set F given by the functions ρ_i. A recent paper of Hambly and Kumagai [HK2] suggests that diffusions on p.c.f.s.s. sets (the most general type of regular fractal which has been studied in detail) have a behaviour a little like this, though it is not likely that the transition density is precisely of the form (3.48).

Spectral properties.

Let X be a FD on a FMS F with diameter $r_0 = 1$. The bounds on the density $p(t, x, y)$ imply that $p(t, ., .)$ has an eigenvalue expansion (see [DaSi, Lemma 2.1]).

Theorem 3.44. *There exist continuous functions φ_i, and λ_i with $0 \leq \lambda_0 \leq \lambda_1 \leq \dots$ such that for each $t > 0$*

$$(3.49) \qquad p(t, x, y) = \sum_{n=0}^{\infty} e^{-\lambda_n t} \varphi_n(x) \varphi_n(y),$$

where the sum in (3.49) is uniformly convergent on $F \times F$.

Remark 3.45. The assumption that X is conservative implies that $\lambda_0 = 0$, while the fact that $p(t, x, y) > 0$ for all $t > 0$ implies that X is irreducible, so that $\lambda_1 > 0$.

A well known argument of Kac (see [Ka, Section 10], and [HS] for the necessary Tauberian theorem) can now be employed to prove that if $N(\lambda) = \#\{\lambda_i : \lambda_i \leq \lambda\}$ then there exists c_i such that

$$(3.50) \qquad c_1 \lambda^{d_s/2} \leq N(\lambda) \leq c_2 \lambda^{d_s/2} \qquad \text{for } \lambda > c_3.$$

So the number of eigenvalues of $\mathcal{L}$ grows roughly as $\lambda^{d_s/2}$. This explains the term *spectral dimension* for d_s.

4. Dirichlet Forms, Markov Processes, and Electrical Networks.

In this chapter I will give an outline of those parts of the theory of Dirichlet forms, and associated concepts, which will be needed later. For a more detailed account of these, see the book [FOT]. I begin with some general introductory remarks.

Let $X = (X_t, t \geq 0, \mathbb{P}^x, x \in F)$ be a Markov process on a metric space F. (For simplicity let us assume X is a Hunt process). Associated with X are its semigroup $(T_t, t \geq 0)$ defined by

$$(4.1) \qquad T_t f(x) = \mathbb{E}^x f(X_t),$$

and its resolvent $(U_\lambda, \lambda > 0)$, given by

$$(4.2) \qquad U_\lambda f(x) = \int_0^\infty T_t f(x) e^{-\lambda t}\, dt = \mathbb{E}^x \int_0^\infty e^{-\lambda s} f(X_s)\, ds.$$

While (4.1) and (4.2) make sense for all functions f on F such that the random variables $f(X_t)$, or $\int e^{-\lambda s} f(X_s)\, ds$, are integrable, to employ the semigroup or resolvent usefully we need to find a suitable Banach space $(B, \| \cdot \|_B)$ of functions on F such that $T_t : B \to B$, or $U_\lambda : B \to B$. The two examples of importance here are

$C_0(F)$ and $L^2(F, \mu)$, where μ is a Borel measure on F. Suppose this holds for one of these spaces; we then have that (T_t) satisfies the semigroup property

$$T_{t+s} = T_t T_s, \quad s, t \geq 0,$$

and (U_λ) satisfies the resolvent equation

$$U_\alpha - U_\beta = (\beta - \alpha) U_\alpha U_\beta, \quad \alpha, \beta > 0.$$

We say (T_t) is *strongly continuous* if $\|T_t f - f\|_B \to 0$ as $t \downarrow 0$. If T_t is strongly continuous then the infinitesimal generator $(\mathcal{L}, \mathcal{D}(L))$ of (T_t) is defined by

(4.3) $$\mathcal{L}f = \lim_{t \downarrow 0} t^{-1}(T_t f - f), \quad f \in \mathcal{D}(\mathcal{L}),$$

where $\mathcal{D}(L)$ is the set of $f \in B$ for which the limit in (4.3) exists (in the space B). The Hille-Yoshida theorem enables one to pass between descriptions of X through its generator $\mathcal{L}$, and its semigroup or resolvent.

Roughly speaking, if we take the analogy between X and a classical mechanical system, $\mathcal{L}$ corresponds to the equation of motion, and T_t or U_λ to the integrated solutions. For a mechanical system, however, there is another formulation, in terms of conservation of energy. The energy equation is often more convenient to handle than the equation of motion, since it involves one fewer differentiation.

For general Markov processes, an "energy" description is not very intuitive. However, for reversible, or symmetric processes, it provides a very useful and powerful collection of techniques. Let μ be a Radon measure on F: that is a Borel measure which is finite on every compact set. We will also assume μ charges every open set. We say that T_t is μ-*symmetric* if for every bounded and compactly supported f, g,

(4.4) $$\int T_t f(x) g(x) \mu(dx) = \int T_t g(x) f(x) \mu(dx).$$

Suppose now (T_t) is the semigroup of a Hunt process and satisfies (4.4). Since $T_t 1 \leq 1$, we have, writing $(\cdot, \cdot)$ for the inner product on $L^2(F, \mu)$, that

$$|T_t f(x)| \leq \left(T_t f^2(x)\right)^{1/2} \left(T_t 1(x)\right)^{1/2} \leq (T_t f^2(x))^{1/2}$$

by Hölder's inequality. Therefore

$$\|T_t f\|_2^2 \leq \|T_t f^2\|_1 = (T_t f^2, 1) = (f^2, T_t 1) \leq (f^2, 1) = \|f\|_2^2,$$

so that T_t is a contraction on $L^2(F, \mu)$.

The definition of the Dirichlet (energy) form associated with (T_t) is less direct than that of the infinitesimal generator: its less intuitive description may be one reason why this approach has until recently received less attention than those based on the resolvent or infinitesimal generator. (Another reason, of course, is the more restrictive nature of the theory: many important Markov processes are not symmetric. I remark here that it is possible to define a Dirichlet form for non-symmetric Markov processes — see [MR]. However, a weaker symmetry condition, the "sector condition", is still required before this yields very much.)

Let F be a metric space, with a locally compact and countable base, and let μ be a Radon measure on F. Set $H = L^2(F, \mu)$.

Definition 4.1. Let $\mathcal{D}$ be a linear subspace of H. A *symmetric form* $(\mathcal{E}, \mathcal{D})$ is a map $\mathcal{E} : \mathcal{D} \times \mathcal{D} \to \mathbb{R}$ such that
(1) $\mathcal{E}$ is bilinear
(2) $\mathcal{E}(f, f) \geq 0$, $\quad f \in \mathcal{D}$.

For $\alpha \geq 0$ define $\mathcal{E}_\alpha$ on $\mathcal{D}$ by $\mathcal{E}_\alpha(f, f) = \mathcal{E}(f, f) + \alpha \|f\|_2^2$, and write

$$\|f\|_{\mathcal{E}_\alpha}^2 = \|f\|_2^2 + \alpha \mathcal{E}(f, f) = \mathcal{E}_\alpha(f, f).$$

Definition 4.2. Let $(\mathcal{E}, \mathcal{D})$ be a symmetric form.
(a) $\mathcal{E}$ is *closed* if $(\mathcal{D}, \| \cdot \|_{\mathcal{E}_1})$ is complete
(b) $(\mathcal{E}, \mathcal{D})$ is *Markov* if for $f \in \mathcal{D}$, if $g = (0 \vee f) \wedge 1$ then $g \in \mathcal{D}$ and $\mathcal{E}(g, g) \leq \mathcal{E}(f, f)$.
(c) $(\mathcal{E}, \mathcal{D})$ is a *Dirichlet form* if $\mathcal{D}$ is dense in $L^2(F, \mu)$ and $(\mathcal{E}, \mathcal{D})$ is a closed, Markov symmetric form.

Some further properties of a Dirichlet form will be of importance:

Definition 4.3. $(\mathcal{E}, \mathcal{D})$ is *regular* if

(4.5) $\qquad\qquad \mathcal{D} \cap C_0(F)$ is dense in $\mathcal{D}$ in $\| \cdot \|_{\mathcal{E}_1}$, and

(4.6) $\qquad\qquad \mathcal{D} \cap C_0(F)$ is dense in $C_0(F)$ in $\| \cdot \|_\infty$.

$\mathcal{E}$ is *local* if $\mathcal{E}(f, g) = 0$ whenever f, g have disjoint support.
$\mathcal{E}$ is *conservative* if $1 \in \mathcal{D}$ and $\mathcal{E}(1, 1) = 0$.
$\mathcal{E}$ is *irreducible* if $\mathcal{E}$ is conservative and $\mathcal{E}(f, f) = 0$ implies that f is constant.

The classical example of a Dirichlet form is that of Brownian motion on $\mathbb{R}^d$:

$$\mathcal{E}_{BM}(f, f) = \tfrac{1}{2} \int_{\mathbb{R}^d} |\nabla f|^2 \, dx, \quad f \in H^{1,2}(\mathbb{R}^d).$$

Later in this section we will look at the Dirichlet forms associated with finite state Markov chains.

Just as the Hille-Yoshida theorem gives a $1 - 1$ correspondence between semigroups and their generators, so we have a $1 - 1$ correspondence between Dirichlet forms and semigroups. Given a semigroup (T_t) the associated Dirichlet form is obtained in a fairly straightforward fashion.

Definition 4.4. (a) The semigroup (T_t) is *Markovian* if $f \in L^2(F, \mu)$, $0 \leq f \leq 1$ implies that $0 \leq T_t f \leq 1$ μ-a.e.
(b) A Markov process X on F is *reducible* if there exists a decomposition $F = A_1 \cup A_2$ with A_i disjoint and of positive measure such that $\mathbb{P}^x(X_t \in A_i \text{ for all } t) = 1$ for $x \in A_i$. X is *irreducible* if X is not reducible.

Theorem 4.5. *([FOT, p. 23]) Let $(T_t, t \geq 0)$ be a strongly continuous μ-symmetric contraction semigroup on $L^2(F, \mu)$, which is Markovian. For $f \in L^2(F, \mu)$ the function $\varphi_f(t)$ defined by*

$$\varphi_f(t) = t^{-1}(f - T_t f, f), \quad t > 0$$

is non-negative and non-increasing. Let

$$\mathcal{D} = \{f \in L^2(F, \mu) : \lim_{t \downarrow 0} \varphi_f(t) < \infty\},$$

$$\mathcal{E}(f, f) = \lim_{t \downarrow 0} \varphi_f(t), \quad f \in \mathcal{D}.$$

Then $(\mathcal{E}, \mathcal{D})$ is a Dirichlet form. If $(\mathcal{L}, \mathcal{D}(\mathcal{L}))$ is the infinitesimal generator of (T_t), then $\mathcal{D}(\mathcal{L}) \subset \mathcal{D}$, $\mathcal{D}(\mathcal{L})$ is dense in $L^2(F, \mu)$, and

$$(4.7) \qquad \mathcal{E}(f, g) = (-\mathcal{L}f, g), \quad f \in \mathcal{D}(\mathcal{L}), g \in \mathcal{D}.$$

As one might expect, by analogy with the infinitesimal generator, passing from a Dirichlet form $(\mathcal{E}, \mathcal{D})$ to the associated semigroup is less straightforward. Since formally we have $U_\alpha = (\alpha - \mathcal{L})^{-1}$, the relation (4.7) suggests that

$$(4.8) \qquad (f, g) = \big((\alpha - \mathcal{L})U_\alpha f, g\big) = \alpha(U_\alpha f, g) + \mathcal{E}(U_\alpha f, g) = \mathcal{E}_\alpha(U_\alpha f, g).$$

Using (4.8), given the Dirichlet form $\mathcal{E}$, one can use the Riesz representation theorem to define $U_\alpha f$. One can verify that U_α satisfies the resolvent equation, and is strongly continuous, and hence by the Hille-Yoshida theorem (U_α) is the resolvent of a semigroup (T_t).

Theorem 4.6. *([FOT, p.18]) Let $(\mathcal{E}, \mathcal{D})$ be a Dirichlet form on $L^2(F, \mu)$. Then there exists a strongly continuous μ-symmetric Markovian contraction semigroup (T_t) on $L^2(F, \mu)$, with infinitesimal generator $(\mathcal{L}, \mathcal{D}(\mathcal{L}))$ and resolvent $(U_\alpha, \alpha > 0)$ such that $\mathcal{L}$ and $\mathcal{E}$ satisfy (4.7) and also*

$$(4.9) \qquad \mathcal{E}(U_\alpha f, g) + \alpha(f, g) = (f, g), \quad f \in L^2(F, \mu), g \in \mathcal{D}.$$

Of course the operations in Theorem 4.5 and Theorem 4.6 are inverses of each other. Using, for a moment, the ugly but clear notation $\mathcal{E} = \text{Thm } 4.5((T_t))$ to denote the Dirichlet form given by Theorem 4.5, we have

$$\text{Thm } 4.6(\text{Thm } 4.5((T_t))) = (T_t),$$

and similarly $\text{Thm } 4.5(\text{Thm } 4.6 \, (\mathcal{E})) = \mathcal{E}$.

Remark 4.7. The relation (4.7) provides a useful computational tool to identify the process corresponding to a given Dirichlet form – at least for those who find it more natural to think of generators of processes than their Dirichlet forms. For example, given the Dirichlet form $\mathcal{E}(f, f) = \int |\nabla f|^2$, we have, by the Gauss-Green formula, for $f, g \in C_0^2(\mathbb{R}^d)$, $(-\mathcal{L}f, g) = \mathcal{E}(f, g) = \int \nabla f . \nabla g = -\int g \Delta f$, so that $\mathcal{L} = \Delta$.

We see therefore that a Dirichlet form $(\mathcal{E}, \mathcal{D})$ give us a semigroup (T_t) on $L^2(F, \mu)$. But does this semigroup correspond to a 'nice' Markov process? In general it need not, but if $\mathcal{E}$ is regular then one obtains a Hunt process. (Recall that a Hunt process $X = (X_t, t \geq 0, \mathbb{P}^x, x \in F)$ is a strong Markov process with cadlag sample paths, which is quasi-left-continuous.)

Theorem 4.8. *([FOT, Thm. 7.2.1.]) (a) Let $(\mathcal{E}, \mathcal{D})$ be a regular Dirichlet form on $L^2(F, \mu)$. Then there exists a μ-symmetric Hunt process $X = (X_t, t \geq 0, \mathbb{P}^x, x \in F)$ on F with Dirichlet form $\mathcal{E}$.*
(b) In addition, X is a diffusion if and only if $\mathcal{E}$ is local.

Remark 4.9. Let $X = (X_t, t \geq 0, \mathbb{P}^x, x \in \mathbb{R}^2)$ be Brownian motion on $\mathbb{R}^2$. Let $A \subset \mathbb{R}^2$ be a polar set, so that

$$\mathbb{P}^x(T_A < \infty) = 0 \text{ for each } x.$$

Then we can obtain a new Hunt process $Y = (X_t \geq 0, \mathbb{Q}^x, x \in \mathbb{R}^2)$ by "freezing" X on A. Set $\mathbb{Q}^x = \mathbb{P}^x$, $x \in A^c$, and for $x \in A$ let $\mathbb{Q}^x(X_t = x, \text{ all } t \in [0, \infty)) = 1$. Then the semigroups (T_t^X), (T_t^Y), viewed as acting on $L^2(\mathbb{R}^2)$, are identical, and so X and Y have the same Dirichlet form.

This example shows that the Hunt process obtained in Theorem 4.8 will not, in general, be unique, and also makes it clear that a semigroup on L^2 is a less precise object than a Markov process. However, the kind of difficulty indicated by this example is the only problem — see [FOT, Thm. 4.2.7.]. In addition, if, as will be the case for the processes considered in these notes, all points are non-polar, then the Hunt process is uniquely specified by the Dirichlet form $\mathcal{E}$.

We now interpret the conditions that $\mathcal{E}$ is conservative or irreducible in terms of the process X.

Lemma 4.10. *If $\mathcal{E}$ is conservative then $T_t 1 = 1$ and the associated Markov process X has infinite lifetime.*

Proof. If $f \in \mathcal{D}(\mathcal{L})$ then $0 \leq \mathcal{E}(1 + \lambda f, 1 + \lambda f)$ for any $\lambda \in \mathbb{R}$, and so $\mathcal{E}(1, f) = 0$. Thus $(-\mathcal{L}1, f) = 0$, which implies that $\mathcal{L}1 = 0$ a.e., and hence that $T_t 1 = 1$. $\square$

Lemma 4.11. *If $\mathcal{E}$ is irreducible then X is irreducible.*

Proof. Suppose that X is reducible, and that $F = A_1 \cup A_2$ is the associated decomposition of the state space. Then $T_t 1_{A_1} = 1_{A_1}$, and hence $\mathcal{E}(1_{A_1}, 1_{A_1}) = 0$. As $1 \neq 1_{A_1}$ in $L^2(F, \mu)$ this implies that $\mathcal{E}$ is not irreducible. $\square$

A remarkable property of the Dirichlet form $\mathcal{E}$ is that there is an equivalence between certain Sobolev type inequalities involving $\mathcal{E}$, and bounds on the transition density of the associated process X. The fundamental connections of this kind were found by Varopoulos [V1]; [CKS] provides a good account of this, and there is a very substantial subsequent literature. (See for instance [Co] and the references therein).

We say $(\mathcal{E}, \mathcal{D})$ satisfies a *Nash inequality* if

$$(4.10) \qquad \|f\|_1^{4/\theta} \left(\delta\|f\|_2^2 + \mathcal{E}(f, f)\right) \geq c\|f\|_2^{2+4/\theta}, \quad f \in \mathcal{D}.$$

This inequality appears awkward at first sight, and also hard to verify. However, in classical situations, such as when the Dirichlet form $\mathcal{E}$ is the one connected with the Laplacian on $\mathbb{R}^d$ or a manifold, it can often be obtained from an isoperimetric inequality.

In what follows we fix a regular conservative Dirichlet form $(\mathcal{E}, \mathcal{D})$. Let (T_t) be the associated semigroup on $L^2(F, \mu)$, and $X = (X_t, t \geq 0, \mathbb{P}^x)$ be the Hunt process associated with $\mathcal{E}$.

Theorem 4.12. *([CKS, Theorem 2.1]) (a) Suppose $\mathcal{E}$ satisfies a Nash inequality with constants c, δ, θ. Then there exists $c' = c'(c, \theta)$ such that*

$$(4.11) \qquad \|T_t\|_{1 \to \infty} \leq c' e^{\delta t} t^{-\theta/2}, \quad t > 0.$$

(b) If (T_t) satisfies (4.11) with constants c', δ, θ then $\mathcal{E}$ satisfies a Nash inequality with constants $c'' = c''(c', \theta)$, δ, and θ.

Proof. I sketch here only (a). Let $f \in \mathcal{D}(\mathcal{L})$. Then writing $f_t = T_t f$, and

$$g_{th} = h^{-1}(f_{t+h} - f_t) - T_t \mathcal{L} f,$$

we have $\|g_{th}\|_2 \leq \|g_{0h}\|_2 \to 0$ as $h \to 0$. It follows that $(d/dt)f_t$ exists in $L^2(F, \mu)$ and that

$$\frac{d}{dt} f_t = T_t \mathcal{L} f = \mathcal{L} T_t f.$$

Set $\varphi(t) = (f_t, f_t)$. Then

$$h^{-1}\big(\varphi(t+h) - \varphi(t)\big) - 2(T_t \mathcal{L} f, T_t f) \quad = (g_{th}, f_t + f_{t+h}) + (T_t \mathcal{L} f, f_{t+h} - f_t),$$

and therefore φ is differentiable, and for $t > 0$

$$(4.12) \qquad \varphi'(t) = 2(\mathcal{L} f_t, f_t) = -2\mathcal{E}(f_t, f_t).$$

If $f \in L^2(F, \mu)$, $T_t f \in \mathcal{D}(\mathcal{L})$ for each $t > 0$. So (4.12) extends from $f \in \mathcal{D}(\mathcal{L})$ to all $f \in L^2(F, \mu)$.

Now let $f \geq 0$, and $\|f\|_1 = 1$: we have $\|f_t\|_1 = 1$. Then by (4.10), for $t > 0$,

$$(4.13) \qquad \varphi'(t) = -2\mathcal{E}(f_t, f_t)) \leq 2\delta\|f_t\|_2^2 - c\|f_t\|_2^{2+4/\theta} = 2\delta\varphi(t)^2 - c\varphi(t)^{1+2/\theta}.$$

Thus φ satisfies a differential inequality. Set $\psi(t) = e^{-2\delta t}\varphi(t)$. Then

$$\psi'(t) \leq -2c\psi(t)^{1+2/\theta} e^{4\delta t/\theta} \leq -2c\psi(t)^{1+2/\theta}.$$

If ψ_0 is the solution of $\psi_0' = -c\psi_0^{1+2/\theta}$ then for some $a \in \mathbb{R}$ we have, for $c_\theta = c_\theta(c, \theta)$,

$$\psi_0(t) = c_\theta(t + a)^{-\theta/2}.$$

If ψ_0 is defined on $(0, \infty)$, then $a \geq 0$, so that

$$\psi_0(t) \leq c_\theta t^{-\theta/2}, \quad t > 0.$$

It is easy to verify that ψ satisfies the same bound – so we deduce that

$$(4.14) \qquad \|T_t f\|_2^2 = e^{2\delta t}\psi(t) \leq c_\theta e^{2\delta t} t^{-\theta/2}, \quad f \in L^2_+, \quad \|f\|_1 = 1.$$

Now let $f, g \in L^2_+(F, \mu)$ with $\|f\|_1 = \|g\|_1 = 1$. Then

$$(T_{2t}f, g) = (T_t f, T_t g) \leq \|T_t f\|_2 \|T_t g\|_2 \leq c_\theta^2 e^{\delta 2t} t^{-\theta/2}.$$

Taking the supremum over g, it follows that $\|T_{2t}f\|_\infty \le c_\theta^2 e^{\delta 2t} t^{-\theta/2}$, that is, replacing $2t$ by t, that

$$\|T_t\|_{1\to\infty} \le c_\theta^2 e^{\delta t} t^{-\theta/2}. \qquad \square$$

Remark 4.13. In the sequel we will be concerned with only two cases: either $\delta = 0$, or $\delta = 1$ and we are only interested in bounds for $t \in (0, 1]$. In the latter case we can of course absorb the constant $e^{\delta t}$ into the constant c.

This theorem gives bounds in terms of contractivity properties of the semigroup (T_t). If T_t has a 'nice' density $p(t, x, y)$, then $\|T_t\|_{1\to\infty} = \sup_{x,y} p(t, x, y)$, so that (4.11) gives global upper bounds on $p(t, \cdot, \cdot)$, of the kind we used in Chapter 3. To derive these, however, we need to know that the density of T_t has the necessary regularity properties.

So let $F, \mathcal{E}, T_t$ be as above, and suppose that (T_t) satisfies (4.11). Write $P_t(x, \cdot)$ for the transition probabilities of the process X. By (4.11) we have, for $A \in \mathcal{B}(F)$, and writing $c_t = ce^{\delta t} t^{-\theta/2}$,

$$P_t(x, A) \le c_t \mu(A) \quad \text{for } \mu\text{-a.a. } x.$$

Since F has a countable base (A_n), we can employ the arguments of [FOT, p.67] to see that

(4.15) $$P_t(x, A_n) \le c_t \mu(A_n), \quad x \in F - N_t,$$

where the set N_t is "properly exceptional". In particular we have $\mu(N_t) = 0$ and

$$\mathbb{P}^x(X_s \in N_t \text{ or } X_{s-} \in N_t \text{ for some } s \ge 0) = 0$$

for $x \in F - N_t$. From (4.15) we deduce that $P_t(x, \cdot) \ll \mu$ for each $x \in F - N_t$. If $s > 0$ and $\mu(B) = 0$ then $P_s(y, B) = 0$ for μ-a.a. y, and so

$$P_{t+s}(x, B) = \int P_s(x, dy) P_t(y, B) = 0, \quad x \in F - N_t.$$

So $P_{t+s}(x, .) \ll \mu$ for all $s \ge 0$, $x \in F - N_t$. So taking a sequence $t_n \downarrow 0$, we obtain a single properly exceptional set $N = \cup_n N_{t_n}$ such that $P_t(x, \cdot) \ll \mu$ for all $t \ge 0$, $x \in F - N$. Write $F' = F - N$: we can reduce the state space of X to F'.

Thus we have for each t, x a density $\tilde{p}(t, x, \cdot)$ of $P_t(x, \cdot)$ with respect to μ. These can be regularised by integration.

Proposition 4.14. (See [Y, Thm. 2]) There exists a jointly measurable transition density $p(t, x, y)$, $t > 0$, $x, y \in F' \times F'$, such that

$$P_t(x, A) = \int_A p(t, x, y) \mu(dy) \text{ for } x \in F', \quad t > 0, A \in \mathcal{B}(F),$$

$$p(t, x, y) = p(t, y, x) \text{ for all } x, y, t,$$

$$p(t + s, x, z) = \int p(s, x, y) p(t, y, z) \mu(dy) \quad \text{for all } x, z, t, s.$$

Corollary 4.15. *Suppose $(\mathcal{E}, \mathcal{D})$ satisfies a Nash inequality with constants c, δ, θ. Then, for all $x, y \in F'$, $t > 0$,*

$$p(t, x, y) \le c' e^{\delta t} t^{-\theta/2}.$$

We also obtain some regularity properties of the transition functions $p(t, x, \cdot)$. Write $q_{t,x}(y) = p(t, x, y)$.

Proposition 4.16. *Suppose $(\mathcal{E}, \mathcal{D})$ satisfies a Nash inequality with constants c, δ, θ. Then for $x \in F'$, $t > 0$, $q_{t,x} \in \mathcal{D}(\mathcal{L})$, and*

(4.16)
$$\|q_{t,x}\|_2^2 \le c_1 e^{2\delta t} t^{-\theta/2},$$

(4.17)
$$\mathcal{E}(q_{t,x}, q_{t,x}) \le c_2 e^{\delta t} t^{-1-\theta/2}.$$

Proof. Since $q_{t,x} = T_{t/2} q_{t/2,x}$, and $q_{t/2,x} \in L^1$, we have $q_{t,x} \in \mathcal{D}(\mathcal{L})$, and the bound (4.16) follows from (4.14).

Fix x, write $f_t = q_{t,x}$, and let $\varphi(t) = \|f_t\|_2^2$. Then

$$\varphi''(t) = \frac{d}{dt}(2\mathcal{L}f_t, f_t) = 4(\mathcal{L}f_t, \mathcal{L}f_t) \ge 0.$$

So, φ' is increasing and hence

$$0 \le \varphi(t) = \varphi(t/2) + \int_{t/2}^t \varphi'(s)\, ds \le \varphi(t/2) + (t/2)\varphi'(t).$$

Therefore using (4.13),

$$\mathcal{E}(f_t, f_t) = -\tfrac{1}{2}\varphi'(t) \le t^{-1}\varphi(t/2) \le c e^{\delta t} t^{-1-\theta/2}. \qquad \square$$

Traces of Dirichlet forms and Markov Processes.

Let X be a μ-symmetric Hunt process on a LCCB metric space (F, μ), with semigroup (T_t) and regular Dirichlet form $(\mathcal{E}, \mathcal{D})$. To simplify things, and because this is the only case we need, we assume

(4.18)
$$\mathrm{Cap}(\{x\}) > 0 \quad \text{for all } x \in F.$$

It follows that x is regular for $\{x\}$, for each $x \in F$, that is, that

$$\mathbb{P}^x(T_x = 0) = 1, \quad x \in F.$$

Hence ([GK]) X has jointly measurable local times $(L_t^x, x \in F, t \ge 0)$ such that

$$\int_0^t f(X_s)\, ds = \int_F f(x) L_t^x \mu(dx), \quad f \in L^2(F, \mu).$$

Now let ν be a σ-finite measure on F. (In general one has to assume ν charges no set of zero capacity, but in view of (4.18) this condition is vacuous here). Let A_t be the continuous additive functional associated with ν:

$$A_t = \int L_t^a \nu(da),$$

and let $\tau_t = \inf\{s : A_s > t\}$ be the inverse of A. Let G be the closed support of ν. Let $\tilde{X}_t = X_{\tau_t}$: then by [BG, p. 212], $\tilde{X} = (\tilde{X}_t, \mathbb{P}^x, x \in G)$ is also a Hunt process. We call $\tilde{X}$ the *trace* of X on G.

Now consider the following operation on the Dirichlet form $\mathcal{E}$. For $g \in L^2(G, \nu)$ set

(4.19) $$\tilde{\mathcal{E}}(g, g) = \inf\{\mathcal{E}(f, f) : f|_G = g\}.$$

Theorem 4.17. *("Trace theorem": [FOT, Thm. 6.2.1]).*
(a) $(\tilde{\mathcal{E}}, \tilde{\mathcal{D}})$ *is a regular Dirichlet form on* $L^2(G, \nu)$.
(b) $\tilde{X}$ *is ν-symmetric, and has Dirichlet form* $(\tilde{\mathcal{E}}, \tilde{\mathcal{D}})$.

Thus $\tilde{\mathcal{E}}$ is the Dirichlet form associated with $\tilde{X}$: we call $\tilde{\mathcal{E}}$ the *trace* of $\mathcal{E}$ (on G).

Remarks 4.18. 1. The domain $\tilde{\mathcal{D}}$ on $\tilde{\mathcal{E}}$ is of course the set of g such that the infimum in (4.19) is finite. If $g \in \tilde{\mathcal{D}}$ then, as $\mathcal{E}$ is closed, the infimum in (4.19) is attained, by f say. If h is any function which vanishes on G^c, then since $(f + \lambda h)|_G = g$, we have

$$\mathcal{E}(f, f) \le \mathcal{E}(f + \lambda h, f + \lambda h), \quad \lambda \in \mathbb{R}$$

which implies $\mathcal{E}(f, h) = 0$. So, if $f \in \mathcal{D}(\mathcal{L})$, and we choose $h \in \mathcal{D}$, then $(-h, \mathcal{L}f) = 0$, so that $\mathcal{L}f = 0$ a.e. on G^c.

This calculation suggests that the minimizing function f in (4.19) should be the harmonic extension of g to F; that is, the solution to the Dirichlet problem

$$\begin{aligned} f &= g &&\text{on } G \\ \mathcal{L}f &= 0 &&\text{on } G^c. \end{aligned}$$

2. We shall sometimes write

$$\tilde{\mathcal{E}} = \text{Tr}(\mathcal{E}|G)$$

to denote the trace of the Dirichlet form $\mathcal{E}$ on G.
3. Note that taking traces has the "tower property"; if $H \subseteq G \subseteq F$, then

$$\text{Tr}(\mathcal{E}|H) = \text{Tr}\big(\text{Tr}(\mathcal{E}|G) \,\big|\, H\big).$$

We now look at continuous time Markov chains on a finite state space. Let F be a finite set.

Definition 4.19. A *conductance matrix* on F is a matrix $A = (a_{xy})$, $x, y \in F$, which satisfies

$$a_{xy} \geq 0, \quad x \neq y,$$
$$a_{xy} = a_{yx},$$
$$\sum_y a_{xy} = 0.$$

Set $a_x = \sum_{y \neq x} a_{xy} = -a_{xx}$. Let $E_A = \{\{x, y\} : a_{xy} > 0\}$. We say that A is *irreducible* if the graph (F, E_A) is connected.

We can interpret the pair (F, A) as an electrical network: a_{xy} is the conductance of the wire connecting the nodes x and y. The intuition from electrical circuit theory is on occasion very useful in Markov Chain theory —for more on this see [DS].

Given (F, A) as above, define the Dirichlet form $\mathcal{E} = \mathcal{E}_A$ with domain $C(F) = \{f : F \to \mathbb{R}\}$ by

(4.20) $$\mathcal{E}(f, g) = \tfrac{1}{2} \sum_{x,y} a_{xy} \big(f(x) - f(y)\big)\big(g(x) - g(y)\big).$$

Note that, writing $f_x = f(x)$ etc.,

$$\mathcal{E}(f, g) = \tfrac{1}{2} \sum_x \sum_{y \neq x} a_{xy}(f_x - f_y)(g_x - g_y)$$
$$= \sum_x \sum_{y \neq x} a_{xy} f_x g_x - \sum_x \sum_{y \neq x} a_{xy} f_x g_y$$
$$= -\sum_x a_{xx} f_x g_x - \sum_x \sum_{y \neq x} a_{xy} f_x g_y$$
$$= -\sum_x \sum_y a_{xy} f_x g_y = -f^T A g.$$

In electrical terms, (4.20) gives the energy dissipation in the circuit (F, A) if the nodes are held at potential f. (A current $I_{xy} = a_{xy}\big(f(y) - f(x)\big)$ flows in the wire connecting x and y, which has energy dissipation $I_{xy}\big(f(y) - f(x)\big) = a_{xy}\big(f(y) - f(x)\big)^2$. The sum in (4.20) counts each edge twice). We can of course also use this interpretation of Dirichlet forms in more general contexts.

(4.20) gives a 1-1 correspondence between conductance matrices and conservative Dirichlet forms on $C(F)$. Let μ be any measure on F which charges every point.

Proposition 4.20. (a) *If A is a conductance matrix, then $\mathcal{E}_A$ is a regular conservative Dirichlet form.*
(b) *If $\mathcal{E}$ is a conservative Dirichlet form on $L^2(F, \mu)$ then $\mathcal{E} = \mathcal{E}_A$ for a conductance matrix A.*
(c) *A is irreducible if and only if $\mathcal{E}$ is irreducible.*

Proof. (a) It is clear from (4.20) that $\mathcal{E}$ is a bilinear form, and that $\mathcal{E}(f, f) \geq 0$. If $g = 0 \vee (1 \wedge f)$ then $|g_x - g_y| \leq |f_x - f_y|$ for all x, y, so since $a_{xy} \geq 0$ for $x \neq y$, $\mathcal{E}$ is Markov. Since $\mathcal{E}(f, f) \leq c(A, \mu)\|f\|_2^2$, $\|.\|_{\mathcal{E}_1}$ is equivalent to $\|.\|_2$, and so $\mathcal{E}$ is closed. It is clear from this that $\mathcal{E}$ is regular.

(b) As $\mathcal{E}$ is a symmetric bilinear form there exists a symmetric matrix A such that $\mathcal{E}(f,g) = -f^T A g$. Let $f = f_{\alpha\beta} = \alpha 1_x + \beta 1_y$; then

$$\mathcal{E}(f,f) = -\alpha^2 a_{xx} - 2\alpha\beta a_{xy} - \beta^2 a_{yy}.$$

Taking $\alpha = 1$, $\beta = 0$ it follows that $a_{xx} \le 0$. The Markov property of $\mathcal{E}$ implies that $\mathcal{E}(f_{01}, f_{01}) \le \mathcal{E}(f_{\alpha 1}, f_{\alpha 1})$ if $\alpha < 0$. So

$$0 \le -\alpha^2 a_{xx} - 2\alpha a_{xy},$$

which implies that $a_{xy} \ge 0$ for $x \ne y$. Since $\mathcal{E}$ is conservative we have $0 = \mathcal{E}(f,1) = -f^T A 1$ for all f. So $A1 = 0$, and therefore $\sum_y a_{xy} = 0$ for all x.

(c) is now evident. $\qquad\square$

Example 4.21. Let μ be a measure on F, with $\mu(\{x\}) = \mu_x > 0$ for $x \in F$. Let us find the generator L of the Markov process associated with $\mathcal{E} = \mathcal{E}_A$ on $L^2(F,\mu)$. Let $z \in F$, $g = 1_z$, and $f \in L^2(F,\mu)$. Then

$$\mathcal{E}(f,g) = -g^T A f = -\sum_y a_{zy} f(y) = \sum_y a_{zy}(f(z) - f(y)).$$

and using (4.7) we have, writing $(\cdot,\cdot)_\mu$ for the inner product on $L^2(F,\mu)$,

$$\mathcal{E}(f,g) = (-Lf,g)_\mu = -\mu_z Lf(z).$$

So,

$$(4.21) \qquad Lf(z) = \sum_{x \ne z}(a_{xz}/\mu_z)(f(x) - f(z)).$$

Note from (4.21) that (as we would expect from the trace theorem), changing the measure μ changes the jump rates of the process, but not the jump probabilities.

Electrical Equivalence.

Definition 4.22. Let (F,A) be an electrical network, and $G \subset F$. If B is a conductance matrix on G, and

$$\mathcal{E}_B = \mathrm{Tr}(\mathcal{E}_A|G)$$

we will say that the networks (F,A) and (G,B) are *(electrically) equivalent on G*.

In intuitive terms, this means that an electrician who is able only to access the nodes in G (imposing potentials, or feeding in currents etc.) would be unable to distinguish from the response of the system between the networks (F,A) and (G,B).

Definition 4.23. (Effective resistance). Let G_0, G_1 be disjoint subsets of F. The effective resistance between G_0 and G_1, $R(G_0,G_1)$ is defined by

$$(4.22) \qquad R(G_0,G_1)^{-1} = \inf\{\mathcal{E}(f,f) : f|_{B_0} = 0, f|_{B_1} = 1\}.$$

This is finite if (F,A) is irreducible.

If $G = \{x, y\}$, then from these definitions we see that (F, A) is equivalent to the network (G, B), where $B = (b_{xy})$ is given by

$$b_{xy} = b_{yx} = -b_{xx} = -b_{yy} = R(x, y)^{-1}.$$

Let (F, A) be an irreducible network, and $G \subseteq F$ be a proper subset. Let $H = G^c$, and for $f \in C(F)$ write $f = (f_H, f_G)$ where f_H, f_G are the restrictions of f to H and G respectively. If $g \in C(G)$, then if $\tilde{\mathcal{E}} = \mathrm{Tr}(\mathcal{E}_A | G)$,

$$\tilde{\mathcal{E}}(g, g) = \inf \left\{ (f_H^T, g^T) A \begin{pmatrix} f_H \\ g \end{pmatrix}, \quad f_H \in C(H) \right\}.$$

We have, using obvious notation

$$(4.23) \qquad (f_H^T, g^T) A \begin{pmatrix} f_H \\ g \end{pmatrix} = f_H^T A_{HH} f_H + 2 f_H^T A_{HG} g + g^T A_{GG} g.$$

The function f_H which minimizes (4.23) is given by $f_H = A_{HH}^{-1} A_{HG} g$. (Note that as A is irreducible, 0 cannot be an eigenvalue of A_{HH}, so A_{HH}^{-1} exists). Hence

$$(4.24) \qquad \tilde{\mathcal{E}}(g, g) = g^T (A_{GG} - A_{GH} A_{HH}^{-1} A_{HG}) g,$$

so that $\tilde{\mathcal{E}} = \mathcal{E}_B$, where B is the conductivity matrix

$$(4.25) \qquad B = A_{GG} - A_{GH} A_{HH}^{-1} A_{HG}.$$

Example 4.24. ($\Delta - Y$ transform). Let $G = \{x_0, x_1, x_2\}$ and B be the conductance matrix defined by,

$$b_{x_0 x_1} = \alpha_2, \quad b_{x_1 x_2} = \alpha_0, \quad b_{x_2 x_0} = \alpha_1.$$

Let $F = G \cup \{y\}$, and A be the conductance matrix defined by

$$a_{x_i x_j} = 0, \quad i \neq j,$$
$$a_{x_i y} = \beta_i, \quad 0 \leq i \leq 2.$$

If the α_i and β_i are strictly positive, and we look just at the edges with positive conductance the network (G, B) is a triangle, while (F, A) is a Y with y at the centre. The $\Delta - Y$ transform is that (F, A) and (G, B) are equivalent if and only if

$$(4.26) \qquad \begin{aligned} \alpha_0 &= \frac{\beta_1 \beta_2}{\beta_0 + \beta_1 + \beta_2}, \\ \alpha_1 &= \frac{\beta_2 \beta_0}{\beta_0 + \beta_1 + \beta_2}, \\ \alpha_2 &= \frac{\beta_0 \beta_1}{\beta_0 + \beta_1 + \beta_2}. \end{aligned}$$

Equivalently, if $S = \alpha_0 \alpha_1 + \alpha_1 \alpha_2 + \alpha_2 \alpha_0$, then

$$(4.27) \qquad \beta_i = \frac{S}{\alpha_i}, \quad 0 \leq i \leq 2.$$

This can be proved by elementary, but slightly tedious, calculations. The $\Delta - Y$ transform can be of great use in reducing a complicated network to a more simple one, though there are of course networks for which it is not effective.

Proposition 4.25. *(See [Ki5]). Let (F, A) be an irreducible electric network, and $R(x, y) = R(\{x\}, \{y\})$ be the 2-point effective resistances. Then R is a metric on F.*

Proof. We define $R(x, x) = 0$. Replacing f by $1 - f$ in (4.22), it is clear that $R(x, y) = R(y, x)$, so it just remains to verify the triangle inequality. Let x_0, x_1, x_2 be distinct points in F, and $G = \{x_0, x_1, x_2\}$.

Using the tower property of traces mentioned above, it is enough to consider the network (G, B), where B is defined by (4.25). Let $\alpha_0 = b_{x_1 x_2}$, and define α_1, α_2 similarly. Let $\beta_0, \beta_1, \beta_2$ be given by (4.27); using the $\Delta - Y$ transform it is easy to see that

$$R(x_i, x_j) = \beta_i^{-1} + \beta_j^{-1}, \quad i \neq j.$$

The triangle inequality is now immediate. $\qquad \square$

Remark 4.26. There are other ways of viewing this, and numerous connections here with linear algebra, potential theory, etc. I will not go into this, except to mention that (4.25) is an example of a Schur complement (see [Car]), and that an alternative viewpoint on the resistance metric is given in [Me6].

The following result gives a connection between resistance and crossing times.

Theorem 4.27. *Let (F, A) be an electrical network, let μ be a measure on F which charges every point, and let $(X_t, t \geq 0)$ be the continuous time Markov chain associated with $\mathcal{E}_A$ on $L^2(F, \mu)$. Write $T_x = \inf\{t > 0 : X_t = x\}$. Then if $x \neq y$,*

(4.28) $$E^x T_y + E^y T_x = R(x, y)\mu(F).$$

Remark. In view of the simplicity of this result, it is rather remarkable that its first appearance (which was in a discrete time context) seems to have been in 1989, in [CRRST]. See [Tet] for a proof in a more accessible publication.

Proof. A direct proof is not hard, but here I will derive the result from the trace theorem. Fix x, y, let $G = \{x, y\}$, and let $\tilde{\mathcal{E}} = \mathcal{E}_B = \mathrm{Tr}(\mathcal{E}|G)$. If $R = R(x, y)$, then we have, from the definitions of trace and effective resistance,

$$B = \begin{pmatrix} -R^{-1} & R^{-1} \\ R^{-1} & -R^{-1} \end{pmatrix}.$$

Let $\nu = \mu|_G$; the process $\tilde{X}_t$ associated with $(\tilde{\mathcal{E}}, L^2(G, v))$ therefore has generator given by

$$\tilde{L}f(z) = (R\mu_z)^{-1} \sum_{w \neq z} (f(w) - f(z)).$$

Writing $\tilde{T}_x, \tilde{T}_y$ for the hitting times associated with $\tilde{X}$ we therefore have

$$E^x \tilde{T}_y + E^y \tilde{T}_x = R(\mu_x + \mu_y).$$

We now use the trace theorem. If $f(x) = 1_z(x)$ then the occupation density formula implies that

$$\mu_z L_t^z = \int_0^t 1_z(X_s)\,ds = |\{s \le t : X_s = z\}|.$$

So

$$A_t = \int_0^t 1_G(X_s)\,ds,$$

and thus if $S = \inf\{t \ge T_y : X_t = x\}$ and $\widetilde{S}$ is defined similarly, we have

$$\widetilde{S} = \int_0^S 1_G(X_s)\,ds.$$

However by Doeblin's theorem for the stationary measure of a Markov Chain

(4.29) $$\mu(G) = (\mathbb{E}^x S)^{-1} \mathbb{E}^x \int_0^S 1_G(X_s)\,ds\,\mu(F).$$

Rearranging, we deduce that

$$\begin{aligned}
\mathbb{E}^x S &= \mathbb{E}^x T_y + \mathbb{E}^y T_x \\
&= \big(\mu(F)/\mu(G)\big)\mathbb{E}^x \widetilde{S} \\
&= \big(\mu(F)/\mu(G)\big)\big(\mathbb{E}^x \widetilde{T}_y + \mathbb{E}^y \widetilde{T}_x\big) = R\mu(F). \qquad \square
\end{aligned}$$

Corollary 4.28. Let $H \subset F$, $x \notin H$. Then

$$\mathbb{E}^x T_H \le R(x, H)\mu(F).$$

Proof. If H is a singleton, this is immediate from Theorem 4.27. Otherwise, it follows by considering the network (F', H') obtained by collapsing all points in H into one point, h, say. (So $F' = (F - H) \cup \{h\}$, and $a'_{zh} = \sum_{y \in H} a_{zy}$). $\square$

Remark. This result is actually older than Theorem 4.27 – see [Tel].

5. Geometry of Regular Finitely Ramified Fractals.

In Section 2 I introduced the Sierpinski gasket, and gave a direct "hands on" construction of a diffusion on it. Two properties of the SG played a crucial role: its symmetry and scale invariance, and the fact that it is finitely ramified. In this section we will introduce some classes of sets which preserve some of these properties, and such that a similar construction has a chance of working. (It will not always do so, as we will see).

There are two approaches to the construction of a family of well behaved regular finitely ramified fractals. The first, adopted by Lindstrøm [L1], and most of the mathematical physics literature, is to look at fractal subsets of $\mathbb{R}^d$ obtained by generalizations of the construction of the Cantor set. However when we come to study processes on F the particular embedding of F in $\mathbb{R}^d$ plays only a small role, and some quite natural sets (such as the "cut square" described below) have no simple embedding. So one may also choose to adapt an abstract approach, defining a collection of well behaved fractal metric spaces. This is the approach of Kigami [Ki2], and is followed in much of the subsequent mathematical literature on general fractal spaces. ("Abstract" fractals may also be defined as quotient spaces of product spaces – see [Kus2]).

The question of embedding has lead to confusion between mathematicians and physicists on at least one (celebrated) occasion. If G is a graph then the natural metric on G for a mathematician is the standard graph distance $d(x, y)$, which gives the length of the shortest path in G between x and y. Physicists call this the *chemical distance*. However, physicists, thinking in terms of the graph G being a model of a polymer, in which the individual strands are tangled up, are interested in the Euclidean distance between x and y in some embedding of G in $\mathbb{R}^d$. Since they regard each path in G as being a random walk path in $\mathbb{Z}^d$, they generally use the metric $d'(x, y) = d(x, y)^{1/2}$.

In this section, after some initial remarks on self-similar sets in $\mathbb{R}^d$, I will introduce the largest class of regular finitely ramified fractals which have been studied in detail. These are the *pc.f.s.s. sets* of Kigami [Ki2], and in what follows I will follow the approach of [Ki2] quite closely.

Definition 5.1. A map $\psi : \mathbb{R}^d \to \mathbb{R}^d$ is a *similitude* if there exists $\alpha \in (0, 1)$ such that $|\psi(x) - \psi(y)| = \alpha|x - y|$ for all x, $y \in \mathbb{R}^d$. We call α the *contraction factor* of ψ.

Let $M \geq 1$, and let $\psi_1, \ldots, \psi_M$ be similitudes with contraction factors α_i. For $A \subset \mathbb{R}^d$ set

$$(5.1) \qquad \Psi(A) = \bigcup_{i=1}^{M} \psi_i(A).$$

Let $\Psi^{(n)}$ denote the n-fold composition of Ψ.

Definition 5.2. Let $\mathcal{K}$ be the set of non-empty compact subsets of $\mathbb{R}^d$. For $A \subset \mathbb{R}^d$ set $\delta_\varepsilon(A) = \{x : |x - a| \leq \varepsilon \text{ for some } a \in A\}$. The *Hausdorff metric* d on $\mathcal{K}$ is defined by

$$d(A, B) = \inf \{\varepsilon > 0 : A \subset \delta_\varepsilon(B) \text{ and } B \subset \delta_\varepsilon(A)\}.$$

Lemma 5.3. *(See [Fe, 2.10.21]). (a) d is a metric on $\mathcal{K}$.*
(b) $(\mathcal{K}, d)$ is complete.
(c) If $K_N = \{K \in \mathcal{K} : K \subset \overline{B}(0, N)\}$ then K_N is compact in $\mathcal{K}$.

Theorem 5.4. *Let $(\psi_1, \ldots, \psi_M)$ be as above, with $\alpha_i \in (0, 1)$ for each $1 \leq i \leq M$. Then there exists a unique $F \in \mathcal{K}$ such that $F = \Psi(F)$. Further, if $G \in \mathcal{K}$ then $\Psi^n(G) \to F$ in d. If $G \in \mathcal{K}$ satisfies $\Psi(G) \subset G$ then $F = \cap_{n=0}^{\infty} \Psi^{(n)}(G)$.*

Proof. Note that $\Psi : \mathcal{K} \to \mathcal{K}$. Set $\alpha = \max_i \alpha_i < 1$. If A_i, $B_i \in \mathcal{K}$, $1 \leq i \leq M$ note that

$$d(\cup_{i=1}^{M} A_i, \cup_{i=1}^{M} B_i) \leq \max_i d(A_i, B_i).$$

(This is clear since if $\varepsilon > 0$ and $B_i \subset \delta_\varepsilon(A_i)$ for each i, then $\cup B_i \subset \delta_\varepsilon(\cup A_i)$). Thus

$$d\big(\Psi(A), \Psi(B)\big) \leq \max_i d\big(\psi_i(A), \psi_i(B)\big)$$
$$= \max_i \alpha_i \, d(A, B) = \alpha d(A, B).$$

So Ψ is a contraction on $\mathcal{K}$, and therefore has a unique fixed point. For the final assertion, note that if $\Psi(G) \subset G$, then $\Psi^{(n)}(G)$ is decreasing. So $\cap_n \Psi^{(n)}(G)$ is non-empty, and must equal F. $\qquad\square$

Examples 5.5. The fractal sets described in Section 2 can all be defined as the fixed point of a map Ψ of this kind.

1. *The Sierpinski gasket.* Let $\{a_1, a_2, a_3\}$ be the 3 corners of the unit triangle, and set

$$(5.2) \qquad \psi_i(x) = a_i + \tfrac{1}{2}(x - a_i), \quad x \in \mathbb{R}^2, \quad 1 \leq i \leq 3.$$

2. *The Vicsek Set.* Let $\{a_1, \ldots, a_4\}$ be the 4 corners of the unit square, let $M = 5$, let $a_5 = (\tfrac{1}{2}, \tfrac{1}{2})$, and let

$$(5.3) \qquad \psi_i(x) = a_i + \tfrac{1}{3}(x - a_i), \quad 1 \leq i \leq 5.$$

It is possible to calculate the dimension of the limiting set F from $(\psi_1, \ldots, \psi_M)$. However an "non-overlap" condition is necessary.

Definition 5.6. $(\psi_1, \ldots, \psi_M)$ satisfies the *open set condition* if there exists an open set U such that $\psi_i(U)$, $1 \leq i \leq M$, are disjoint, and $\Psi(U) \subset U$. Note that, since $\Psi(\overline{U}) \subset \overline{U}$, then the fixed point F of Ψ satisfies $F = \cap \Psi^{(n)}(\overline{U})$.

For the Sierpinski gasket, if H is the convex hull of $\{a_1, a_2, a_3\}$, then one can take $U = \text{int}(H)$.

Theorem 5.7. *Let $(\psi_1, \ldots, \psi_M)$ satisfy the open set condition, and let F be the fixed point of Ψ. Let β be the unique real such that*

$$(5.4) \qquad \sum_{i=1}^{M} \alpha_i^\beta = 1.$$

Then $\dim_H(F) = \beta$, and $0 < \mathcal{H}^\beta(F) < \infty$.

Proof. See [Fa2, p. 119].

Remark. If $\alpha_i = \alpha$, $1 \leq i \leq M$, then (5.4) simplifies to $M\alpha^\beta = 1$, so that

$$(5.5) \qquad \beta = \frac{\log M}{\log \alpha^{-1}}.$$

We now wish to set up an abstract version of this, so that we can treat fractals without necessarily needing to consider their embeddings in $\mathbb{R}^d$. Let (F, d) be a compact metric space, let $I = I_M = \{1, \ldots, M\}$, and let

$$\psi_i : F \to F, \quad 1 \leq i \leq M$$

be continuous injections. We wish the copies $\psi_i(F)$ to be strictly smaller than F, and we therefore assume that there exists $\delta > 0$ such that

$$(5.6) \qquad d\big(\psi_i(x), \psi_i(y)\big) \leq (1-\delta)d(x,y), \quad x,y \in F, \qquad i \in I_M.$$

Definition 5.8. $(F, \psi_i, 1 \leq i \leq M)$ is a *self-similar structure* if (F, d) is a compact metric space, ψ_i are continuous injections satisfying (5.6) and

$$(5.7) \qquad F = \bigcup_{i=1}^{M} \psi_i(F).$$

Let $(F, \psi_i, 1 \leq i \leq M)$ be a self-similar structure. We can use iterations of the maps ψ_i to give the 'address' of a point in F. Introduce the word spaces

$$\mathbb{W}_n = I^n, \quad \mathbb{W} = I^{\mathbb{N}}.$$

We endow $\mathbb{W}$ with the usual product topology. For $w \in \mathbb{W}_n$, v in $\mathbb{W}_n$ or $\mathbb{W}$, let $w \cdot v = (w_1, \ldots, w_n, v_1, \ldots)$, and define the left shift σ on $\mathbb{W}$ (or $\mathbb{W}_n$) by

$$\sigma w = (w_2, \ldots).$$

For $w = (w_1, ..., w_n) \in \mathbb{W}_n$ define

$$(5.8) \qquad \psi_w = \psi_{w_1} \circ \psi_{w_2} \circ \ldots \circ \psi_{w_n}.$$

It is clear from (5.7) that for each $n \geq 1$,

$$F = \bigcup_{w \in \mathbb{W}_n} \psi_w(F).$$

If $a = (a_1, \ldots, a_M)$ is a vector indexed by I, we write

$$(5.9) \qquad a_w = \prod_{i=1}^{n} a_{w_i}, \quad w \in \mathbb{W}_n.$$

Write $A_w = \psi_w(A)$ for $w \in \cup_n \mathbb{W}_n$, $A \subset F$. If $n \geq 1$, and $w \in \mathbb{W}$ (or $\mathbb{W}_m$ with $m \geq n$) write

$$(5.10) \qquad w|n = (w_1, \ldots, w_n) \in \mathbb{W}_n.$$

Lemma 5.9. *For each $w \in \mathbb{W}$, there exists a $x_w \in F$ such that*

$$(5.11) \qquad \bigcap_{n=1}^{\infty} \psi_{w|n}(F) = \{x_w\}.$$

Proof. Since $\psi_{w|(n+1)}(F) = \psi_{w|n}\big(\psi_{w_{n+1}}(F)\big) \subset \psi_{w|n}(F)$, the sequence of sets in (5.11) is decreasing. As ψ_i are continuous, $\psi_{w|n}(F)$ are compact, and therefore $A = \cap_n F_{w|n}$ is non-empty. But as $\text{diam}(F_{w|n}) \leq (1-\delta)^n \text{diam}(F)$, we have $\text{diam}(A) = 0$, so that A consists of a single point. $\qquad \square$

Lemma 5.10. *There exists a unique map* $\pi : \mathbb{W} \to F$ *such that*

$$(5.12) \qquad \pi(i \cdot w) = \psi_i\big(\pi(w)\big), \quad w \in \mathbb{W}, \quad i \in I.$$

π *is continuous and surjective.*

Proof. Define $\pi(w) = x_w$, where x_w is defined by (5.11). Let $w \in \mathbb{W}$. Then for any n,

$$\pi(i \cdot w) \in F_{(i \cdot w)|n} = F_{i \cdot (w|n-1)} = \psi_i(F_{w|n-1}).$$

So $\pi(i \cdot w) \in \cap_m \psi_i(F_m) = \{\psi_i(x_w)\}$, proving (5.12). If π' also satisfies (5.12) then $\pi'(v \cdot w) = \psi_v\big(\pi'(w)\big)$ for $v \in \mathbb{W}_n$, $w \in \mathbb{W}$, $n \geq 1$. Then $\pi'(w) \in F_{w|n}$ for any $n \geq 1$, so $\pi' = \pi$.

To prove that π is surjective, let $x \in F$. By (5.7) there exists $w_1 \in I_M$ such that $x \in F_{w_1} = \psi_{w_1}(F) = \cup_{w_2=1}^{M} F_{w_1 w_2}$. So there exists w_2 such that $x \in F_{w_1 w_2}$, and continuing in this way we obtain a sequence $w = (w_1, w_2, \ldots) \in \mathbb{W}$ such that $x \in F_{w|n}$ for each n. It follows that $x = \pi(w)$.

Let U be open in F, and $w \in \pi^{-1}(U)$. Then $F_{w|n} \cap U^c$ is a decreasing sequence of compact sets with empty intersection, so there exists m with $F_{w|m} \subset U$. Hence $V = \{v \in \mathbb{W} : v|m = w|m\} \subset \pi^{-1}(U)$, and since V is open in $\mathbb{W}$, $\pi^{-1}(U)$ is open. Thus π is continuous. □

Remark 5.11. It is easy to see that (5.12) implies that

$$(5.13) \qquad \pi(v \cdot w) = \psi_v\big(\pi(w)\big), \quad v \in \mathbb{W}_n, \quad w \in \mathbb{W}.$$

Lemma 5.12. *For* $x \in F$, $n \geq 0$ *set*

$$N_n(x) = \bigcup \{F_w : w \in \mathbb{W}_n, x \in F_w\}.$$

Then $\{N_n(x), n \geq 1\}$ *form a base of neighbourhoods of* x.

Proof. Fix x and n. If $v \in \mathbb{W}_n$ and $x \notin F_v$ then, since F_v is compact, $d(x, F_v) = \inf\{d(x, y) : y \in F_v\} > 0$. So, as $\mathbb{W}_n$ is finite, $d(x, N_n(x)^c) = \min\{d(x, F_v) : x \notin F_v, v \in \mathbb{W}_n\} > 0$. So $x \in \text{int}(N_n(x))$. Since $\text{diam} F_w \leq (1-\delta)^n \text{diam}(F)$ for $w \in \mathbb{W}_n$ we have $\text{diam} N_n(x) \leq 2(1-\delta)^n \text{diam}(F)$. So if $U \ni x$ is open, $N_n(x) \subset U$ for all sufficiently large n. □

The definition of a self-similar structure does not contain any condition to prevent overlaps between the sets $\psi_i(F)$, $i \in I_M$. (One could even have $\psi_1 = \psi_2$ for example). For sets in $\mathbb{R}^d$ the open set condition prevents overlaps, but relies on the existence of a space in which the fractal F is embedded. A general, abstract, non-overlap condition, in terms of dimension, is given in [KZ1]. However, for finitely ramified sets the situation is somewhat simpler.

For a self-similar structure $\mathcal{S} = (F, \psi_i, i \in I_M)$ set

$$B = B(\mathcal{S}) = \bigcup_{i,j,i \neq j} F_i \cap F_j.$$

As one might expect, we will require $B(S)$ to be finite. However, this on its own is not sufficient: we will require a stronger condition, in terms of the word space $\mathbb{W}$. Set

$$\Gamma = \pi^{-1}\big(B(S)\big),$$

$$P = \bigcup_{n=1}^{\infty} \sigma^n(\Gamma).$$

Definition 5.13. A self-similar structure (F, ψ) is *post critically finite*, or p.c.f., if P is finite. A metric space (F, d) is a *p.c.f.s.s. set* if there exists a p.c.f. self-similar structure $(\psi_i, 1 \leq i \leq M)$ on F.

Remarks 5.14. 1. As this definition is a little impenetrable, we will give several examples below. The definition is due to Kigami [Ki2], who called Γ the *critical set* of S, and P the *post critical set*.

2. The definition of a self-similar structure given here is slightly less general than that given in [Ki2]. Kigami did not impose the constraint (5.6) on the maps ψ_i, but made the existence and continuity of π an axiom.

3. The initial metric d on F does not play a major role. On the whole, we will work with the natural structure of neighbourhoods of points provided by the self-similar structure and the sets $F_w, w \in \mathbb{W}_n, n \geq 0$.

Examples 5.15. 1. *The Sierpinski gasket.* Let a_1, a_2, a_3 be the corners of the unit triangle in $\mathbb{R}^d$, and let

$$\psi_i(x) = a_i + \tfrac{1}{2}(x - a_i), \quad x \in \mathbb{R}^2, \quad 1 \leq i \leq 3.$$

Write G for the Sierpinski gasket; it is clear that $(G, \psi_1, \psi_2, \psi_3)$ is a self-similar structure. Writing $\dot{s} = (s, s, \ldots)$, we have

$$\pi(\dot{s}) = a_s, \quad 1 \leq s \leq 3.$$

So

$$B(S) = \left\{ \tfrac{1}{2}(a_3 + a_1), \quad \tfrac{1}{2}(a_1 + a_2), \quad \tfrac{1}{2}(a_2 + a_3) \right\},$$
$$\Gamma = \left\{ (1\dot{3}), (3\dot{1}), (1\dot{2}), (2\dot{1}), (2\dot{3}), (3\dot{2}) \right\},$$

and

$$P = \sigma(\Gamma) = \left\{ (\dot{1}), (\dot{2}), (\dot{3}) \right\}.$$

2. *The cut square.* This is an example of a p.c.f.s.s. set which has no convenient embedding in Euclidean space. (Though of course such an embedding can certainly be found).

Start with the unit square $C_0 = [0,1]^2$. Now make 'cuts' along the line $L_1 = \{(\tfrac{1}{2}, y) : 0 < y < \tfrac{1}{2}\}$, and the 3 similar lines (L_2, L_3, L_4 say) obtained from L_1 by rotation. So the set C_1 consists of C_0, but with the points in the line segment $(\tfrac{1}{2}, y-), (\tfrac{1}{2}, y+)$, viewed as distinct, for $0 < y < \tfrac{1}{2}$. (And similarly for the 3 similar sets obtained by rotation). Alternatively, C_1 is the closure of $A = C_0 - \cup_{i=1}^4 L_i$ in the geodesic metric d_A defined in Section 2. One now repeats this construction on each of the 4 squares of side $\tfrac{1}{2}$ which make up C_1 to obtain successively $C_2, C_3, \ldots$; the cut square C is the limit.

This is a p.c.f.s.s. set; one has $M = 4$, and if $a_1, \ldots, a_4$ are the 4 corners of $[0,1]^2$, then the maps ψ_i agree at all points with irrational coordinates with the maps $\varphi_i(x) = a_i + \frac{1}{2}(x - a_i)$. We have

$$B = \left\{ (0, \tfrac{1}{2}), (\tfrac{1}{2}, \tfrac{1}{2}), (1, \tfrac{1}{2}), (\tfrac{1}{2}, 0), (\tfrac{1}{2}, 1) \right\}$$
$$\Gamma = \left\{ (1\dot{2}), (2\dot{1}), (2\dot{3}), (3\dot{2}), (3\dot{4}), (4\dot{3}), (4\dot{1}), (1\dot{4}), (1\dot{3}), (3\dot{1}), (2\dot{4}), (4\dot{2}) \right\},$$

so that

$$P = \left\{ (\dot{1}), (\dot{2}), (\dot{3}), (\dot{4}) \right\}.$$

Note also that $\pi(1\dot{2}) = \pi(2\dot{1})$, and $\pi(1\dot{3}) = \pi(3\dot{1}) = \pi(2\dot{4}) = \pi(4\dot{2}) = z$, the centre of the square.

In both the examples above we had $P = \{(\dot{s}), s \in I_M\}$, and $P = \sigma^n P$ for all $n \geq 1$. However P can take a more complicated form if the sets $\psi_i(F)$, $\psi_j(F)$ overlap at points which are sited at different relative positions in the two sets.

3. *Sierpinski gasket with added triangle.* (See [Kum2]). We describe this set as a subset of $\mathbb{R}^2$. Let $\{a_1, a_2, a_3\}$ be the corners of the unit triangle in $\mathbb{R}^2$, and let $\psi_i(x) = \frac{1}{2}(x - a_i) + a_i$, $1 \leq i \leq 3$. Let $a_4 = \frac{1}{3}(a_1 + a_2 + a_3)$ be the centre of the triangle, and let $\psi_4(x) = a_4 + \frac{1}{4}(x - a_4)$. Of course (ψ_1, ψ_2, ψ_3) gives the Sierpinski gasket, but $\Psi = (\psi_1, \psi_2, \psi_3, \psi_4)$ still satisfies the open set condition, and if $F = F(\Psi)$ is the fixed point of Ψ then $(F, \psi_1, \ldots, \psi_4)$ is a self-similar structure. Writing b_1, b_2, b_3 for the mid-points of (a_2, a_3), (a_3, a_1), (a_1, a_2) respectively, and $c_i = \frac{1}{2}(a_i + b_i)$, $1 \leq i \leq 3$, we have

$$B = \{b_1, b_2, b_3, c_1, c_2, c_3\},$$

$\pi^{-1}(b_1) = \{(2\dot{3}), (3\dot{2})\}$, while $\pi^{-1}(c_1) = \{(12\dot{3}), (13\dot{2}), (4\dot{1})\}$, with similar expressions for $\pi^{-1}(b_j)$, $\pi^{-1}(c_j)$, $j = 2, 3$. So $\#(\Gamma) = 15$, and

$$\sigma(\Gamma) = \left\{ (\dot{1}), (\dot{2}), (\dot{3}), (2\dot{3}), (3\dot{2}), (3\dot{1}), (1\dot{3}), (1\dot{2}), (2\dot{1}) \right\},$$
$$\sigma^2(\Gamma) = \{(\dot{1}), (\dot{2}), (\dot{3})\}.$$

Then $P = \sigma(\Gamma)$ consists of 9 points in $\mathbb{W}$, and $\#(\pi(P)) = 6$.

Fig. 5.1 : Sierpinski gasket with added triangle.

4. *(Rotated triangle).* Let a_i, b_i, ψ_i, $1 \leq i \leq 3$, be as above. Let $\lambda \in (0, 1)$, and let $p_1 = \lambda b_2 + (1 - \lambda)b_3$, with p_2, p_3 defined similarly. Evidently $\{p_1, p_2, p_3\}$ is an

equilateral triangle; let ψ_4 be the similitude such that $\psi_4(a_i) = p_i$. Let $F = F(\Psi)$ be the fixed point of Ψ. If H is the convex hull of $\{a_1, a_2, a_3\}$, then $\Psi(H) \subset H$, so clearly F is finitely ramified, and

$$B = \{b_1, b_2, b_3, p_1, p_2, p_3\}.$$

Fig. 5.2 : Rotated triangle with $\lambda = 2/3$.

As before, $\pi^{-1}(b_1) = \{(2\dot{3}), (3\dot{2})\}$. Let $y_1 = \psi_1^{-1}(p_1)$; then y_1 lies on the line segment connecting a_2 and a_3. If $A = \pi^{-1}(y_1)$ then A consists of one or two points, according to whether λ is a dyadic rational or not. Let $A = \{v, w\}$, where $v = w$ if $\lambda \notin \mathbb{D}$. Note that for each element $u \in A$, we have, writing $u = (u_1, u_2, \ldots)$, that $u_k \in \{2, 3\}$, $k \geq 1$. Then $\pi^{-1}(p_1) = \{(4\dot{1}), (1 \cdot v), (1 \cdot w)\}$. If $\theta : \mathbb{W} \to \mathbb{W}$ is defined by $\theta(w) = w'$, where $w_i' = w_i + 1 \pmod 3$, and

$$A_n = \{(\dot{1}), \sigma^n v, \sigma^n w\},$$

then $\sigma^n(\Gamma) = A_n \cup \theta(A_n) \cup \theta^2(A_n)$.

(a) $\lambda = \frac{1}{2}$ gives Example 3 above.

(b) If λ is irrational, then $P = \cup_{n \geq 1} \sigma^n(\Gamma)$ is infinite. This example therefore shows that the "p.c.f." condition in Definition 5.13 is strictly stronger than the requirement that the set F be finitely ramified and self-similar.

(c) Let $\lambda = \frac{2}{3}$. Then $v = w = (\dot{2}\dot{3})$. Therefore B consists of p_1 and b_1, with their rotations, and $\sigma(L)$ consists of $(2\dot{3})$, $(3\dot{2})$, $(4\dot{1})$, $(123\dot{2}\dot{3})$ and their "rotations" by θ. Hence

$$P = \left\{(\dot{1}), (\dot{2}), (\dot{3}), (2\dot{3}), (3\dot{2}), (3\dot{1}), (1\dot{3}), (1\dot{2}), (2\dot{1})\right\}.$$

So $\lambda = \frac{2}{3}$ does give a p.c.f.s.s. set.

(d) In general, as is clear from the examples above, while F is finitely ramified for any $\lambda \in (0, 1)$, F is a p.c.f.s.s. set if and only if $\lambda \in \mathbb{Q} \cap (0, 1)$.

Fig. 5.3 : Rotated triangle with $\lambda = 0.721$.

We now introduce some more notation.

Definition 5.16. Let $(F, \psi_1, \ldots, \psi_M)$ be a p.c.f.s.s. set. Set for $n \geq 0$,

$$P^{(n)} = \{w \in \mathbb{W} : \sigma^n w \in P\},$$
$$V^{(n)} = \pi(P^{(n)}).$$

Any set of the form F_w, $w \in \mathbb{W}_n$, we call an *n-complex*, and any set of the form $\psi_w(V^{(0)}) = V_w^{(0)}$ we call a *n-cell*.

Lemma 5.17. (a) Let $x \in V^{(n)}$. Then $x = \psi_w(y)$, where $y \in V^{(0)}$ and $w \in \mathbb{W}_n$.
(b) $V^{(n)} = \cup_{w \in \mathbb{W}_n} V_w^{(0)}$.

Proof. (a) From the definition, $x = \pi(w \cdot v)$, for $w \in \mathbb{W}_n$, $v \in \mathbb{W}$. Then if $y = \pi(v)$, $y \in V^{(0)}$, and by (5.13), $x = \pi(w \cdot v) = \psi_w(y)$.
(b) Let $x \in V_w^{(0)}$. Then $x = \psi_w\big(\pi(v)\big)$, where $v \in P$. Hence $x = \pi(w \cdot v)$, and since $w \cdot v \in P^{(n)}$, $x \in V^{(n)}$. The other inclusion follows from (a). $\qquad\square$

We think of $V^{(0)}$ as being the "boundary" of the set F. The set F consists of the union of M^n n-complexes F_w (where $w \in \mathbb{W}_n$), which intersect only at their boundary points.

Lemma 5.18. (a) If $w, v \in \mathbb{W}_n$, $w \neq v$, then $F_w \cap F_v = V_w^{(0)} \cap V_v^{(0)}$.
(b) If $n \geq 0$, $\pi^{-1}\big(\pi(P^{(n)})\big) = \pi^{-1}(V^{(n)}) = P^{(n)}$.

Proof. (a) Let $n \geq 1$, $v, w \in \mathbb{W}_n$, and $x \in F_w \cap F_v$. So $x = \pi(w \cdot u) \neq \pi(v \cdot u')$ for $u, u' \in \mathbb{W}$. Suppose first that $w_1 \neq v_1$. Then as $F_w \subset F_{w_1}$, we have $x \in F_{w_1} \cap F_{v_1} \subset B$. So $w \cdot u$, $v \cdot u' \in \Gamma$, and thus $u = \sigma^{n-1}\sigma(w \cdot u) \in P$. Therefore $\pi(u) \in V^{(0)}$, and $x = \psi_w\big(\pi(u)\big) \in V_w^{(0)}$. If $w_1 = v_1$ then let k be the largest integer such that $w|k = v|k$. Applying $\psi_{w|k}^{-1}$ we can then use the argument above.

(b) It is elementary that $P^{(n)} \subset \pi^{-1}\big(\pi(P^{(n)})\big)$. Let $n = 0$ and $w \in \pi^{-1}\big(\pi(P)\big)$. Then there exists $v \in P$ such that $\pi(w) = \pi(v)$. As $v \in P$, $v \in \sigma^m(\Gamma)$ for some $m \geq 1$. Hence there exists $u \in \mathbb{W}_m$ such that $u \cdot v \in \pi^{-1}(B)$. However $\pi(u \cdot w) = \psi_u\big(\pi(w)\big) = \pi(u \cdot v) \in B$, and thus $u \cdot v \in \sigma$. Hence $v \in P$.

If $n \geq 1$, and $\pi(w) \in \pi(P^{(n)}) = V^{(n)}$, then $\pi(w) \in V_v^{(0)}$ for some $v \in \mathbb{W}_n$. So $\pi(w) \in V_v^{(0)} \cap F_{w|n} = V_v^{(0)} \cap V_{w|n}^{(0)}$ by (a). Therefore $\pi(w) \in V_{w|n}^{(0)}$, and thus $\pi(w) = \psi_{w|n}\big(\pi(v)\big)$, where $v \in P$. So $\pi(w) = \pi(w|n \cdot v)$, and thus $\pi(\sigma^n w) = \pi(v)$. By the case $n = 0$ above $\sigma^n w \in P$, and hence $w \in P^{(n)}$. $\qquad\square$

Remark 5.19. Note we used the fact that $\pi(v \cdot w) = \pi(v \cdot w')$ implies $\pi(w) = \pi(w')$, which follows from the fact that ψ_v is injective.

Lemma 5.20. *Let* $s \in \{1, \ldots, M\}$. *Then* $\pi(\dot{s})$ *is in exactly one n-complex, for each* $n \geq 1$.

Proof. Let $n = 1$, and write $x_s = \pi(\dot{s})$. Plainly $x_s \in F_s$; suppose $x_s \in F_i$ where $i \neq s$. Then $x_s = \psi_i\big(\pi(w)\big)$ for some $w \in \mathbb{W}$. Since $x_s = \psi_s^k(x_s)$ for any $k \geq 1$, $x_s = \psi_s^k\big(\pi(i \cdot w)\big) = \pi(s^k \cdot i \cdot w)$, where $s^k = (s, s, \ldots, s) \in \mathbb{W}_k$. Since $x_s \in F_i \cap F_s \subset B$, $\pi^{-1}(x_s) \in C$. But therefore $s^k \cdot i \cdot w \in C$ for each $k \geq 1$, and since $i \neq s$, C is infinite, a contradiction.

Now let $n \geq 2$, and suppose $x_s = \pi(\dot{s}) \in F_w$, where $w \in \mathbb{W}_n$ and $w \neq s^n$. Let $0 \leq k \leq n - 1$ be such that $w = s^k \cdot \sigma^k w$, and $w_{k+1} \neq s$. Then applying ψ_s^{-k} to F_{s^k} we have that $x_s \in F_{\sigma^k w} \cap F_{s^{n-k}}$, which contradicts the case $n = 1$ above. $\qquad\square$

Let $(F, \psi_1, \ldots, \psi_M, \pi)$ be a p.c.f.s.s. set. For $x \in F$, let

$$m_n(x) = \#\{w \in \mathbb{W}_n : x \in F_w\}$$

be the n-multiplicity of x, that is the number of distinct n-complexes containing x. Plainly, if $x \notin \cup_n V^{(n)}$, then $m_n(x) = 1$ for all n. Note also that $m.(x)$ is increasing.

Proposition 5.21. *For all* $x \in F$, $n \geq 1$,

$$m_n(x) \leq M\#(P).$$

Proof. Suppose $x \in F_{w^1} \cap \ldots \cap F_{w^k}$, where w^i, $1 \leq i \leq k$ are distinct elements of $\mathbb{W}_n$. Suppose first that $w_1^i \neq w_1^j$ for some $i \neq j$. Then $x \in B$, and therefore there exist $v^1, \ldots, v^k \in \mathbb{W}$ such that $\pi(w^l \cdot v^l) = x$, $1 \leq l \leq k$. Hence $w^l \cdot v^l \in \Gamma$ for each l, and so $\#(\Gamma) \geq k$. But $\#(P) \geq M^{-1}\#(\Gamma)$, and thus $k \leq M\#(P)$.

If all the w^l contain a common initial string v, then applying ψ_v^{-1} we can use the argument above. $\qquad\square$

Nested Fractals and Affine Nested fractals.

Nested fractals were introduced by Lindstrøm [L1], and affine nested fractals (ANF) by [FHK]. These are of p.c.f.s.s. sets, but have two significant additional properties:

(1) They are embedded in Euclidean space,
(2) They have a large symmetry group.

I will first present the definition of an ANF, and then relate it to that for p.c.f.s.s. sets. Let $\psi_1, \ldots, \psi_M$ be similitudes in $\mathbb{R}^d$, and let F be the associated compact set. Writing ψ_i also for the restrictions of ψ_i to F, $(F, \psi_1, \ldots, \psi_M)$ is a self similar structure. Let $\mathbb{W}$, π, $V^{(0)}$, etc. be as above. For $x, y \in V^{(0)}$ let $g_{xy} : \mathbb{R}^d \to \mathbb{R}^d$ be reflection in the hyperplane which bisects the line segment connecting x and y. As each ψ_i is a contraction, it has a unique fixed point, z_i say. Let $\overline{V} = \{z_1, \ldots, z_M\}$ be the set of fixed points. Call $x \in \overline{V}$ an *essential fixed point* if there exists $y \in \overline{V}$, and $i \neq j$ such that $\psi_i(x) = \psi_j(y)$. Write $\overline{V}^{(0)}$ for the set of essential fixed points. Set also

$$\overline{V}^{(n)} = \bigcup_{w \in W_n} \overline{V}^{(0)}.$$

Definition 5.22. $(F, \psi_1, \ldots, \psi_M)$ is an *affine nested fractal* if $\psi_1, \ldots, \psi_M$ satisfy the open set condition, $\#(\overline{V}^{(0)}) \geq 2$, and

(A1) (Connectivity) For any i, j there exists a sequence of 1-cells $V_{i_0}^{(0)}, \ldots, V_{i_k}^{(0)}$ such that $i_0 = i$, $i_k = j$ and $\overline{V}_{i_{r-1}}^{(0)} \cap \overline{V}_{i_r}^{(0)} \neq \emptyset$ for $1 \leq r \leq k$.

(A2) (Symmetry) For each $x, y \in \overline{V}^{(0)}$, $n \geq 0$, g_{xy} maps n cells to n cells.

(A3) (Nesting) If $w, v \in W_n$ and $w \neq v$ then

$$F_w \cap F_v = \overline{V}_w^{(0)} \cap \overline{V}_v^{(0)}.$$

In addition $(F, \psi_1, \ldots, \psi_M)$ is a *nested fractal* if the ψ_i all have the same contraction factor.

If ψ_i has contraction factor α_i, then by (5.4) $\dim_H(F) = \beta$, where β solves

$$(5.14) \qquad \sum_{i=1}^{M} \alpha_i^\beta = 1.$$

If $\alpha_i = \alpha$, so that F is a nested fractal, then

$$(5.15) \qquad \dim_H(F) = \frac{\log M}{\log(1/\alpha)}.$$

Following Lindstrøm we will call M the *mass scale factor*, and $1/\alpha$ the *length scale factor*, of the nested fractal F.

Lemma 5.23. *Let $(F, \psi_1, \ldots, \psi_M)$ be an affine nested fractal. Write z_i for the fixed point of ψ_i. Then $z_i \notin F_j$ for any $j \neq i$.*

Proof. Suppose that $z_1 \in F_2$. Then by Definition 5.22(A3) $F_1 \cap F_2 = \overline{V}_1^{(0)} \cap \overline{V}_2^{(0)}$, so $z_1 \in \overline{V}_2^{(0)}$, and $z_1 = \psi_2(z_i)$, for some $z_i \in \overline{V}^{(0)}$. We cannot have $i = 2$, as $\psi_2(z_2) = z_2 \neq z_1$. Also, if $i = 1$ then ψ_2 would fix both z_1 and z_2, so could not be a contraction. So let $i = 3$. Therefore for any $k \geq 0$, $i \geq 0$,

$$z_1 = \psi_1^k \circ \psi_2 \circ \psi_3^i(z_3) \in F_{1^k \cdot 2 \cdot 3^i}.$$

Write $D_n = \{w \in W_n : z_1 \in F_w\}$: by the above $\#(D_n) \geq n$. Let U be the open set given by the open set condition. Since $F \subset \overline{U}$ we have $z_i \in \overline{U}$ for each i. So $z_1 \in \overline{U}_w$ for each $w \in D_n$, while the open set condition implies that the sets $\{U_w, w \in D_n\}$ are disjoint. So z_1 is on the boundary of at least n disjoint open sets. If (as is true for nested fractals) all these sets are congruent then a contradiction is almost immediate.

For the general case of affine nested fractals we need to work a little harder to obtain the same conclusion. Let $a > 0$ be such that

$$|B(z_i, 1) \cap U| > a \quad \text{for each } i.$$

Let α_i, $1 \leq i \leq M$ be the contraction factors of the ψ_i. Recall the notation $\alpha_w = \Pi_{i=1}^n \alpha_{w_i}$, $w \in W_n$. Set $\delta = \min_{w \in D_n} \alpha_w$, and let $\beta = \min_i \alpha_i$. For each $w \in D_n$ let $w' = w \cdot 1 \ldots 1$ be chosen so that $\beta\delta < \alpha_{w'} \leq \delta$. Then $z_1 \in F_{w'} \subset \overline{U}_{w'}$, for each $w \in D_n$, and the sets $\{U_{w'}, w \in D_n\}$ are still disjoint. (Since $\Psi(U) \subset U$ we have $U_{w'} \subset U_w$ for each $w \in D_n$).

Now if $w \in D_n$ then if j is such that $z_1 = \psi_{w'}(z_j)$

$$|B(z_1, \delta) \cap U_{w'}| = \alpha_{w'}^d |B(z_j, \delta/\alpha_{w'}) \cap U| \geq (\beta\delta)^d |B(z_j, 1) \cap U| \geq a(\beta\delta)^d.$$

So

$$c_d \delta^d = |B(z_1, \delta)| \geq \sum_{w \in D_n} |B(z_1, \delta) \cap U_{w'}| \geq na(\beta\delta)^d.$$

Choosing n large enough this gives a contradiction. $\qquad\square$

Proposition 5.24. *Let $(F, \psi_1, \ldots, \psi_M)$ be an affine nested fractal. Write z_i for the fixed point of ψ_i. Then $(F, \psi_1, \ldots, \psi_M)$ is a p.c.f.s.s. set, and*
(a) $\overline{V}^{(0)} = V^{(0)}$,
(b) $P = \left\{ (\dot{s}) : z_s \in \overline{V}^{(0)} \right\}$.
(c) If $z \in V^{(0)}$ then z is in exactly one n-complex for each $n \geq 1$.
(d) Each 1-complex contains at most one element of $V^{(0)}$.

Proof. It is clear that $(F, \psi_1, \ldots, \psi_M)$ is a self-similar structure. Relabelling the ψ_i, we can assume $\overline{V}^{(0)} = \{z_1, \ldots, z_k\}$ where $2 \leq k \leq M$. We begin by calculating B, Γ and P. It is clear from (A3) that

$$B = \bigcup_{s \neq t} (\overline{V}_s^{(0)} \cap \overline{V}_t^{(0)}).$$

Let $w \in \Gamma$. Then $\pi(w) \in B$, so (as $\pi(w) \in F_{w_1}$) $\pi(w) \in \overline{V}^{(0)}_{w_1}$, and therefore $\pi(\sigma w) \in \overline{V}^{(0)}$. Say $\pi(\sigma w) = z_s$, where $s \in \{1, .., k\}$. Then since $z_s \in F_{w_2}$, by Lemma 5.23 we must have $w_2 = s$. So $\psi_s(\pi(\sigma^2 w)) = \pi(s \cdot \sigma^2 w) = \pi(\sigma w) = z_s$, and therefore $\pi(\sigma^2 w) = z_s$. So $w_3 = s$, and repeating we deduce that $\sigma w = (\dot{s})$. Therefore $\{\sigma w, w \in \Gamma\} = \{(\dot{s}), 1 \le s \le k\}$. This proves (b); as P is finite $(F, \psi_1, \ldots, \psi_M)$ is a p.c.f.s.s. set. (a) is immediate, since $\pi(P) = V^{(0)} = \{\pi(\dot{s})\} = \overline{V}^{(0)}$.
(c) This is now immediate from (a), (b) and Lemma 5.23.
(d) Suppose F_i contains z_s and z_t, where $s \ne t$. Then one of s, t is distinct from i – suppose it is s. Then $z_s \in F_s \cap F_i$, which contradicts (c). $\qquad\square$

Remarks 5.25. 1. Of the examples considered above, the SG is a nested fractal and the SG with added triangle is an ANF. The cut square is not an ANF, since if it were, the maps $\psi_i : \mathbb{R}^d \to \mathbb{R}^d$ would preserve the plane containing its 4 corners, and then the nesting axiom fails. The rotated triangle fails the symmetry axiom unless $\lambda = 1/2$. The Vicsek set defined in Section 2 is a nested fractal, but the Sierpinski carpet fails the nesting axiom.

2. The simplest examples of p.c.f.s.s. sets, and nested fractals can be a little misleading. Note the following points:
(a) Proposition 5.24(c) fails for p.c.f.s.s. sets. See for example the SG with added triangle, where $V^{(0)}$ contains the points $\{b_1, b_2, b_3\}$ as well as the corners $\{a_1, a_2, a_3\}$, and each of the points b_i lies in 2 distinct 1-cells.
(b) This example also shows that for a general p.c.f.s.s. set it is possible to have $F - V^{(0)}$ disconnected even if F is connected.
(c) Let $V_i^{(0)}$ and $V_j^{(0)}$ be two distinct 1-cells in a p.c.f.s.s. set. Then one can have $\#(V_i^{(0)} \cap V_j^{(0)}) \ge 2$. (The cut square is an example of this). For nested fractals, I do not know whether it is true that

$$(5.16) \qquad \#(V_i^{(0)} \cap V_j^{(0)}) \le 1 \quad \text{if } i \ne j.$$

In [FHK, Prop. 2.2(4)] it is asserted that (5.16) holds for affine nested fractals, quoting a result of J. Murai: however, the result of Murai was proved under stronger hypotheses. While much of the work on nested fractals has assumed that (5.16) holds, this difficulty is not a serious one, since only minor modifications to the definitions and proofs in the literature are needed to handle the general case.

3. The symmetry hypothesis (A2) is very strong. We have

$$(5.17) \qquad g_{xy} : V^{(0)} \to V^{(0)} \quad \text{for all} \quad x \ne y, \qquad x, y \in V^{(0)}.$$

The question of which sets $V^{(0)}$ satisfy (5.17) leads one into questions concerning reflection groups in $\mathbb{R}^d$. It is easy to see that $V^{(0)}$ satisfies (5.17) if $V^{(0)}$ is a regular planar polygon, a d-dimensional tetrahedron or a d-dimensional simplex. (That is, the set $V^{(0)} = \{e_i, -e_i, 1 \le i \le d\} \subset \mathbb{R}^d$, where $e_i = (\delta_{1i}, \ldots, \delta_{di})$. I have been assured by two experts in this area that these are the only possibilities, and my web page see $(\text{http://www.math.ubc.ca/})$ contains a letter from G. Maxwell with a sketch of a proof of this fact.

Note that the cube in $\mathbb{R}^3$ fails to satisfy (5.17).

4. Note also that if F is a nested fractal in $\mathbb{R}^d$, and $V^{(0)} \subset H$ where H is a k-dimensional subspace, one does not necessarily have $F \subset H$. This is the case of the Koch curve, for example. (See [L1, p. 39]).

Example 5.26. (Lindstrøm snowflake). This nested fractal is the "classical example", used in [L1] as an illustration of the axioms. It may be defined briefly as follows. Let z_i, $1 \le i \le 6$ be the vertices of a regular hexagon in $\mathbb{R}^2$, and let $z_7 = \frac{1}{6}(z_1 + \dots z_6)$ be the centre. Set

$$\psi_i(x) = z_i + \tfrac{1}{3}(x - z_i), \qquad 1 \le i \le 7.$$

It is easy to verify that this set satisfies the axioms (A1)–(A3) above.

Fig. 5.4. Lindstrøm snowflake.

Measures on p.c.f.s.s. sets.

The structure of these sets makes it easy to define measures which have good properties relative to the maps ψ_i. We begin by considering measures on $\mathbb{W}$. Let $\theta = (\theta_1, \dots, \theta_M)$ satisfy

$$\sum_{i=1}^{M} \theta_i = 1, \qquad 0 < \theta_i < 1 \quad \text{for each} \quad i \in I_M.$$

Recall the notation $\theta_w = \prod_{i=1}^{n} \theta_{w_i}$ for $w \in \mathbb{W}_n$. We define the measure $\tilde{\mu}_\theta$ on $\mathbb{W}$ to be the natural product measure associated with the vector θ. More precisely, let $\xi_n : \mathbb{W} \to I_M$ be defined by $\xi_n(w) = w_n$; then $\tilde{\mu}_\theta$ is the measure which makes (ξ_n) i.i.d. random variables with distribution given by $\mathbb{P}(\xi_n = r) = \theta_r$. Note that for any $n \ge 1$, $w \in \mathbb{W}_n$,

$$(5.18) \qquad \tilde{\mu}_\theta \left(\{v \in \mathbb{W} : v|n = w\}\right) = \prod_{i=1}^{n} \theta_{w_i}.$$

Definition 5.27. Let $\mathcal{B}(F)$ be the σ-field of subsets of F generated by the sets $\{F_w, w \in \mathbb{W}_n, n \ge 1\}$. (By Lemma 5.12 this is the Borel σ-field). For $A \in \mathcal{B}(F)$, set

$$\mu(A) = \tilde{\mu}\left(\pi^{-1}(A)\right).$$

Then for $w \in \mathbb{W}_n$

$$(5.19) \qquad \mu_\theta(F_w) = \tilde{\mu}_\theta\left(\pi^{-1}(F_w)\right) = \tilde{\mu}_\theta\left(\{v : v|n = w\}\right) = \theta_w = \prod_{i=1}^{n} \theta_{w_i}.$$

In contexts when θ is fixed we will write μ for μ_θ.

Remark. If $(F, \psi_1, \ldots, \psi_M)$ is a nested fractal, then the sets $\psi_i(F)$, $1 \le i \le M$ are congruent, and it is natural to take $\theta_i = M^{-1}$. More generally, for an ANF, the 'natural' θ is given by

$$\theta_i = \alpha_i^\beta,$$

where β is defined by (5.4).

The following Lemma summarizes the self-similarity of μ in terms of the space $L^1(F, \mu)$.

Lemma 5.28. Let $f \in L^1(F, \mu)$. Then for $n \ge 1$

$$(5.20) \qquad \int_F f \, d\mu = \sum_{w \in \mathbb{W}_n} \theta_w \int (f \circ \psi_w) \, d\mu, \qquad n \ge 1.$$

Proof. It is sufficient to prove (5.20) in the case $n = 1$: the general case then follows by iteration. Write $G = F - V^{(0)}$. Note that $G_v \cap G_w = \emptyset$ if $v, w \in \mathbb{W}_n$ and $v \ne w$. As μ is non-atomic we have $\mu(F_w) = \mu(G_w)$ for any $w \in \mathbb{W}_n$. Let $f = 1_{G_w}$ for some $w \in \mathbb{W}_n$. Then $f \circ \psi_i = 0$ if $i \ne w_1$, and $f \circ \psi_{w_1} = 1_{G_{\sigma w}}$. Thus

$$\int (f \circ \psi_i) \, d\mu = \mu(G_{\sigma w}) = \theta_{w_1}^{-1} \mu(G_w) = \theta_{w_1}^{-1} \int f \, d\mu,$$

proving (5.20) for this particular f. The equality then extends to L^1 by a standard argument. $\qquad\qquad \square$

We will also need related measures on the sets $V^{(n)}$. Let $N_0 = \#V^{(0)}$. Fix θ and set

$$(5.21) \qquad \mu_n(x) = N_0^{-1} \sum_{w \in \mathbb{W}_n} \theta_w 1_{V_w^{(0)}}(x), \qquad x \in V^{(n)}.$$

Lemma 5.29. μ_n is a probability measure on $V^{(n)}$ and

$$\text{wlim}_{n \to \infty} \mu_n = \mu_\theta.$$

Proof. Since $\#V_w^{(0)} = N_0$ we have

$$\mu_n(V^{(n)}) = \sum_{x \in V^{(n)}} N_0^{-1} \sum_{w \in \mathbb{W}_n} \theta_w 1_{V_w^{(0)}}(x) = \sum_{w \in \mathbb{W}_n} \theta_w = 1,$$

proving the first assertion.

We may regard μ_n as being derived from μ by shifting the mass on each n-complex F_w to the boundary $V_w^{(0)}$, with an equal amount of mass being moved to each point. (So a point $x \in V_w^{(0)}$ obtains a contribution of θ_w from each n-complex it belongs to). So if $f : F \to \mathbb{R}$ then

(5.22)
$$\left| \int_F f d\mu - \int_F f d\mu_n \right| \le \max_{w \in W_n} \sup_{x,y \in F_w} |f(x) - f(y)|$$

It follows that $\mu_n \overset{w}{\to} \mu_\theta$. $\qquad\square$

Symmetries of p.c.f.s.s. sets.

Definition 5.30. Let $\mathcal{G}$ be a group of continuous bijections from F to F. We call $\mathcal{G}$ a *symmetry group* of F if
(1) $g : V^{(0)} \to V^{(0)}$ for all $g \in \mathcal{G}$.
(2) For each $i \in I$, $g \in \mathcal{G}$ there exists $j \in I$, $g' \in \mathcal{G}$ such that

(5.23)
$$g \circ \psi_i = \psi_j \circ g'.$$

Note that if g, h satisfy (5.23) then

$$(g \circ h) \circ \psi_i = g \circ (h \circ \psi_i) = g \circ (\psi_j \circ h') = (g \circ \psi_j) \circ h'$$
$$= (\psi_k \circ g') \circ h' = \psi_k \circ g'',$$

for some $j, k \in I$, $g', h', g'' \in \mathcal{G}$. The calculation above also shows that if $\mathcal{G}_1$ and $\mathcal{G}_2$ are symmetry groups then the group generated by $\mathcal{G}_1$ and $\mathcal{G}_2$ is also a symmetry group. Write $\mathcal{G}(F)$ for the largest symmetry group of F. If $\mathcal{G}$ is a symmetry group, and $g \in \mathcal{G}$ write $\tilde{g}(i)$ for the unique element $j \in I$ such that (5.23) holds.

Lemma 5.31. *Let $g \in \mathcal{G}$. Then for each $n \ge 0$, $w \in W_n$, there exist $v \in W_n$, $g' \in \mathcal{G}$ such that $g \circ \psi_w = \psi_v \circ g'$. In particular $g : V^{(n)} \to V^{(n)}$.*

Proof. The first assertion is just (5.23) if $n = 1$. If $n \ge 1$, and the assertion holds for all $v \in W_n$ then if $w = i \cdot v \in W_{n+1}$ then

$$g \circ \psi_w = g \circ \psi_i \circ \psi_v = \psi_j \circ g' \circ \psi_v = \psi_j \circ \psi_{v'} \circ g'',$$

for $j \in I$, $g', g'' \in \mathcal{G}$. $\qquad\square$

Proposition 5.32. *Let $(F, \psi_1, \ldots, \psi_M)$ be an ANF. Let $\mathcal{G}_1$ be the set of isometries of $\mathbb{R}^d$ generated by reflections in the hyperplanes bisecting the line segments $[z_i, z_j]$, $i \ne j$, $z_i, z_j \in V^{(0)}$. Let $\mathcal{G}_0$ be the group generated by $\mathcal{G}_1$. Then $\mathcal{G}_R = \{g|_F : g \in \mathcal{G}_0\}$ is a symmetry group of F.*

Proof. If $g \in \mathcal{G}_1$ then $g : V^{(n)} \to V^{(n)}$ for each n and hence also $g : F \to F$. Let $i \in I$: by the symmetry axiom (A2) $g(V_i^{(0)}) = V_j^{(0)}$ for some $j \in I$. For each of the possible forms of $V^{(0)}$ given in Remark 5.25(3), the symmetry group of $V^{(0)}$ is generated by the reflections in $\mathcal{G}_1$. So, there exists $g' \in \mathcal{G}_0$ such that $g \circ \psi_i = \psi_j \circ g'$. Thus (5.23) is verified for each $g \in \mathcal{G}_1$, and it follows that (5.23) holds for all $g \in \mathcal{G}_0$. $\qquad\square$

Remark 5.33. In [BK] the collection of 'p.c.f. morphisms' of a p.c.f.s.s. set was introduced. These are rather different from the symmetries defined here since the definition in [BK] involved 'analytic' as well as 'geometric' conditions.

Connectivity Properties.

Definition 5.34. Let F be a p.c.f.s.s. set. For $n \geq 0$, define a graph structure on $V^{(n)}$ by taking $\{x,y\} \in \mathbf{E}_n$ if $x \neq y$, and $x,y \in V_w^{(0)}$ for some $w \in \mathbb{W}_n$.

Proposition 5.35. *Suppose that $(V^{(1)}, \mathbf{E}_1)$ is connected. Then $(V^{(n)}, \mathbf{E}_n)$ is connected for each $n \geq 2$, and F is pathwise connected.*

Proof. Suppose that $(V^{(n)}, \mathbf{E}_n)$ is connected, where $n \geq 1$. Let $x, y \in V^{(n+1)}$. If $x, y \in V_w^{(1)}$ for some $w \in \mathbb{W}_n$, then, since $(V^{(1)}, \mathbf{E}_1)$ is connected, there exists a path $\psi_w^{-1}(x) = z_0, z_1, \ldots, z_k = \psi_w^{-1}(y)$ in $(V^{(1)}, \mathbf{E}_1)$ connecting $\psi_w^{-1}(x)$ and $\psi_w^{-1}(y)$. We have $z_{i-1}, z_i \in V_{w_i}^{(0)}$ for some $w_i \in \mathbb{W}_1$, for each $1 \leq i \leq k$. Then if $z_i' = \psi_w(z_i)$, $z_{i-1}', z_i' \in F_{w_i \cdot w}$ and so $\{z_{i-1}', z_i'\} \in \mathbf{E}_{n+1}$. Thus x, y are connected by a path in $(V^{(n+1)}, \mathbf{E}_{n+1})$.

For general $x, y \in V^{(n+1)}$, as $(V^{(n)}, \mathbf{E}_n)$ is connected there exists a path $y_0, \ldots, y_m$ in $(V^{(n)}, \mathbf{E}_n)$ such that $\{y_{i-1}, y_i\} \in \mathbf{E}_n$ and x, y_0, and y, y_m, lie in the same $n+1$-cell. Then, by the above, the points $x, y_0, y_1, \ldots, y_m, y$ can be connected by chains of edges in $\mathbf{E}_{n+1}$.

To show that F is path-connected we actually construct a continuous path $\gamma : [0,1] \to F$ such that $F = \{\gamma(t), t \in [0,1]\}$. Let $x_0, \ldots, x_N$ be a path in $(V^{(1)}, \mathbf{E}_1)$ which is "space-filling", that is such that $V^{(1)} \subset \{x_0, \ldots, x_N\}$. Define $\gamma(i/N) = x_i$, $A_1 = \{i/N, 0 \leq i \leq N\}$. Now $x_0, x_1 \in V_w^{(0)}$, for some $w \in \mathbb{W}_1$. Let $x_0 = y_0, y_1, \ldots, y_m = x_1$ be in a space-filling path in $(V_w^{(1)}, \mathbf{E}_2)$. Define $\gamma(k/Nm) = y_k$, $0 \leq k \leq m$. Continuing in this way we fill each of the sets $V_w^{(1)}$, $w \in \mathbb{W}_1$, and so can define $A_2 \subset [0,1]$ such that $A_1 \subset A_2$, and $\gamma(t)$, $t \in A_2$ is a space filling path in the graph $(V^{(2)}, \mathbf{E}_2)$. Repeating this construction we obtain an increasing sequence (A_n) of finite sets such that $\gamma(t)$, $t \in A_n$ is a space filling path in $(V^{(n)}, \mathbf{E}_n)$, and $\cup_n A_n$ is dense in $[0,1]$. If $t \in A_n$, and $t' < t < t''$ are such that $(t', t'') \cap A_n = \{t\}$, then $\gamma(s)$ is in the same n-complex as $\gamma(t)$ for $s \in (t', t'')$. So, if $t \in [0,1] - A$, and $s_n, t_n \in A_n$ are chosen so that $s_n < t < t_n$, $(s_n, t_n) \cap A_n = \emptyset$, then the points $\gamma(u)$, $u \in A \cap (s, t)$ all lie in the same n-complex. So defining $\gamma(t) = \lim_n \gamma(t_n)$, we have that the limit exists, and γ is continuous. The construction of γ also gives that γ is space filling; if $w \in \mathbb{W}$ then for any $n \geq 1$ a section of the path, $\gamma(s)$, $a_n \leq s \leq b_n$, $s \in A_n$, fills $V_{w|n}^{(0)}$.

It follows immediately from the existence of γ that F is pathwise connected. $\square$

Remark. This proof returns to the roots of the subject – the original papers of Sierpinski [Sie1, Sie2] regarded the Sierpinski gasket and Sierpinski carpet as "curves".

Corollary 5.36. *Any ANF is pathwise connected.*

Remark 5.37. If F is a p.c.f.s.s. set, and the graph $(V^{(1)}, \mathbf{E}_1)$ is not connected, then it is easy to see that F is not connected.

For the case of ANFs, we wish to examine the structure of the graphs $(V^{(n)}, \mathbf{E}_n)$ a little more closely. Let $(F, \psi_1, \ldots, \psi_M)$ be an ANF. Then let

$$a = \min\left\{ |x - y| : x, y \in V^{(0)}, x \neq y \right\},$$

and set

$$E_0' = \big\{\{x,y\} \in V^{(0)} : |x - y| = a\big\},$$

$$E_n' = \Big\{\{x,y\} \in E_n : x = \psi_w(x'), y = \psi_w(y') \quad \text{for some}$$

$$w \in W_n, \{x',y'\} \in E_0'\Big\}, n \geq 1.$$

Proposition 5.38. *Let F be an ANF.*
(a) *Let $x,y,z \in V^{(0)}$ be distinct points. Then there exists a path in $(V^{(0)}, E_0')$ connecting x and y and not containing z.*
(b) *Let $x,y \in V^{(0)}$. There exists a path in $(V^{(1)}, E_1')$ connecting x,y which does not contain any point in $V^{(0)} - \{x,y\}$.*
(c) *Let $x,y,x',y' \in V^{(0)}$ with $|x - y| = |x' - y'|$. Then there exists $g \in \mathcal{G}_R$ such that $g(x') = x$, $g(y') = y$.*

Proof. If $\#\big(V^{(0)}\big) = 2$ then $E_0 = E_0'$, so (a) is vacuous and (b) is immediate from Corollary 5.36. So suppose $\#\big(V^{(0)}\big) \geq 3$.
(a) Since (see Remark 5.25(3)) $V^{(0)}$ is either a d-dimensional tetrahedron, or a d-dimensional simplex, or a regular polygon, this is evident. (For a proof which does not use this fact, see [L1, p. 34–35]).
(b) This now follows from (a) by the same kind of argument as that given in Proposition 5.35.
(c) Write $g[x,y]$ for the reflection in the hyperplane bisecting the line segment $[x,y]$. Let $g_1 = g[y,y']$, and $z = g_1(x')$. Then if $z = x$ we are done. Otherwise note that $|x - y| = |x' - y'| = |z - y|$, so if $g_2 = g[x,z]$ then $g_2(y) = y$. Hence $g_1 \circ g_2$ works. $\square$

Metrics on Nested Fractals.

Nested fractals, and ANFs, are subsets of $\mathbb{R}^d$, and so of course are metric spaces with respect to the Euclidean metric. Also, p.c.f.s.s. sets have been assumed to be metric spaces. However, these metrics do not necessarily have all properties we would wish for, such as the mid-point property that was used in Section 3. We saw in Section 2 that the geodesic metric on the Sierpinski gasket was equivalent to the Euclidean metric, but for a general nested fractal there may be no path of finite length between distinct points. (It is easy to construct examples). It is however, still possible to construct a geodesic metric on a ANF.

For simplicity, we will just treat the case of nested fractals. Let $(F, (\psi_i)_{i=1}^M)$ be a nested fractal, with length scale factor L. Write $d_n(x,y)$ for the natural graph distance in the graph $(V^{(n)}, E_n)$. Fix $x_0, y_0 \in V^{(0)}$ such that $\{x_0, y_0\} \in E_0'$, and let $a_n = d_n(x_0, y_0)$, and b_0 be the maximum distance between points in $(V^{(0)}, E_0')$.

Lemma 5.39. *If $x,y \in V^{(0)}$ then $a_n \leq d_n(x,y) \leq b_0 a_n$.*

Proof. Since x,y are connected by a path of length at most b_0 in $(V^{(0)}, E_0')$, the upper bound is evident. Fix x,y, and let $k = d_n(x,y)$. If $\{x,y\} \in E_0'$ then $d_n(x,y) = d_n(x_0,y_0) = a_n$, so suppose $\{x,y\} \notin E_0'$. Choose $y' \in V^{(0)}$ such that $\{x,y'\} \in E_0'$, let H be the hyperplane bisecting $[y,y']$ and let g be reflection in H. Write A, A' for the components of $\mathbb{R}^d - H$ containing y, y' respectively. As $|x - y'| < |x - y|$ we have $x \in A'$. Let $x = z_0, z_1, \ldots, z_k = y$ be the shortest path in $(V^{(n)}, E_n)$ connecting x and y. Let $j = \min\{i : z_i \in A\}$, and write $z_i' = z_i$ if $i < j$, $z_i' = g(z_i)$

if $i \geq j$. Then z_i', $0 \leq i \leq k$ is a path in $(V^{(n)}, \mathbf{E}_n)$ connecting x and y', and so $d_n(x, y) = k \geq d_n(x, y') = a_n$. $\qquad \square$

Lemma 5.40. *Let $x, y \in V^{(n)}$. Then for $m \geq 0$*

$$(5.24) \qquad a_m d_n(x, y) \leq d_{n+m}(x, y) \leq b_0 a_m d_n(x, y).$$

In particular

$$(5.25) \qquad a_n a_m \leq a_{n+m} \leq b_0 a_n a_m, \quad n \geq 0, m \geq 0.$$

Proof. Let $k = d_n(x, y)$, and let $x = z_0, z_1, \ldots, z_k = y$ be a shortest path connecting x and y in $(V^{(n)}, \mathbf{E}_n)$. Then since by Lemma 5.39 $d_m(z_{i-1}, z_i) \leq b_0 a_m$, the upper bound in (5.24) is clear.

For the lower bound, let $r = d_{n+m}(x, y)$, and let $(z_i)_{i=0}^{r}$ be a shortest path in $(V^{(n+m)}, \mathbf{E}_{n+m})$ connecting x, y. Let $0 = i_0, i_1, \ldots, i_s = r$ be successive disjoint hits by this path on $V^{(n)}$. (Recall the definition from Section 2: of course it makes sense for a deterministic path as well as a process). We have $s = d_n(x, y) \geq a_n$. Then since $z_{i_{j-1}}, z_{i_j}$ lie in the same n-cell, $i_j - i_{j-1} = d_m(z_{i_{j-1}}, z_{i_j}) \geq a_m$, by Lemma 5.39. So $r = \sum_{j=1}^{s}(i_j - i_{j-1}) \geq a_n a_m$. $\qquad \square$

Corollary 5.41. *There exists $\gamma \in [L, b_0 a_1]$ such that*

$$(5.26) \qquad b_0^{-1} \gamma^n \leq a_n \leq \gamma^n.$$

Proof. Note that $\log(b_0 a_n)$ is a subadditive sequence, and that $\log a_n$ is superadditive. So by the general theory of these sequences there exist θ_0, θ_1 such that

$$\theta_0 = \lim_{n \to \infty} n^{-1} \log(b_0 a_n) = \inf_{n \geq 0} n^{-1} \log(b_0 a_n),$$

$$\theta_1 = \lim_{n \to \infty} n^{-1} \log(a_n) = \sup_{n \geq 0} n^{-1} \log(a_n).$$

So $\theta_0 = \theta_1$, and setting $\gamma = e^{\theta_0}$, (5.26) follows.

To obtain bounds on γ note first that as $a_n \leq b_0 a_1 a_{n-1}$ we have $\gamma \leq b_0 a_1$. Also,

$$|x_0 - y_0| \leq a_n L^{-n} |x_0 - y_0|,$$

so $\gamma \geq L$. $\qquad \square$

Definition 5.42. We call $d_c = \log \gamma / \log L$ the *chemical exponent* of the fractal F, and γ the *shortest path scaling factor*.

Theorem 5.43. *There exists a metric d_F on F with the following properties.*
(a) There exists $c_1 < \infty$ such that for each $n \geq 0$, $w \in \mathbf{W}_n$,

$$(5.27) \qquad d_F(x, y) \leq c_1 \gamma^{-n} \text{ for } x, y \in F_w,$$

and

$$(5.28) \qquad d_F(x, y) \geq c_2 \gamma^{-n} \text{ for } x \in V^{(n)}, y \in N_n(x)^c.$$

(b) d_F induces the same topology on F as the Euclidean metric.

(c) d_F has the midpoint property.

(d) The Hausdorff dimension of F with respect to the metric d_F is

$$d_f(F) = \frac{\log M}{\log \gamma}. \tag{5.29}$$

Proof. Write $V = \cup_n V^{(n)}$. By Lemma 5.41 for $x, y \in V$ we have

$$b_0^{-1} \gamma^m d_n(x, y) \le d_{n+m}(x, y) \le b_0 \gamma^m d_n(x, y). \tag{5.30}$$

So $(\gamma^{-m} d_{n+m}(x, y), m \ge 0)$ is bounded above and below. By a diagonalization argument we can therefore find a subsequence $n_k \to \infty$ such that

$$d_F(x, y) = \lim_{k \to \infty} \gamma^{-n_k} d_{n_k}(x, y) \text{ exists for each } x, y \in V.$$

So, if $x, y \in V_w^{(0)}$ where $w \in \mathbb{W}_n$ then

$$c_0^{-1} \gamma^{-n} \le d_F(x, y) \le c_0 \gamma^{-n}. \tag{5.31}$$

It is clear that d_F is a metric on V.

Let $n \ge 0$ and $y \in V^{(n)}$. For $m = n - 1, n - 2, \ldots, 0$ choose inductively $y_m \in V^{(m)}$ such that y_m is in the same m-cell as $y_{m+1}, \ldots, y_n$. Then

$$d_{m+1}(y_m, y_{m+1}) \le \max\{d_1(x', y') : x', y' \in V^{(1)}\} = c < \infty.$$

So by (5.30) $d_n(y_m, y_{m+1}) \le b_0 \gamma^{n-(m+1)} c$, and therefore

$$d(y_k, y) \le c \sum_{i=k}^{\infty} \gamma^{-i-1} = c' \gamma^{-k}.$$

So if $x, y \in V$ are in the same k-cell, choosing x_k in the same way we have

$$d_F(x, y) \le d_F(x, x_k) + d_F(x_k, y_k) + d_F(y_k, y) \le c_1 \gamma^k, \tag{5.32}$$

since $d_k(x_k, y_k) \le b_0$. Thus d_F is uniformly continuous on $V \times V$, and so extends by continuity to a metric d_F on F. (a) is immediate from (5.31).

If $x, y \in V^{(n)}$ and $x \ne y$ then $d_F(x, y) \ge b_0^{-1} \gamma^{-n}$. This, together with (5.30), implies (b).

If $x, y \in V^{(n)}$ then there exists $z \in V^{(n)}$ such that

$$|d_n(u, z) - \tfrac{1}{2} d_n(x, y)| \le 1, \quad u = x, y.$$

So the metrics d_n have an approximate midpoint property: (c) follows by an easy limiting argument.

Let μ be the measure on F associated with the vector $\theta = (M^{-1}, \ldots, M^{-1})$. Thus $\mu(F_w) = M^{-|w|}$ for each $w \in \mathbb{W}_n$. Since we have $\text{diam}_{d_F}(F_w) \asymp \gamma^{-|w|}$, it follows that, writing $d_f = \log M / \log \gamma$,

$$c_5 r^{d_f} \le \mu(B_{d_F}(x, r)) \le c_6 r^{d_f}, \quad x \in F$$

and the conclusion then follows from Corollary 2.8. $\qquad\square$

Remark 5.44. The results here on the metric d_F are not the best possible. The construction here used a subsequence, and did not give a procedure for finding the scale factor γ. See [BS], [Kum2], [FHK], [Ki6] for more precise results.

6. Renormalization on Finitely Ramified Fractals.

Let $(F, \psi_1, \dots, \psi_M)$ be a p.c.f.s.s. set. We wish to construct a sequence $Y^{(n)}$ of random walks on the sets $V^{(n)}$, nested in a similar fashion to the random walks on the Sierpinski gasket considered in Section 2. The example of the Vicsek set shows that, in general, some calculation is necessary to find such a sequence of walks. As the random walks we treat will be symmetric, we will find it convenient to use the theory of Dirichlet forms, and ideas from electrical networks, in our proofs.

Fix a p.c.f.s.s. set $\big(F, (\psi_i)_{i=1}^M\big)$, and a Bernoulli measure $\mu = \mu_\theta$ on F, where each $\theta_i > 0$. We also choose a vector $r = (r_1, \dots, r_M)$ of positive "weights": loosely speaking r_i is the size of the set $\psi_i(F) = F_i$, for $1 \le i \le M$. We call r a *resistance vector*.

Definition 6.1. Let $\mathbb{D}$ be the set of Dirichlet forms $\mathcal{E}$ defined on $C\big(V^{(0)}\big)$. From Section 4 we have that each element $\mathcal{E} \in \mathbb{D}$ is of the form $\mathcal{E}_A$, where A is a conductance matrix. Let also $\mathbb{D}_1$ be the set of Dirichlet forms on $C\big(V^{(1)}\big)$.

We consider two operations on $\mathbb{D}$:

(1) Replication – i.e. extension of $\mathcal{E} \in \mathbb{D}$ to a Dirichlet form $\mathcal{E}^R \in \mathbb{D}_1$.
(2) Decimation/Restriction/Trace. Reduction of a form $\mathcal{E} \in \mathbb{D}_1$ to a form $\tilde{\mathcal{E}} \in \mathbb{D}$.

Note. In Section 4, we defined a Dirichlet form $(\mathcal{E}, \mathcal{D})$ with domain $\mathcal{D} \subset L^2(F, \mu)$. But for a finite set F, as long as μ charges every point in the set it plays no role in the definition of $\mathcal{E}$. We therefore will find it more convenient to define $\mathcal{E}$ on $C(F) = \{f : F \to \mathbb{R}\}$.

Definition 6.2. Given $\mathcal{E} \in \mathbb{D}$, define for $f, g \in C\big(V^{(1)}\big)$,

$$(6.2) \qquad \mathcal{E}^R(f, g) = \sum_{i=1}^M r_i^{-1} \mathcal{E}(f \circ \psi_i, g \circ \psi_i).$$

(Note that as $\psi_i : V^{(0)} \to V^{(1)}$, $f \circ \psi_i \in C\big(V^{(0)}\big)$.) Define $R : \mathbb{D} \to \mathbb{D}_1$ by $R(\mathcal{E}) = \mathcal{E}^R$.

Lemma 6.3. Let $\mathcal{E} = \mathcal{E}_A$, and let

$$(6.3) \qquad a_{xy}^R = \sum_{i=1}^M 1_{\big(x \in V_i^{(0)}\big)} 1_{\big(y \in V_i^{(0)}\big)} r_i^{-1} a_{\psi_i^{-1}(x), \psi_i^{-1}(y)}.$$

Then

$$(6.4) \qquad \mathcal{E}^R(f, g) = \tfrac{1}{2} \sum_{x,y} a_{xy}^R \big(f(x) - f(y)\big)\big(g(x) - g(y)\big).$$

$A^R = (a_{xy}^R)$ is a conductance matrix, and $\mathcal{E}^R$ is the associated Dirichlet form.

Proof. As the maps ψ_i are injective, it is clear that $a_{xy}^R \geq 0$ if $x \neq y$, and $a_{xx}^R \leq 0$. Also $a_{xy}^R = a_{yx}^R$ is immediate from the symmetry of A. Writing $x_i = \psi_i^{-1}(x)$ we have

$$\sum_{y \in V^{(1)}} a_{xy}^R = \sum_i r_i^{-1} 1_{V_i^{(0)}}(x) \sum_{y \in V^{(1)}} 1_{V_i^{(0)}}(y) a_{\psi_i^{-1}(x), \psi_i^{-1}(y)}$$

$$= \sum_i r_i^{-1} 1_{V_i^{(0)}}(x) \sum_{y \in V_i^{(0)}} a_{x,y} = 0,$$

so A^R is a conductance matrix.

To verify (6.4), it is sufficient by linearity to consider the case $f = g = \delta_z$, $z \in V^{(1)}$. Let $B = \{i \in \mathbb{W}_1 : z \in V_i^{(0)}\}$. If $i \notin B$, then $f \circ \psi_i(x) = 0$, since $\psi_i(x)$ cannot equal z. If $i \in B$, then $f \circ \psi_i(x) = \delta_{z_i}(x)$, where $z_i = \psi_i^{-1}(z)$. So,

$$\mathcal{E}(f \circ \psi_i, f \circ \psi_i) = \mathcal{E}(\delta_{z_i}, \delta_{z_i}) = -a_{z_i z_i}.$$

Thus

$$\mathcal{E}^R(f,f) = -\sum_{i \in B} r_i^{-1} a_{z_i z_i} = -\sum_{i=1}^M r_i^{-1} 1_{V_i^{(0)}}(z) a_{\psi_i^{-1}(z), \psi_i^{-1}(z)} = -a_{zz}^R,$$

while

$$\tfrac{1}{2} \sum_{x,y} a_{xy}^R \big(f(x) - f(y) \big)^2 = -f^T A^R f = -a_{zz}^R.$$

So (6.4) is verified. $\qquad\square$

The most intuitive explanation of the replication operation is in terms of electrical networks. Think of $V^{(0)}$ as an electric network. Take M copies of $V^{(0)}$, and rescale the ith one by multiplying the conductance of each wire by r_i^{-1}. (This explains why we called r a resistance vector). Now assemble these to form a network with nodes $V^{(1)}$, using the ith network to connect the nodes in $V_i^{(0)}$. Then $\mathcal{E}^R$ is the Dirichlet form corresponding to the network $V^{(1)}$.

As we saw in the previous section, for $x, y \in V^{(1)}$ there may in general be more than one 1-cell which contains both x and y: this is why the sum in (6.3) is necessary. If x and y are connected by k wires, with conductivities $c_1, \ldots, c_k$ then this is equivalent to connection by one wire of conductance $c_1 + \ldots + c_k$.

Remark 6.4. The replication of conductivities defined here is not the same as the replication of transition probabilities discussed in Section 2. To see the difference, consider again the Sierpinski gasket. Let $V^{(0)} = \{z_1, z_2, z_3\}$, and y_3 be the midpoint of $[z_1, z_2]$, and define y_1, y_2 similarly. Let A be a conductance matrix on $V^{(0)}$, and write $a_{ij} = a_{z_i z_j}$. Take $r_1 = r_2 = r_3 = 1$. While the continuous time Markov Chains $X^{(0)}$, $X^{(1)}$ associated with $\mathcal{E}_A$ and $\mathcal{E}_A^R$ will depend on the choice of a measure on $V^{(0)}$ and $V^{(1)}$, their discrete time skeletons that is, the processes $X^{(i)}$

sampled at their successive jump times do not – see Example 4.21. Write $Y^{(i)}$ for these processes. We have

$$\mathbb{P}^{y_s}\left(Y_1^{(1)} \in \{z_2, y_1\}\right) = \frac{a_{12} + a_{31}}{2a_{12} + a_{31} + a_{23}}.$$

On the other hand, if we replicate probabilities as in Section 2,

$$\mathbb{P}^{y_s}\left(Y_1^{(1)} \in \{z_2, y_1\}\right) = \mathbb{P}^{y_s}\left(Y_1^{(1)} \in \{z_1, y_2\}\right) = \tfrac{1}{2};$$

in general these expressions are different. So, even when we confine ourselves to symmetric Markov Chains, replication of conductivities and transition probabilities give rise to different processes.

Since the two replication operations are distinct, it is not surprising that the dynamical systems associated with the two operations should have different behaviours. In fact, the simple symmetric random walk on $V^{(0)}$ is stable fixed point if we replicate conductivities, but an unstable one if we replicate transition probabilities.

The second operation on Dirichlet forms, that of restriction or trace, has already been discussed in Section 4.

Definition 6.5. For $\mathcal{E} \in \mathbb{D}_1$ let

(6.5) $$T(\mathcal{E}) = Tr(\mathcal{E}|V^{(0)}).$$

Define $\Lambda : \mathbb{D} \to \mathbb{D}$ by $\Lambda(\mathcal{E}) = T\big(R(\mathcal{E})\big)$. Note that Λ is homogeneous in the sense that if $\theta > 0$,

$$\Lambda(\theta\mathcal{E}) = \theta\Lambda(\mathcal{E}).$$

Example 6.6. (The Sierpinski gasket). Let A be the conductance matrix corresponding to the simple random walk on $V^{(0)}$, so that

$$a_{xy} = 1, \quad x \neq y, \quad a_{xx} = -2.$$

Then A^R is the network obtained by joining together 3 symmetric triangular networks. If $\Lambda(\mathcal{E}_A) = \mathcal{E}_B$, then B is the conductance matrix such that the networks $(V^{(1)}, A^R)$ and $(V^{(0)}, B)$ are electrically equivalent on $V^{(0)}$. The simplest way to calculate B is by the $\Delta - Y$ transform. Replacing each of the triangles by an (upside down) Y, we see from Example 4.24 that the branches in the Y each have conductance 3. Thus $(V^{(1)}, A^R)$ is equivalent to a network consisting of a central triangle of wires of conductance $3/2$, and branches of conductance 3. Applying the transform again, the central triangle is equivalent to a Y with branches of conductance $9/2$. Thus the whole network is equivalent to a Y with branches of conductance $9/5$, or a triangle with sides of conductance $3/5$.

Thus we deduce

$$\Lambda(\mathcal{E}_A) = \mathcal{E}_B, \qquad \text{where} \quad B = \tfrac{3}{5}A.$$

The example above suggests that to find a decimation invariant random walk we need to find a Dirichlet form $\mathcal{E} \in \mathbb{D}$ such that for some $\lambda > 0$

$$(6.6) \qquad\qquad \Lambda(\mathcal{E}) = \lambda\mathcal{E}.$$

Thus we wish to find an eigenvector for the map Λ on $\mathbb{D}$. Since however (as we will see shortly) Λ is non-linear, this final formulation is not particularly useful. Two questions immediately arise: does there always exist a non-zero $(\mathcal{E}, \lambda)$ satisfying (6.6) and if so, is this solution (up to constant multiples) unique? We will abuse terminology slightly, and refer to an $\mathcal{E} \in \mathbb{D}$ such that (6.6) holds as a *fixed point* of Λ. (In fact it is a fixed point of Λ defined on a quotient space of $\mathcal{D}$.)

Example 6.7. (*"abc gaskets"* – see [HHW1]).

Let m_1, m_2, m_3 be integers with $m_i \geq 1$. Let z_1, z_2, z_3 be the corners of the unit triangle in $\mathbb{R}^2$, H be the closed convex hull of $\{z_1, z_2, z_3\}$. Let $M = m_1 + m_2 + m_3$, and let ψ_i, $1 \leq i \leq M$ be similitudes such that (writing for convenience $\psi_{M+j} = \psi_j$, $1 \leq j \leq M$) $H_i = \psi_i(H) \subset H$, and the M triangles H_i are arranged round the edge of H, such that each triangle H_i touches only H_{i-1} and H_{i+1}. (H_1 touches H_M and H_2 only). In addition, let $z_1 \in H_1$, $z_2 \in H_{m_3+1}$, $z_3 \in H_{m_3+m_1+1}$. So there are $m_3 + 1$ triangles along the edge $[z_1, z_2]$, and $m_1 + 1$, $m_2 + 1$ respectively along $[z_2, z_3]$, $[z_3, z_1]$. We assume that ψ_i are rotation-free. Note that the triangles H_2 and H_M do not touch, unless $m_1 = m_2 = m_3 = 1$.

Let F be the fractal obtained by Theorem 5.4 from $(\psi_1, \ldots, \psi_M)$. To avoid unnecessarily complicated notation we write ψ_i for both ψ_i and $\psi_i|_F$.

Figure 6.1: *abc* gasket with $m_1 = 4$, $m_2 = 3$, $m_3 = 2$.

It is easy to check that $(F, \psi_1, \ldots, \psi_M)$ is a p.c.f.s.s. set. Write $r = 1$, $s = m_3+1$, $t = m_3 + m_1 + 1$. We have $\pi(i\dot{s}) = \pi((i+1)\dot{r})$ for $1 \leq i \leq m_3$, $\pi(i\dot{t}) = \pi((i+1)\dot{s})$ for $m_3 + 1 \leq i \leq m_3 + m_1$, $\pi(i\dot{r}) = \pi((i+1)\dot{t})$ for $m_3 + m_1 + 1 \leq i \leq M - 1$, and $\pi(M\dot{r}) = \pi(1\dot{t})$. The set $B = \cup(H_i \cap H_j)$ consists of these points. Hence

$$P = \{(\dot{r}), (\dot{s}), (\dot{t})\}, \quad V^{(0)} = \{z_1, z_2, z_3\}.$$

While it is easier to define F in $\mathbb{R}^2$, rather than abstractly, doing so has the misleading consequence that it forces the triangles H_i to be of different sizes. However, we will view F as an abstract metric space in which all the triangles H_i are of equal size, and so we will take $r_i = 1$ for $1 \leq i \leq M$.

We now study the renormalization map Λ. If $\mathcal{E} = \mathcal{E}_A \in \mathbb{D}$, then A is specified by the conductivities

$$\alpha_1 = a_{z_2,z_3}, \quad \alpha_2 = a_{z_3,z_1}, \quad \alpha_3 = a_{z_1,z_2}.$$

Let $f : \mathbb{R}^3 \to \mathbb{R}^3$ be the renormalization map acting on $(\alpha_1, \alpha_2, \alpha_3)$. (So if $A = A(\alpha)$ then $\Lambda(\mathcal{E}) = \mathcal{E}_{A(f(\alpha))}$.)

It is easier to compute the action of the renormalization map on the variables β_i given by the $\triangle-Y$, transform. So let $\varphi : (0,\infty)^3 \to (0,\infty)^3$ be the $\triangle-Y$ map given in Example 4.24. Note that φ is bijective. Let $\beta = \varphi(\alpha)$ be the $Y-$conductivities, and write $\tilde{\beta} = (\tilde{\beta}_1, \tilde{\beta}_2, \tilde{\beta}_3)$ for the renormalized $Y-$conductivities: then $\tilde{\beta} = \varphi(f(\alpha))$.

Applying the $\triangle - Y$ transform on each of the small triangles, we obtain a network with nodes $z_1, z_2, z_3, y_1, y_2, y_3$, where $\{z_i, y_i\}$ has conductivity β_i, and if $i \neq j$ $\{y_i, y_j\}$ has conductivity β_i, and if $i \neq j$, $\{y_i, y_i\}$ has conductivity

$$\gamma_k = \frac{\beta_i \beta_j}{(\beta_i + \beta_j) m_k},$$

where $k = k(i,j)$ is such that $k \in \{1, 2, 3\} - \{i, j\}$.

Apply the $\triangle - Y$ transform again to $\{y_1, y_2, y_3\}$, to obtain a Y, with conductivities $\delta_1, \delta_2, \delta_3$, in the branches where

$$\delta_i \gamma_i = S = \gamma_1 \gamma_2 + \gamma_2 \gamma_3 + \gamma_3 \gamma_1, \qquad 1 \leq i \leq 3.$$

Then

(6.7)
$$\tilde{\beta}_1^{-1} = \beta_1^{-1} + \delta_1^{-1} = \beta_1^{-1} + \frac{\beta_2 \beta_3}{(\beta_2 + \beta_3) m_1 S}.$$

Suppose that $\alpha \in (0,\infty)^3$ is such that $\varphi(\alpha) = \lambda \alpha$ for some $\lambda > 0$. Then since $\varphi(\theta \alpha) = \theta \varphi(\alpha)$ for any $\theta > 0$, we deduce that $\tilde{\beta} = \varphi(f(\alpha)) = \lambda \beta$. So, from (6.7),

$$\lambda^{-1} \beta_1^{-1} = \beta_1^{-1} + \frac{\beta_2 \beta_2}{(\beta_2 + \beta_3) m_1 S},$$

which implies that $\lambda^{-1} > 1$. Writing $T = \beta_1 \beta_2 \beta_3 / S$, and $\theta = T\lambda(1 - \lambda)^{-1}$, we therefore have

$$m_1(\beta_2 + \beta_3) = \theta,$$

and (as S, T are symmetric in the β_i) we also obtain two similar equations. Hence

(6.8)
$$\beta_2 + \beta_3 = \theta/m_1, \quad \beta_3 + \beta_1 = \theta/m_2, \quad \beta_1 + \beta_2 = \theta/m_3,$$

which has solution

(6.9)
$$2\beta_1 = \theta(m_2^{-1} + m_3^{-1} - m_1^{-1}), \quad \text{etc.}$$

Since, however we need the $\beta_i > 0$, we deduce that a solution to the conductivity renormalization problem exists only if m_i^{-1} satisfy the triangle condition, that is that

(6.10) $\qquad m_2^{-1} + m_3^{-1} > m_1^{-1}, \quad m_3^{-1} + m_1^{-1} > m_2^{-1}, \quad m_1^{-1} + m_2^{-1} > m_3^{-1}.$

If (6.10) is satisfied, then (6.9) gives β_i such that the associated $\alpha = \varphi^{-1}(\beta)$ does satisfy the eigenvalue problem.

In the discussion above we looked for strictly positive α such that $\varphi(\alpha) = \lambda\alpha$. Now suppose that just one of the α_i, α_3 say, equals 0. Then while z, and z_2 are only connected via z_3 in the network $V^{(0)}$ they are connected via an additional path in the network $V^{(1)}$. So, $\varphi(\alpha)_3 > 0$, and α cannot be a fixed point. If now $\alpha_1 > 0$, and $\alpha_2 = \alpha_3 = 0$ then we obtain $\varphi(\alpha)_2 = \varphi(\alpha)_3 = 0$. So $\alpha = (1,0,0)$ satisfies $\varphi(\alpha) = \lambda\alpha$ for some $\lambda > 0$. Similarly $(0,1,0)$ and $(0,0,1)$ are also fixed points. Note that in these cases the network $(V^{(0)}, A(\alpha))$ is not connected.

The example of the abc gaskets shows that, even if fixed points exist, they may correspond to a reducible (ie non-irreducible) $\mathcal{E} \in \mathbb{D}$. The random walks (and limiting diffusion) corresponding to such a fixed point will be restricted to part of the fractal F. We therefore wish to find a *non-degenerate fixed point* of (6.6), that is an $\mathcal{E}_A \in \mathbb{D}$ such that the network $(V^{(0)}, A)$ is connected.

Definition 6.8. Let $\mathbb{D}^i$ be the set of $\mathcal{E} \in \mathbb{D}_0$ such that $\mathcal{E}$ is irreducible -- that is the network $(V^{(0)}, A)$ is connected. Call $\mathcal{E} \in \mathbb{D}$ *strongly irreducible* if $\mathcal{E} = \mathcal{E}_A$ and $a_{xy} > 0$ for all $x \neq y$. Write $\mathbb{D}^{si}$ for the set of strongly irreducible Dirichlet forms on $V^{(0)}$.

The existence problem therefore takes the form:

Problem 6.9. (Existence). Let $(F, \psi_1, ..., \psi_M)$ be a p.c.f.s.s. set and let $r_i > 0$. Does there exist $\mathcal{E} \in \mathbb{D}^i$, $\lambda > 0$, such that

(6.12 $\qquad\qquad\qquad\qquad \Lambda(\mathcal{E}) = \lambda\mathcal{E}?$

Before we pose the uniqueness question, we need to consider the role of symmetry. Let $(F, (\psi_i))$ be a p.c.f.s.s. set, and let $\mathcal{H}$ be a symmetry group of F.

Definition 6.10. $\mathcal{E} \in \mathbb{D}$ is *$\mathcal{H}$-invariant* if for each $h \in \mathcal{H}$

$$\mathcal{E}(f \circ h, g \circ h) = \mathcal{E}(f, g), \quad f, g \in C(V^{(0)}).$$

r is *$\mathcal{H}$-invariant* if $r_{\tilde{h}(i)} = r_i$ for all $h \in \mathcal{H}$. (Here $\tilde{h}$ is the bijection on I associated with h).

Lemma 6.11. (a) Let $\mathcal{E} = \mathcal{E}_A$. Then $\mathcal{E}$ is $\mathcal{H}$-invariant if and only if:

(6.13) $\qquad\qquad\qquad a_{h(x)\,h(y)} = a_{xy}$ for all $x, y \in V^{(0)}$, $h \in \mathcal{H}$.

(b) Let $\mathcal{E}$ and r be $\mathcal{H}$-invariant. Then $\Lambda\mathcal{E}$ is $\mathcal{H}$-invariant.

Proof. (a) This is evident from the equation $\mathcal{E}(1_x, 1_y) = -a_{xy}$.

(b) Let $f \in C(V^{(1)})$. Then if $h \in \mathcal{H}$,

$$\mathcal{E}^R(f \circ h, f \circ h) = \sum_i r_i^{-1} \mathcal{E}(f \circ h \circ \psi_i, f \circ h \circ \psi_i)$$

$$= \sum_i r_i^{-1} \mathcal{E}(f \circ \psi_{\bar{h}(i)} \circ h, f \circ \psi_{\bar{h}(i)} \circ h,)$$

$$= \sum_i r_{\bar{h}(i)}^{-1} \mathcal{E}(f \circ \psi_{\bar{h}(i)}, f \circ \psi_{\bar{h}(i)}) = \mathcal{E}^R(f, f).$$

If $g \in C(V^{(0)})$ then writing $\tilde{\mathcal{E}} = \Lambda(\mathcal{E})$, if $f|_{V^{(0)}} = g$ then as $f \circ h|_{V^{(0)}} = g \circ h$,

$$\tilde{\mathcal{E}}(g \circ h, g \circ h) \le \mathcal{E}^R(f \circ h, f \circ h) = \mathcal{E}^R(f, f),$$

and taking the infimum over f, we deduce that for any $h \in \mathcal{H}$, $\tilde{\mathcal{E}}(g \circ h, g \circ h) \le \tilde{\mathcal{E}}(g, g)$. Replacing g by $g \circ h$ and h by h^{-1} we see that equality must hold. □

If the fractal F has a non-trivial symmetry group $\mathcal{G}(F)$ then it is natural to restrict our attention to $\mathcal{G}(F)$-symmetric diffusions. We can now pose the uniqueness problem.

Problem 6.12. (Uniqueness). Let $(F, (\psi_i))$ be a p.c.f.s.s. set, let $\mathcal{H}$ be a symmetry group of F, and let r be $\mathcal{H}$-invariant. Is there at most one $\mathcal{H}$-invariant $\mathcal{E} \in \mathbb{D}^i$ such that $\Lambda(\mathcal{E}) = \lambda \mathcal{E}$?

(Unless otherwise indicated, when I refer to fixed points for nested fractals, I will assume they are invariant under the symmetry group $\mathcal{G}_R$ generated by the reflections in hyperplanes bisecting the lines $[x, y]$, $x, y \in V^{(0)}$).

The following example shows that uniqueness does not hold in general.

Example 6.13. (Vicsek sets – see [Me3].) Let $(F, \psi_i, 1 \le i \le 5)$ be the Vicsek set – see Section 2. Write $\{z_1, z_2, z_3, z_4,\}$ for the 4 corners of the unit square in $\mathbb{R}^2$. For $\alpha, \beta, \gamma > 0$ let $A(\alpha, \beta, \gamma)$ be the conductance matrix given by

$$a_{12} = a_{23} = a_{34} = a_{41} = \alpha, \quad \alpha_{13} = \beta, \quad a_{24} = \gamma,$$

where $a_{ij} = a_{z_i z_j}$. If $\mathcal{H}$ is the group on F generated by reflections in the lines $[z_1, z_3]$ and $[z_2, z_4]$ then A is clearly $\mathcal{H}$-invariant. Define $\tilde{\alpha}, \tilde{\beta}, \tilde{\gamma}$ by

$$\Lambda(\mathcal{E}_A) = \mathcal{E}_{A(\tilde{\alpha}, \tilde{\beta}, \tilde{\gamma})}.$$

Then several minutes calculation with equivalent networks shows that

(6.14)
$$\tilde{\alpha} = \frac{\alpha(\alpha + \beta)(\alpha + \gamma)}{5\alpha^2 + 3\alpha\beta + 3\alpha\gamma + \beta\gamma},$$
$$\tilde{\beta} = \tfrac{1}{3}(\alpha + \beta) - \tilde{\alpha},$$
$$\tilde{\gamma} = \tfrac{1}{3}(\alpha + \gamma) - \tilde{\alpha}.$$

If $(1, \beta, \gamma)$ is a fixed point then $(\tilde{\alpha}, \tilde{\beta}, \tilde{\gamma}) = (\theta, \theta\beta, \theta\gamma)$ for some $\theta \ge 0$, so that $\tilde{\beta} = \tilde{\alpha}\beta$, $\tilde{\gamma} = \tilde{\alpha}\gamma$. So $\tilde{\alpha} = \tfrac{1}{3}$, and this implies that $\beta\gamma = 1$. We therefore have that $(1, \beta, \beta^{-1})$ is a fixed point (with $\lambda = \tfrac{1}{3}$) for any $\beta \in (0, \infty)$ Thus for the group $\mathcal{H}$ uniqueness does not hold.

However if we replace $\mathcal{H}$ by the group $\mathcal{G}_R = \mathcal{G}(F)$, generated by all the symmetries of the square then for $\mathcal{E}_A$ to be $\mathcal{G}_R$-invariant we have to have $\beta = \gamma$. So in this case we obtain

(6.15)
$$\tilde{\alpha}(\alpha,\beta) = \frac{\alpha(\alpha+\beta)^2}{5\alpha^2 + 6\alpha\beta + \beta^2},$$
$$\tilde{\beta}(\alpha,\beta) = \tfrac{1}{3}(\alpha+\beta) - \tilde{\alpha}.$$

This has fixed points $(0,\beta)$, $\beta > 0$, and (α,α), $\alpha > 0$. The first are degenerate, the second not, so in this case, as we already saw in Section 2, uniqueness does hold for Problem 6.12.

This example also shows that Λ is in general non-linear.

As these examples suggest, the general problem of existence and uniqueness is quite hard. For all but the simplest fractals, explicit calculation of the renormalization map Λ is too lengthy to be possible without computer assistance – at least for 20th century mathematicians. Lindstrøm [L1] proved the existence of a fixed point $\mathcal{E} \in \mathbb{D}^{si}$ for nested fractals, but did not treat the question of uniqueness. After the appearance of [L1], the uniqueness of a fixed point for Lindstrøm's canonical example, the snowflake (Example 5.26) remained open for a few years, until Green [Gre] and Yokai [Yo] proved uniqueness by computer calculations.

The following analytic approach to the uniqueness problem, using the theory of quadratic forms, has been developed by Metz and Sabot – see [Me2-Me5, Sab1, Sab2]. Let M_+ be set of symmetric bilinear forms $Q(f,g)$ on $C(V^{(0)})$ which satisfy

$$Q(1,1) = 0,$$
$$Q(f,f) \geq 0 \quad \text{for all } f \in C(V^{(0)}).$$

For $Q_1, Q_2 \in M_+$ we write $Q_1 \geq Q_2$, if $Q_2 - Q_1 \in M_+$ or equivalently if $Q_2(f,f) \geq Q_1(f,f)$ for all $f \in C(V^{(0)})$. Then $\mathbb{D} \subset M_+$; it turns out that we need to consider the action of Λ on M_+, and not just on $\mathbb{D}$. For $Q \in M_+$, the replication operation is defined exactly as in (6.2)

(6.16)
$$Q^R(f,g) = \sum_{i=1}^{M} r_i^{-1} Q(f \circ \psi_i, g \circ \psi_i), \quad f, g \in C(V^{(1)}).$$

The decimation operation is also easy to extend to M_+ :

$$T(Q^R)(g,g) = \inf\{Q^R(f,f) : f \in C(V^{(0)}), f|_{V^{(0)}} = g\};$$

we can write $T(Q^R)$ in matrix terms as in (4.24). We set $\Lambda(Q) = T(Q^R)$.

Lemma 6.14. *The map Λ on M_+ satisfies:*
(a) $\Lambda : M_+ \to M_+$, and is continuous on $\text{int}(M_+)$.
(b) $\Lambda(Q_1 + Q_2) \geq \Lambda(Q_1) + \Lambda(Q_2)$.
(c) $\Lambda(\theta Q) = \theta \Lambda(Q)$

Proof. (a) is clear from the formulation of the trace operation in matrix terms.

Since the replication operation is linear, we clearly have $Q^R = Q_1^R + Q_2^R$, $(\theta Q)^R = \theta Q^R$. (c) is therefore evident, while for (b), if $g \in C(V^{(0)})$,

$$
\begin{aligned}
T(Q^R)(g,g) &= \inf\{Q_1^R(f,f) + Q_2^R(f,f) : f|_{V^{(0)}} = g\} \\
&\geq \inf\{Q_1^R(f,f) : f|_{V^{(0)}} = g\} + \inf\{Q_2^R(f,f) : f|_{V^{(0)}} = g\} \\
&= T(Q_1^R)(g,g) + T(Q_2^R)(g,g). \qquad\square
\end{aligned}
$$

Note that for $\mathcal{E} \in \mathbb{D}^i$, we have $\mathcal{E}(f,f) = 0$ if only if f is constant.

Definition 6.15. For $\mathcal{E}_1, \mathcal{E}_2 \in \mathbb{D}^i$ set

$$
\begin{aligned}
m(\mathcal{E}_1/\mathcal{E}_2) &= \sup\{\alpha \geq 0; \ \mathcal{E}_1 - \alpha\mathcal{E}_2 \in \mathbb{M}_+\}. \\
&= \inf\{\frac{\mathcal{E}_1(f,f)}{\mathcal{E}_2(f,f)} : f \text{ non constant}\}.
\end{aligned}
$$

Similarly let

$$
M(\mathcal{E}_1/\mathcal{E}_2) = \sup\{\frac{\mathcal{E}_1(f,f)}{\mathcal{E}_2(f,f)} : f \text{ non constant}\}.
$$

Note that

$$
(6.18) \qquad\qquad M(\mathcal{E}_1/\mathcal{E}_2) = m(\mathcal{E}_2/\mathcal{E}_1)^{-1}.
$$

Lemma 6.16. (a) For $\mathcal{E}_1, \mathcal{E}_2 \in \mathbb{D}^i$, $0 < m(\mathcal{E}_1, \mathcal{E}_2) < \infty$.
(b) If $\mathcal{E}_1$, $\mathcal{E}_2 \in \mathbb{D}^i i$ then $m(\mathcal{E}_1/\mathcal{E}_2) = M(\mathcal{E}_1/\mathcal{E}_2)$ if and only if $\mathcal{E}_2 = \lambda\mathcal{E}_1$ for some $\lambda > 0$.
(c) If $\mathcal{E}_1, \mathcal{E}_2, \mathcal{E}_3 \in \mathbb{D}^i$ then

$$
m(\mathcal{E}_1/\mathcal{E}_3) \geq m(\mathcal{E}_1/\mathcal{E}_2)\, m(\mathcal{E}_2/\mathcal{E}_3),
$$

$$
M(\mathcal{E}_1/\mathcal{E}_3) \leq M(\mathcal{E}_1/\mathcal{E}_2)\, M(\mathcal{E}_2/\mathcal{E}_3).
$$

Proof. (a) This follows from the fact that $\mathcal{E}_i$ are irreducible, and so vanish only on the subspace of constant functions.
(b) is immediate from the definition of m and M.
(c) We have

$$
m(\mathcal{E}_1/\mathcal{E}_3) = \inf_f \frac{\mathcal{E}_1(f,f)}{\mathcal{E}_2(f,f)} \frac{\mathcal{E}_2(f,f)}{\mathcal{E}_3(f,f)} \geq m(\mathcal{E}_1/\mathcal{E}_2)m(\mathcal{E}_2/\mathcal{E}_3);
$$

while the second assertion is immediate from (6.18). $\qquad\square$

Definition 6.17. Define

$$
d_H(\mathcal{E}_1, \mathcal{E}_2) = \log\frac{M(\mathcal{E}_1\mathcal{E}_2)}{m(\mathcal{E}_1\mathcal{E}_2)}, \quad \mathcal{E}_1, \mathcal{E}_2 \in \mathbb{D}^i.
$$

Let $p\mathbb{D}^i$ be the projective space $\mathbb{D}^i/\sim$, where $\mathcal{E}_1 \sim \mathcal{E}_2$ if $\mathcal{E}_1 = \lambda\mathcal{E}_2$. d_H is called *Hilbert's projective metric* – see [Nus], [Me4].

Proposition 6.18. *(a)* $d_H(\mathcal{E}_1, \mathcal{E}_2) = 0$ *if and only if* $\mathcal{E}_1 = \lambda \mathcal{E}_2$ *for some* $\lambda > 0$.
(b) d_H *is a pseudo-metric on* $\mathbb{D}^i$, *and a metric on* $p\,\mathbb{D}^i$.
(c) If $\mathcal{E}, \mathcal{E}_0, \mathcal{E}_1 \in \mathbb{D}^i$ *then for* $\alpha_0, \alpha_1 > 0$,

$$d_H(\mathcal{E}, \alpha_0\,\mathcal{E}_0 + \alpha_1 \mathcal{E}_1) \leq \max(d_H(\mathcal{E}, \mathcal{E}_0), d_H(\mathcal{E}, \mathcal{E}_1)).$$

In particular open balls in d_H *are convex.*
(d) $(p\,\mathbb{D}^i, d_H)$ *is complete.*

Proof. (a) is evident from Lemma 6.17(b). To prove (b) note that $d_H(\mathcal{E}_1, \mathcal{E}_2) \geq 0$, and that $d_H(\mathcal{E}_1, \mathcal{E}_2) = d_H(\mathcal{E}_2, \mathcal{E}_1)$ from (6.18). The triangle inequality is immediate from Lemma 6.17(c). So d_H is a pseudo metric on $\mathbb{D}^i$.
   To see that d_H is a metric on $p\,\mathbb{D}^i$, note that

$$m(\lambda \mathcal{E}_1 / \mathcal{E}_2) = \lambda m(\mathcal{E}_1 / \mathcal{E}_2), \quad \lambda > 0,$$

from which it follows that $d_H(\lambda \mathcal{E}_1, \mathcal{E}_2) = d_H(\mathcal{E}_1, \mathcal{E}_2)$ and thus d_H is well defined on $p\,\mathbb{D}^i$. The remaining properties are now immediate from those of d_H on $\mathbb{D}^i$.
(c) Replacing $\mathcal{E}_1$ by $\big(m(\mathcal{E}_1/\mathcal{E}_0)/m(\mathcal{E}/\mathcal{E}_1)\big)\mathcal{E}_1$ we can suppose that

$$m(\mathcal{E}/\mathcal{E}_0) = m(\mathcal{E}/\mathcal{E}_1) = m.$$

Write $M_i = M(\mathcal{E}/\mathcal{E}_i)$. Then if $\mathcal{F} = \alpha_0\,\mathcal{E}_0 + \alpha_1 \mathcal{E}_1$,

$$M(\mathcal{E}/\mathcal{F}) = \inf_f \frac{\alpha_0 \mathcal{E}_0(f, f) + \alpha_1 \mathcal{E}_1(f, f)}{\mathcal{E}(f, f)}$$
$$\geq \alpha_0 m(\mathcal{E}/\mathcal{E}_0) + \alpha_1 m(\mathcal{E}/\mathcal{E}_1) = \alpha_0 + \alpha_1.$$

Similarly $M(\mathcal{E}/\mathcal{F}) \leq \alpha_0 M_0 + \alpha_1 M_1$. Therefore

$$\exp d_H(\mathcal{E}, \mathcal{F}) \leq (\alpha_0/(\alpha_0 + \alpha_1))(M_0/m) + (\alpha_1/(\alpha_0 + \alpha_1))(M_1/m)$$
$$\leq \max(M_0/m, M_1/m).$$

   It is immediate that if $\mathcal{E}_i \in B(\mathcal{E}, r)$ then $d_H(\mathcal{E}, \lambda \mathcal{E}_0 + (1 - \lambda)\mathcal{E}_1) < r$, so that $B(\mathcal{E}, r)$ is convex. For (d) see [Nus, Thm. 1.2]. □

Theorem 6.19. *Let* $\mathcal{E}_1, \mathcal{E}_2 \in \mathbb{D}^i$. *Then*

(6.19) $$m(\Lambda(\mathcal{E}_1), \Lambda(\mathcal{E}_2)) \geq m(\mathcal{E}_1, \mathcal{E}_2),$$

(6.20) $$M(\Lambda(\mathcal{E}_1), \Lambda(\mathcal{E}_2)) \leq M(\mathcal{E}_1, \mathcal{E}_2).$$

In particular Λ *is non-expansive in* d_H :

(6.21) $$d_H(\Lambda(\mathcal{E}_1), \Lambda(\mathcal{E}_2)) \leq d_H(\mathcal{E}_1, \mathcal{E}_2).$$

Proof. Suppose $\alpha < m(\mathcal{E}_1, \mathcal{E}_2)$. Then $Q = \mathcal{E}_1 - \alpha \mathcal{E}_2 \in \mathbb{M}_+$, and $Q(f, f) > 0$, for all non-constant $f \in C(V^{(0)})$. So by Lemma 6.14

$$\Lambda(\mathcal{E}_1) = \Lambda(Q + \alpha \mathcal{E}_2) \geq \Lambda(Q) + \alpha \Lambda(\mathcal{E}_2),$$

and since $\Lambda(Q) \geq 0$, this implies that $\Lambda(\mathcal{E}_1) - \alpha\Lambda(\mathcal{E}_2) \geq 0$. So $\alpha < m(\Lambda(\mathcal{E}_1), \Lambda(\mathcal{E}_2))$, and thus $m(\mathcal{E}_1, \mathcal{E}_2) \leq m(\Lambda(\mathcal{E}_1), \Lambda(\mathcal{E}_2))$, proving (6.19). (6.20) and (6.21) then follow immediately from (6.19), and the definition of d_H. $\qquad\square$

A strict inequality in (6.21) would imply the uniqueness of fixed points. Thus the example of the Vicsek set above shows that strict inequality cannot hold in general. So this Theorem gives us much less than we might hope. Nevertheless, we can obtain some useful information.

Corollary 6.20. *(See [HHW1, Cor. 3.7]) Suppose $\mathcal{E}_1, \mathcal{E}_2$ are fixed points satisfying $\Lambda(\mathcal{E}_i) = \lambda_i \mathcal{E}_i$, $i = 1, 2$. Then $\lambda_1 = \lambda_2$.*

Proof. From (6.19)

$$m(\mathcal{E}_1/\mathcal{E}_2) \leq m\big(\Lambda(\mathcal{E}_1)/\Lambda(\mathcal{E}_2)\big) = (\lambda_1/\lambda_2)m(\mathcal{E}_1/\mathcal{E}_2),$$

so that $\lambda_1 \geq \lambda_2$. Interchanging $\mathcal{E}_1$ and $\mathcal{E}_2$ we obtain $\lambda_1 = \lambda_2$. $\qquad\square$

We can also deduce the existence of $\mathcal{H}$-invariant fixed points.

Proposition 6.21. *Let $\mathcal{H}$ be a symmetry group of F. If Λ has a fixed point $\mathcal{E}_1$ in $\mathbb{D}^i$ then Λ has an $\mathcal{H}$-invariant fixed point in $\mathbb{D}^i$.*

Proof. Let $A = \{\mathcal{E} \in \mathbb{D}^i : \mathcal{E}$ is $\mathcal{H}$-invariant.$\}$. (It is clear from Lemma 6.11 that A is non-empty). Then by Lemma 6.11(b) $\Lambda : A \to A$. Let $\mathcal{E}_0 \in A$, and write $r = d_H(\mathcal{E}_1, \mathcal{E}_0)$, $B = B_{d_H}(\mathcal{E}_1, 2r)$. By Theorem 6.20 $\Lambda : B \to B$. So $\Lambda : A \cap B \to A \cap B$. Each of A, B is convex (A is convex as the sum of two $\mathcal{H}$-invariant forms is $\mathcal{H}$-invariant, B by Proposition 6.18(c)), and so $A \cap B$ is convex. Since Λ is a continuous function on a convex space, by the Brouwer fixed point theorem Λ has a fixed point $\mathcal{E}' \in A \cap B$, and $\mathcal{E}'$ is $\mathcal{H}$-invariant. $\qquad\square$

We will not make use of the following result, but is useful for understanding the general situation.

Corollary 6.22. *Suppose Λ has two distinct fixed points $\mathcal{E}_1$ and $\mathcal{E}_2$ (with $\mathcal{E}_1 \neq \lambda\mathcal{E}_2$ for any λ). Then Λ has uncountably many fixed points.*

Proof. (Note that the example of the Vicsek set shows that $\frac{1}{2}(\mathcal{E}_1 + \mathcal{E}_2)$ is not necessarily a fixed point). Let $\mathbb{F} \subset \mathbb{D}^i$ be the set of fixed points. Let $\mathcal{E}_0, \mathcal{E}_1 \in \mathbb{F}$; multiplying $\mathcal{E}_1$ by a scalar we can take $m(\mathcal{E}_0, \mathcal{E}_1) = 1$. Write $R = d_H(\mathcal{E}_0, \mathcal{E}_1)$. If $\mathcal{E}_\lambda = \lambda\mathcal{E}_1 + (1 - \lambda)\mathcal{E}_0$ then as in Proposition 6.19(c)

$$\exp d_{\mathcal{H}}(\mathcal{E}_\lambda, \mathcal{E}_0) \leq (1 - \lambda) + \lambda M(\mathcal{E}_1, \mathcal{E}_0)$$

and so

$$d_{\mathcal{H}}(\mathcal{E}_{1/2}, \mathcal{E}_0) \leq \log((1 + e^R)/2).$$

Thus there exists δ, depending only on R, such that

$$A = \{\mathcal{E} \in \mathbb{D}^i : \mathcal{E} \in B(\mathcal{E}_0, (1 - \delta)R) \bigcap B(\mathcal{E}_1, (1 - \delta)R)\}$$

is non-empty. Since Λ preserves A, Λ has a fixed point in A. $\mathbb{F}$ thus has the property:

if $\mathcal{E}_1, \mathcal{E}_2$ are distinct elements of $\mathbb{F}$ then there exists $\mathcal{E}_3 \in \mathbb{F}$

such that $0 < d_{\mathcal{H}}(\mathcal{E}_3, \mathcal{E}_1) < d_{\mathcal{H}}(\mathcal{E}_2, \mathcal{E}_1)$.

As $\mathbb{F}$ is closed (since Λ is continuous) we deduce that $\mathbb{F}$ is perfect, and therefore uncountable. □

This if as far as we will go in general. For nested fractals the added structure – symmetry and the embedding in $\mathbb{R}^d$, enables us to obtain stronger results. If $(F, (\psi_i))$ is a nested fractal, or an ANF, we only consider the set $\mathbb{D}^i \cap \{\mathcal{E} : \mathcal{E}$ is $\mathcal{G}_R$-invariant$\}$, so that in discussing the existence and uniqueness of fixed points we will be considering only $\mathcal{G}_R$-invariant ones.

Let $(F, (\psi_i))$ be a nested fractal, write $\mathcal{G} = \mathcal{G}_R$ and let $\mathcal{E}_A$ be a ($\mathcal{G}$-invariant) Dirichlet form on $C(V^{(0)})$. $\mathcal{E}_A$ is determined by the conductances on the equivalence classes of edges in $(V^{(0)}, \mathcal{E}_0)$ under the action of $\mathcal{G}$. By Proposition 5.38(c) if $|x - y| = |x' - y'|$ then the edges $\{x, y\}$ and $\{x', y'\}$ are equivalent, so that $A_{xy} = A_{x'y'}$.

List the equivalence classes in order of increasing Euclidean distance, and write $\alpha_1, \alpha_2 ..., \alpha_k$ for the common conductances of the edges. Since $\tilde{A} = \Lambda(A)$ is also $\mathcal{G}$-invariant, Λ induces a map $\Lambda' : \mathbb{R}_+^k \to \mathbb{R}_+^k$ such that, using obvious notation, $\Lambda(A(\alpha)) = A(\Lambda'(\alpha))$.

Set $\mathbb{D}^* = \{\alpha : \alpha_1 \geq \alpha_2 \geq ... \geq \alpha_k > 0\}$. Clearly we have $\mathbb{D}^* \subset \mathbb{D}^{si}$. We have the following existence theorem for nested fractals.

Theorem 6.23. *(See [L1, p. 48]). Let $(F, (\psi_i))$ be a nested fractal (or an ANF). Then Λ has a fixed point in $\mathbb{D}^*$.*

Proof. Let $\mathcal{E}_A \in \mathbb{D}^*$, and let $\alpha_1, ... \alpha_k$ be the associated conductivities. Let $(Y_t, t \geq 0, \mathbb{Q}^x, x \in V^{(0)})$ be the continuous time Markov chain associated with $\mathcal{E}_A$, and let $(\widehat{Y}_n, n \geq 0, \mathbb{Q}^x, x \in V^{(0)})$ be the discrete time skeleton of Y.

Let $E_0^{(1)}, ..., E_0^{(k)}$ be the equivalence classes of edges in $(V^{(0)}, E_0)$, so that $A_{xy} = \alpha_j$ if $\{x, y\} \in E_0^{(j)}$. Then if $\{x, y\} \in E_0^{(j)}$,

$$\mathbb{Q}^x(\widehat{Y}_1 = y) = \frac{\alpha_j}{\sum_{y \neq x} A_{xy}}.$$

As $c_1 = \sum_{y \neq x} A_{xy}$ does not depend on x (by the symmetry of $V^{(0)}$) the transition probabilities of $\widehat{Y}$ are proportional to the α_j.

Now let $R(A)$ be the conductivity matrix on $V^{(1)}$ attained by replication of A. Let $(X_t, t \geq 0, \mathbb{P}^x, x \in V^{(1)})$ and $(\widehat{X}_n, n \geq 0, \mathbb{P}^x, x \in V^{(1)})$ be the associated Markov Chains. Let $T_0, T_1, ...$ be successive disjoint hits (see Definition 2.14) on $V^{(0)}$ by $\widehat{X}_n$.

Write $\tilde{A} = \Lambda(A)$, and $\tilde{\alpha}$ for the edge conductivities given by A. Using the trace theorem,

$$\mathbb{P}^x(\widehat{X}_{T_1} = y) = \tilde{\alpha}_j / c_1 \quad \text{if } \{x, y\} \in E_0^{(j)}.$$

Now let $x_1, y_1, y_2 \in V^{(0)}$, with $|x - y_1| < |x - y_2|$. We will prove that

(6.23) $$\mathbb{P}^{x_1}(\widehat{X}_{T_1} = y_2) < \mathbb{P}^{x_1}(\widehat{X}_{T_1} = y_1).$$

Let H be the hyperplane bisecting $[y_1, y_2]$, let g be reflection in H, and $x_2 = g(x_1)$. Let

$$T = \min\{n \geq 0 : \widehat{X}_n \in V^{(0)} - \{x_1\}\},$$

so that $T_1 = T$ $\mathbb{P}^{x_1}$-almost surely. Set

$$f_n(x) = \mathbb{E}^x 1_{(T \le n)}(1_{y_1}(\widehat{X}_T) - 1_{y_2}(\widehat{X}_T)).$$

Let $p(x,y)$, $x,y \in V^{(0)}$ be the transition probabilities of $\widehat{X}$. Then

(6.24
$$f_{n+1}(x) = 1_A(x) f_0(x) + 1_{A^c}(x) \sum_y p(x,y) f_n(y).$$

Let $J_{12} = \{x \in V^{(1)} : |x - y_1| \le |x - y_2|\}$, and define J_{21} analogously. We prove by induction that f_n satisfies

(6.25a) $$f_n(x) \ge 0, \qquad x \in J_{12},$$
(6.25b) $$f_n(x) + f_n(g(x)) \ge 0, \qquad x \in J_{12}.$$

Since $f_0 = 1_{y_1} - 1_{y_2}$, and $y_1 \in J_{12}$, f_0 satisfies (6.25). Let $x \in A^c \cup J_{12}$ and suppose f_n satisfies (6.25). If $p(x,y) > 0$, and $y \in J_{12}^c$, then x, y are in the same 1-cell so if $y' = g(y)$, y' is also in the same 1-cell as x_1 and $|x - y'| \le |x - y|$. So (since $\mathcal{E}_A \in \mathbb{D}^*$), $p(x,y') \ge p(x,y)$ and using (6.25b), as $f_n(y') \ge 0$,

$$p(x,y)f_n(y) + p(x,y')f_n(y') \ge p(x,y)(f_n(y) + f_n(g(y))) \ge 0.$$

Then by (6.24), $f_{n+1}(x) \ge 0$. A similar argument implies that f_{n+1} satisfies (6.25b).

So (f_n) satisfies (6.25) for all n, and hence its limit f_∞ does. Thus $f_\infty(x_1) = \mathbb{P}^x(\widehat{X}_T = y_1) - \widehat{\mathbb{P}}(\widehat{X}_T = y_2) \ge 0$, proving (6.23).

From (6.23) we deduce that $\tilde{\alpha}_1 \ge \tilde{\alpha}_2 \ge ... \ge \tilde{\alpha}_k$, so that $\Lambda : \mathbb{D}^* \to \mathbb{D}^*$. As $\Lambda'(\theta \alpha) = \theta \Lambda'(\alpha)$, we can restrict the action of Λ' to the set

$$\{\alpha \in \mathbb{R}_+^k : \alpha_1 \ge ... \ge \alpha_k \ge 0, \sum \alpha_i = 1\}.$$

This is a closed convex set, so by the Brouwer fixed point theorem, Λ' has a fixed point in $\mathbb{D}^*$. $\qquad \square$

Remark 6.24. The proof here is essentially the same as that in Lindstrøm [L1]. The essential idea is a kind of reflection argument, to show that transitions along shorter edges are more probable. This probabilistic argument yields (so far) a stronger existence theorem for nested fractals than the analytic arguments used by Sabot [Sab1] and Metz [Me7]. However, the latter methods are more widely applicable.

It does not seem easy to relax any of the conditions on ANFs without losing some link in the proof of Theorem 6.23. This proof used in an essential fashion not only the fact that $V^{(0)}$ has a very large symmetry group, but also the Euclidean embedding of $V^{(0)}$ and $V^{(1)}$.

The following uniqueness theorem for nested fractals was proved by Sabot [Sab1]. It is a corollary of a more general theorem which gives, for p.c.f.s.s. sets, sufficient conditions for existence and uniqueness of fixed points. A simpler proof of this result has also recently been obtained by Peirone [Pe].

Theorem 6.25. *Let* $(F, (\psi_i))$ *be a nested fractal. Then* Λ *has a unique* $\mathcal{G}_R$*-invariant non-degenerate fixed point.*

Definition 6.26. Let $\mathcal{E}$ be a fixed point of Λ. The *resistance scaling factor* of $\mathcal{E}$ is the unique $\rho > 0$ such that
$$\Lambda(\mathcal{E}) = \rho^{-1} \mathcal{E}.$$

Very often we will also call ρ the resistance scaling factor of F: in view of Corollary 6.21, ρ will have the same value for any two non-degenerate fixed points.

Proposition 6.27. *Let* $(F, (\psi_i))$ *be a p.c.f.s.s. set, let* (r_i) *be a resistance vector, and let* $\mathcal{E}_A$ *be a non-degenerate fixed point of* Λ. *Then for each* $s \in \{1, ...M\}$ *such that* $\pi(\dot{s}) \in V^{(0)}$,

$$(6.27) \qquad\qquad r_s \rho^{-1} < 1.$$

Proof. Fix $1 \le s \le M$, let $x = \pi(\dot{s})$, and let $f = 1_x \in C(V^{(0)})$. Then
$$\mathcal{E}_A(f, f) = \sum_{y \in V^{(0)}, \, y \neq x} A_{xy} = |A_{xx}|.$$

Let $g = 1_x \in C(V^{(1)})$. As $\Lambda(\mathcal{E}_A) = \rho^{-1} \mathcal{E}_A$,

$$(6.28) \qquad\qquad \rho^{-1} |A_{xx}| = \Lambda(\mathcal{E}_A)(f, f) < \mathcal{E}_A^R(g, g):$$

since g is not harmonic with respect to $\mathcal{E}_A^R$, strict inequality holds in (6.28). By Proposition 5.24(c), x is in exactly one 1-complex. So
$$\mathcal{E}_A^R(g, g) = \sum_i r_i^{-1} \mathcal{E}_A(g \circ \psi_i, g \circ \psi_i) = r_s^{-1} |A_{xx}|,$$

and combining this with (6.28) gives (6.27). $\qquad\qquad\qquad\qquad\qquad\qquad\square$

Since $r_s = 1$ for nested fractals, we deduce

Corollary 6.28. *Let* $(F, (\psi_i))$ *be a nested fractal. Then* $\rho > 1$.

For nested fractals, many properties of the process can be summarized in terms of certain scaling factors.

Definition 6.29. Let $(F, (\psi_i))$ be a nested fractal, and $\mathcal{E}$ be the (unique) non-degenerate fixed point. See Definition 5.22 for the length and mass scale factors L and M. The *resistance scale factor* ρ of F is the resistance scaling factor of $\mathcal{E}$. Let

$$(6.29 \qquad\qquad\qquad \tau = M\rho;$$

we call τ the *time scaling factor.* (In view of the connection between resistances and crossing times given in Theorem 4.27, it is not surprising that τ should have a connection with the space-time scaling of processes on F.)

It may be helpful at this point to draw a rough distinction between two kinds of structure associated with the nested fractal (F, ψ). The quantities introduced in Section 5, such as L, M, the geodesic metric d_F, the chemical exponent γ and the dimension $d_w(F)$ are all *geometric* – that is, they can be determined entirely by a geometric inspection of F. On the other hand, the resistance and time scaling factors ρ and τ are *analytic* or *physical* – they appear in some sense to lie deeper than the geometric quantities, and arise from the solution to some kind of equation on the space. On the Sierpinski gasket, for example, while one obtains $L = \gamma = 2$, and $M = 3$ almost immediately, a brief calculation (Lemma 2.16) is needed to obtain ρ. For more complicated sets, such as some of the examples given in Section 5, the calculation of ρ would be very lengthy.

Unfortunately, while the distinction between these two kinds of constant arises clearly in practice, it does not seem easy to make it precise. Indeed, Corollary 6.20 shows that the geometry does in fact determine ρ: it is not possible to have one nested fractal (a geometric object) with two distinct analytic structures which both satisfy the symmetry and scale invariance conditions.

We have the following general inequalities for the scaling factors.

Proposition 6.30. *Let* $(F, (\psi_i))$, *be a nested fractal with scaling factors* L, M, ρ, τ. *Then*

$$(6.30) \qquad L > 1, \qquad M \geq 2, \qquad M \geq L, \qquad \tau = M\rho \geq L^2.$$

Proof. $L > 1$, $M \geq 2$ follow from the definition of nested fractals. If $\theta = \mathrm{diam}(V^{(0)})$, then, as $V^{(1)}$ consists of M copies of $V^{(0)}$ each of diameter $L^{-1}\theta$, by the connectivity axiom we deduce $ML^{-1}\theta \geq \theta$. Thus $M \geq L$.

To prove the final inequality in (6.30) we use the same strategy as in Proposition 6.27, but with a better choice of minimizing function.

Let $\mathcal{H}$ be the set of functions f of the form $f(x) = Ox + a$, where $x \in \mathbb{R}^d$ and O is an orthogonal matrix. Set $\mathcal{H}_n = \{f|_{V^{(n)}}, f \in \mathcal{H}\}$. Let $\theta = \sup\{\mathcal{E}(f, f) : f \in \mathcal{H}_0\}$: clearly $\theta < \infty$. Choose f to attain the supremum, and let $g \in \mathcal{H}$ be such that $f = g|_{V^{(0)}}$. Then if $f_1 = g|_{V^{(1)}}$

$$\rho^{-1}\theta = \rho^{-1}\mathcal{E}(f, f) = \Lambda(\mathcal{E})(f, f) \leq \mathcal{E}^R(g_1, g_1) = \sum_{i=1}^{M} \mathcal{E}(g_1 \circ \psi_i, g_1 \circ \psi_i).$$

However, $g_1 \circ \psi_i$ is the restriction to $V^{(0)}$ of a function of the form $L^{-1}Ox + a_i$, and so $\mathcal{E}(g \circ \psi_i, g \circ \psi_i) \leq L^{-2}\theta$. Hence $\rho^{-1}\theta \leq ML^{-2}\theta$, proving (6.30). $\qquad\square$

The following comparison theorem provides a technique for bounding ρ in certain situations.

Proposition 6.31. *Let* $(F_1, \{\psi_i, 1 \leq i \leq M_1\})$ *be a p.c.f.s.s. set. Let* $F_0 \subset F_1$, $M_0 \leq M_1$, *and suppose that* $(F_0, \{\psi_i, 1 \leq i \leq M_0\})$ *is also a p.c.f.s.s. set, and that* $V_{F_1}^{(0)} = V_{F_0}^{(0)}$. *Let* $(r_i^{(k)}, 1 \leq i \leq M_k)$ *be resistance vectors for* $k = 0, 1$, *and suppose that* $r_i^{(0)} \geq r_i^{(1)}$ *for* $1 \leq i \leq M_0$. *Let* Λ_k *be the renormalization map for* $(F_k, (\psi_i)_{i=1}^{M_k}, (r_i^{(k)})_{i=1}^{M_k})$. *If* $\mathcal{E}_k$ *are non-degenerate Dirichlet forms satisfying* $\Lambda_k(\mathcal{E}_k) = \rho_k^{-1}\mathcal{E}_k$, $k = 0, 1$, *then* $\rho_1 \leq \rho_0$.

Proof. Since $V_{F_1}^{(0)} \subset V_{F_1}^{(1)}$, we have, writing R_i for the replication maps associated with F_i,

$$R_1 \mathcal{E}(f, f) \geq R_0 \mathcal{E}(f, f), \qquad f \in C(V_{F_1}^{(1)}).$$

So $\Lambda_1(\mathcal{E}) \geq \Lambda_0(\mathcal{E})$ for any $\mathcal{E} \in \mathbb{D}$. If $m = m(\mathcal{E}_1/\mathcal{E}_0)$, then

$$\rho_1^{-1}\mathcal{E}_1 = \Lambda_1(\mathcal{E}_1) \geq \Lambda_1(m\,\mathcal{E}_0) \geq \Lambda_0(m\,\mathcal{E}_0) = m\rho_0^{-1}\mathcal{E}_0 \geq \rho_0^{-1}\mathcal{E}_1,$$

which implies that $\rho_0 \geq \rho_1$. $\qquad\qquad\square$

7. Diffusions on p.c.f.s.s. sets.

Let $\big(F, (\psi_i)\big)$ be a p.c.f.s.s. set, and r_i be a resistance vector. We assume that the graph $(V^{(1)}, \mathbf{E}_1)$ is connected. Suppose that the renormalization map Λ has a non-degenerate fixed point $\mathcal{E}^{(0)} = \mathcal{E}_A$, so that $\Lambda(\mathcal{E}^{(0)}) = \rho^{-1}\mathcal{E}^{(0)}$. Fixing F, r, and $\mathcal{E}_A$, in this section we will construct a diffusion X on F, as a limit of processes on the graphical approximations $V^{(n)}$. In Section 2 this was done probabilistically for the Sierpinski gasket, but here we will use Dirichlet form methods, following [Kus2, Ful, Ki2].

Definition 7.1. For $f \in C(V^{(n)})$, set

$$(7.1) \qquad \mathcal{E}^{(n)}(f, f) = \rho^n \sum_{w \in W_n} r_w^{-1}\mathcal{E}^{(0)}(f \circ \psi_w, f \circ \psi_w).$$

This is the Dirichlet form on $V^{(n)}$ obtained by replication of scaled copies of $\mathcal{E}^{(0)}$, where the scaling associated with the map ψ_w is $\rho^n r_w^{-1}$.

These Dirichlet forms have the following nesting property.

Proposition 7.2. (a) For $n \geq 1$, $Tr(\mathcal{E}^{(n)}|V^{(n-1)}) = \mathcal{E}^{(n-1)}$.
(b) If $f \in C(V^{(n)})$, and $g = f|_{V^{(n-1)}}$ then $\mathcal{E}^{(n)}(f, f) \geq \mathcal{E}^{(n-1)}(g, g)$.
(c) $\mathcal{E}^{(n)}$ is non-degenerate.

Proof. (a) Let $f \in C(V^{(n)})$. Then decomposing $w \in W_n$ into $v \cdot i$, $v \in W_{n-1}$,

$$(7.2) \qquad \mathcal{E}^{(n)}(f, f) = \rho^n \sum_{v \in W_{n-1}} r_v^{-1} \sum_i r_i^{-1}\mathcal{E}^{(0)}(f \circ \psi_v \circ \psi_i, f \circ \psi_v \circ \psi_i)$$

$$= \rho^{n-1} \sum_{v \in W_{n-1}} r_v^{-1}\mathcal{E}^{(1)}(f_v, f_v),$$

where $f_v = f \circ \psi_v \in C(V^{(1)})$. Now let $g \in C(V^{(n-1)})$. If $f|_{V^{(n-1)}} = g$ then $f_v|_{V^{(0)}} = g \circ \psi_v = g_v$. As $\mathcal{E}^{(0)}$ is a fixed point of Λ,

$$(7.3) \qquad \inf\left\{\mathcal{E}^{(1)}(h, h) : h|_{V^{(0)}} = g_v\right\} = \rho \inf\left\{R\mathcal{E}^{(0)}(h, h) : h|_{V^{(0)}} = g_v\right\}$$

$$= \rho\Lambda(\mathcal{E}^{(0)})(g_v, g_v) = \mathcal{E}^{(0)}(g_v, g_v).$$

Summing over $v \in W_{n-1}$ we deduce therefore

$$\inf\left\{\mathcal{E}^{(n)}(f, f) : f|_{V^{(n-1)}} = g\right\} \leq \rho^{n-1} \sum_v r_v^{-1}\mathcal{E}^{(0)}(g, g) = \mathcal{E}^{(n-1)}(g, g).$$

For each $v \in \mathbb{W}_{n-1}$, let $h_v \in C(V^{(1)})$ be chosen to attain the infimum in (7.3). We wish to define $f \in C(V^{(n)})$ such that

$$(7.4) \qquad f \circ \psi_v = h_v, \qquad v \in \mathbb{W}_{n-1}.$$

Let $v \in \mathbb{W}_{n-1}$. We define

$$f(\psi_v(y)) = h_v(y), \qquad y \in V^{(1)}.$$

We need to check f is well-defined; but if v, u are distinct elements of $\mathbb{W}_{n-1}$ and $x = \psi_v(y) = \psi_u(z)$, then $x \in V^{(n-1)}$ by Lemma 5.18, and so y, $z \in V^{(0)}$. Therefore

$$f(\psi_v(y)) = h_v(y) = g_v(y) = g(x) = f(\psi_u(z)),$$

so the definitions of f at x agree. (This is where we use the fact that F is finitely ramified: it allows us to minimize separately over each set of the form $V_v^{(1)}$).
  So

$$\mathcal{E}^{(n)}(f, f) = \mathcal{E}^{(n-1)}(g, g),$$

and therefore $Tr\left(\mathcal{E}^{(n)}|V^{(n-1)}\right) = \mathcal{E}^{(n-1)}$.
(b) is evident from (a).
(c) We prove this by induction. $\mathcal{E}^{(0)}$ is non-degenerate by hypothesis. Suppose $\mathcal{E}^{(n-1)}$ is non-degenerate, and that $\mathcal{E}^{(n)}(f, f) = 0$. From (7.2) we have

$$\mathcal{E}^{(n)}(f, f) = \rho \sum_{v \in \mathbb{W}_1} r_v^{-1} \mathcal{E}^{(n-1)}(f \circ \psi_v, f \circ \psi_v),$$

and so $f \circ \psi_v$ is constant for each $v \in \mathbb{W}_1$. Thus f is constant on each 1-complex, and as $(V^{(1)}, \mathbf{E}_1)$ is connected this implies that f is constant. $\qquad \square$

To avoid clumsy notation we will identify functions with their restrictions, so, for example, if $f \in C(V^{(n)})$, and $m < n$, we will write $\mathcal{E}^{(m)}(f, f)$ instead of $\mathcal{E}^{(m)}\left(f|_{V^{(m)}}, f|_{V^{(m)}}\right)$.

Definition 7.3. Set $V^{(\infty)} = \cup_{n=0}^{\infty} V^{(n)}$. Let $U = \{f : V^{(\infty)} \to \mathbb{R}\}$. Note that the sequence $\left(\mathcal{E}^{(n)}(f, f)\right)_{n=1}^{\infty}$ is non-decreasing. Define

$$\mathcal{D}' = \{f \in U : \sup_n \mathcal{E}^{(n)}(f, f) < \infty\},$$

$$\mathcal{E}'(f, g) = \sup_n \mathcal{E}^{(n)}(f, g); \quad f, g \in \mathcal{D}'.$$

$\mathcal{E}'$ is the initial version of the Dirichlet form we are constructing.

Lemma 7.4. $\mathcal{E}'$ *is a symmetric Markov form on* $\mathcal{D}'$.

Proof. $\mathcal{E}'$ clearly inherits the properties of symmetry, bilinearity, and positivity from the $\mathcal{E}^{(n)}$. If $f \in \mathcal{D}'$, and $g = (0 \vee f) \wedge 1$ then $\mathcal{E}^{(n)}(g, g) \le \mathcal{E}^{(n)}(f, f)$, as the $\mathcal{E}^{(n)}$ are Markov. So $\mathcal{E}'(g, g) \le \mathcal{E}'(f, f)$. $\qquad \square$

What we have done here seems very easy. However, more work is needed to obtain a 'good' Dirichlet form $\mathcal{E}$ which can be associated with a diffusion on F. Note the following scaling result for $\mathcal{E}'$.

Lemma 7.5. *For* $n \geq 1$, $f \in \mathcal{D}'$,

$$(7.5) \qquad \mathcal{E}'(f,f) = \sum_{w \in W_n} \rho^n r_w^{-1} \mathcal{E}'(f \circ \psi_w, f \circ \psi_w).$$

Proof. We have, for $m \geq n$, $f \in \mathcal{D}'$,

$$\mathcal{E}^{(m)}(f,f) = \sum_{w \in W_n} \rho^n r_w^{-1} \mathcal{E}^{(m-n)}(f \circ \psi_w, f \circ \psi_w).$$

Letting $m \to \infty$ it follows, first that $f \circ \psi_w \in \mathcal{D}'$, and then that (7.5) holds. $\qquad \square$

If H is a set, and $f : H \to \mathbb{R}$, we write

$$(7.6) \qquad \mathrm{Osc}(f,B) = \sup_{x,y \in B} |f(x) - f(y)|, \quad B \subset H.$$

Lemma 7.6. *There exists a constant* c_0, *depending only on* $\mathcal{E}$, *such that*

$$\mathrm{Osc}(f, V^{(0)}) \leq c_0 \mathcal{E}^{(0)}(f,f), \qquad f \in C(V^{(0)}).$$

Proof. Let $\widetilde{E}_0 = \{\{x,y\} : A_{xy} > 0\}$. As $\mathcal{E}^{(0)}$ is non-degenerate, $(V^{(0)}, \widetilde{E}_0)$ is connected; let N be the maximum distance between points in this graph. Set $\alpha = \min\{A_{xy}, \{x,y\} \in \widetilde{E}_0\}$. If $x, y \in V^{(0)}$, there exists a chain $x = x_0, x_1, \ldots, x_n = y$ connecting x, y with $n \leq N$, and therefore,

$$\begin{aligned}
|f(x) - f(y)|^2 &\leq \left(\sum_{i=1}^{n} |f(x_i) - f(x_{i-1})| \right)^2 \\
&\leq n \sum_{i=1}^{n} |f(x_i) - f(x_{i-1})|^2 \\
&\leq n\alpha^{-1} \sum_{i=1}^{n} A_{x_{i-1}, x_i} |f(x_i) - f(x_{i-1})|^2 \\
&\leq N\alpha^{-1} \mathcal{E}^{(0)}(f,f). \qquad \square
\end{aligned}$$

Since $V^{(1)}$ consists of M copies of $V^{(0)}$ we deduce a similar result for $V^{(1)}$.

Corollary 7.7. *There exists a constant* $c_1 = c_1(F, r, A)$ *such that*

$$(7.7) \qquad \mathrm{Osc}(f, V^{(1)}) \leq c_1 \mathcal{E}^{(1)}(f,f), \qquad f \in \mathcal{D}'.$$

Proof. For $i \in W_1$, $f \in C(V^{(1)})$,

$$\mathrm{Osc}(f, V_i^{(0)}) = \mathrm{Osc}(f \circ \psi_i, V^{(0)}) \leq c_0 \mathcal{E}^{(0)}(f \circ \psi_i, f \circ \psi_i).$$

So, as $V^{(1)}$ is connected,

$$\begin{aligned}
\mathrm{Osc}(f, V^{(1)}) &\leq \sum_i \mathrm{Osc}(f, V_i^{(0)}) \\
&\leq \sum_i c_0 \mathcal{E}^{(0)}(f \circ \psi_i, f \circ \psi_i) \leq c_1 \mathcal{E}^{(1)}(f,f),
\end{aligned}$$

where c_1 is chosen so that $c_0 \leq c_1 \rho r_i^{-1}$ for each $i \in \mathbb{W}_1$. □

Corollary 7.8. *Let* $w \in \mathbb{W}_n$, *and* $x, y \in V_w^{(1)}$. *Then*

$$\mathrm{Osc}(f, V_w^{(1)}) \leq c_1 r_w \rho^{-n} \mathcal{E}'(f, f), \qquad f \in \mathcal{D}'.$$

Proof. We have $\mathrm{Osc}(f, V_w^{(1)}) = \mathrm{Osc}(f \circ \psi_w, V^{(1)}) \leq c_1 \mathcal{E}^{(1)}(f \circ \psi_w, f \circ \psi_w)$. Since $\mathcal{E}^{(1)} \leq \mathcal{E}'$, and by (7.5)

$$\mathcal{E}'(f \circ \psi_w, f \circ \psi_w) \leq r_w \rho^{-n} \mathcal{E}'(f, f),$$

the result is immediate. □

Definition 7.9. We will call the fixed point $\mathcal{E}^{(0)}$ a *regular fixed point* if

$$(7.8) \qquad\qquad r_i < \rho \qquad \text{for} \quad 1 \leq i \leq M.$$

Proposition 6.27 implies that (7.8) holds for any $s \in \{1, \ldots, M\}$ such that $\pi(\dot{s}) \in V^{(0)}$. In particular therefore, for nested fractals, where every point in $V^{(0)}$ is of this form and r is constant, any fixed point is regular.

It is not hard to produce examples of non-regular fixed points. Consider the Lindstrøm snowflake, but with $r_i = 1$, $1 \leq i \leq 6$, $r_7 = r > 1$. Writing $\rho(r)$ for the resistance scale factor, we have (by Proposition 6.31) that $\rho(r)$ is increasing in r. However, also by Proposition 6.31, $\rho(r) \leq \rho_0$, where ρ_0 is the resistance scale factor of the nested fractal obtained just from ψ_i, $1 \leq i \leq 6$. So if we choose $r_7 > \rho_0$, then as $r_7 > \rho_0 \geq \rho(r_7)$, we have an example of an affine nested fractal with a non-regular fixed point.

From now on we take $\mathcal{E}^{(0)}$ to be a regular fixed point. (See [Kum3] for the general situation). Write $\gamma = \max_i r_i / \rho < 1$. For $x, y \in F$, set $w(x, y)$ to be the longest word w such that $x, y \in F_w$.

Proposition 7.10. *(Sobolev inequality). Let* $f \in \mathcal{D}'$. *Then if* $\mathcal{E}^{(0)}$ *is a regular fixed point*

$$(7.8) \qquad |f(x) - f(y)|^2 \leq c_2 r_{w(x,y)} \rho^{-|w(x,y)|} \mathcal{E}'(f, f), \qquad x, y \in V^{(\infty)}.$$

Proof. Let $x, y \in V^{(n)}$, let $w = w(x, y)$ and let $|w| = m$. We prove (7.8) by a standard kind of chaining argument, similar to those used in continuity results such as Kolmogorov's lemma. (But this argument is deterministic and easier). We may assume $n \geq m$.

Let $u \in \mathbb{W}_n$ be an extension of w, such that $x \in V_u^{(0)}$: such a u certainly exists, as $x \in V_n^{(0)} \cap F_w$. Write $u_k = u|k$ for $m \leq k \leq n$. Now choose a sequence z_k, $m \leq k \leq n$ such that $z_n = x$, and $z_k \in V_{u_k}^{(0)}$ for $k \leq m \leq n-1$. For each $k \in \{m, \ldots, n-1\}$ we have $z_k, z_{k+1} \in V_{u_k}^{(1)}$. So

$$(7.9) \qquad |f(z_n) - f(z_m)| \leq \sum_{k=m}^{n-1} |f(z_{k+1}) - f(z_k)|$$

$$\leq \sum_{k=m}^{n-1} \left(c_1 r_{u_k} \rho^{-k} \mathcal{E}(f, f) \right)^{1/2}$$

$$= \left(c_1 r_w \rho^{-m} \mathcal{E}(f,f)\right)^{1/2} \left(\sum_{k=m}^{n-1} \frac{r_{u_k}}{r_w} \rho^{-k+m}\right)^{1/2}.$$

As $\mathcal{E}$ is a regular fixed point, $\gamma = \max_i r_i/\rho < 1$, so the final sum in (7.9) is bounded by $(\sum_{k=m}^{\infty} \gamma^{k-m})^{1/2} = c_3 < \infty$. Thus we have

$$|f(x) - f(z_m)|^2 \le c_1 c_3 r_w \rho^{-n} \mathcal{E}'(f,f),$$

and as a similar bound holds for $|f(y) - f(z_m)|^2$, this proves (7.8). $\qquad \square$

We have not so far needed a measure on F. However, to define a Dirichlet form we need some L^2 space in which the domain of $\mathcal{E}$ is closed. Let μ be a probability measure on $(F, \mathcal{B}(F))$ which charges every set of the form F_w, $w \in \mathbb{W}_n$. Later we will take μ to be the Bernouilli measure μ_θ associated with a vector of weights $\theta \in (0, \infty)^M$, but for now any measure satisfying the condition above will suffice.

As $\mu(F) = 1$, $C(F) \subset L^2(F, \mu)$. Set

$$\mathcal{D} = \{f \in C(F) : f|_{V^{(\infty)}} \in \mathcal{D}'\}$$
$$\mathcal{E}(f,f) = \mathcal{E}'(f|_{V^{(\infty)}}, f|_{V^{(\infty)}}), \quad f \in \mathcal{D}.$$

Proposition 7.11. $(\mathcal{E}, \mathcal{D})$ *is a closed symmetric form on* $L^2(F, \mu)$.

Proof. Note first that the condition on μ implies that if $f, g \in \mathcal{D}$ then $\|f - g\|_2 = 0$ implies that $f = g$. We need to prove that $\mathcal{D}$ is complete in the norm $\|f\|_{\mathcal{E}_1}^2 = \mathcal{E}(f,f) + \|f\|_2^2$. So suppose (f_n) is Cauchy in $\|\cdot\|_{\mathcal{E}_1}$. Since (f_n) is Cauchy in $\|\cdot\|_2$, passing to a subsequence there exists $\tilde{f} \in L^2(F, \mu)$ such that $f_n \to \tilde{f}$ μ–a.e. Fix $x_0 \in F$ such that $f_n(x_0) \to \tilde{f}(x)$. Then since $f_n - f_m$ is continuous, (7.8) extends to an estimate on the whole of F and so

$$|f_n(x) - f_m(x)| \le |(f_n - f_m)(x) - (f_n - f_m)(x_0)| + |(f_n - f_m)(x_0)|$$
$$\le c_2^{1/2} \mathcal{E}(f_n - f_m, f_n - f_m)^{1/2} + |f_n(x_0) - f_m(x_0)|.$$

So (f_n) is Cauchy in the uniform norm, and thus there exists $f \in C(F)$ such that $f_n \to f$ uniformly.

Let $n \ge 1$. Then as $\mathcal{E}^{(n)}(g, g)$ is a finite sum,

$$\mathcal{E}^{(n)}(f,f) = \lim_{m \to \infty} \mathcal{E}^{(n)}(f_m, f_m) \le \limsup_{m \to \infty} \mathcal{E}(f_m, f_m)$$
$$\le \sup_m \|f_m\|_{\mathcal{E}_1} < \infty.$$

Hence $\mathcal{E}^{(n)}(f,f)$ is bounded, so $f \in \mathcal{D}$. Finally, by a similar calculation, for any $N \ge 1$,

$$\mathcal{E}^{(N)}(f_n - f, f_n - f) \le \lim_{m \to \infty} \mathcal{E}(f_n - f_m, f_n - f_m).$$

So $\mathcal{E}(f_n - f, f_n - f) \to 0$ as $n \to \infty$, and thus $\|f - f_n\|_{\mathcal{E}_1}^2 \to 0$. $\qquad \square$

To show that $(\mathcal{E}, \mathcal{D})$ is a Dirichlet form, it remains to show that $\mathcal{D}$ is dense in $L^2(F, \mu)$. We do this by studying the harmonic extension of a function.

Definition 7.12. Let $f \in C(V^{(n)})$. Recall that $\mathcal{E}^{(n)}(f,f) = \inf\{\mathcal{E}^{(n+1)}(g,g) : g|_{V^{(n)}} = f\}$. Let $\widetilde{H}_{n+1}f \in C(V^{(n+1)})$ be the (unique, as $\mathcal{E}^{(n+1)})$ is non-degenerate) function which attains the infimum.

For $x \in V^{(\infty)}$ set

$$\widehat{H}_n f(x) = \lim_{m \to \infty} \widetilde{H}_m \widetilde{H}_{m-1} \ldots \widetilde{H}_{n+1} f(x);$$

note that (as $\widetilde{H}_{n+1}f = f$ on $V^{(n)}$) this limit is ultimately constant.

Proposition 7.13. Let $\mathcal{E}$ be a regular fixed point.
(a) $\widehat{H}_n f$ has a continuous extension to a function $H_n f \in \mathcal{D} \cap C(F)$, which satisfies

$$\mathcal{E}(H_n f, H_n f) = \mathcal{E}^{(n)}(f,f).$$

(b) If $f, g \in C(F)$

$$(7.10) \qquad \mathcal{E}(H_n f, g) = \mathcal{E}^{(n)}(f,g).$$

Proof. From the definition of $\widetilde{H}_{n+1}$, $\mathcal{E}^{(n+1)}(\widetilde{H}_{n+1}f, \widetilde{H}_{n+1}f) = \mathcal{E}^{(n)}(f,f)$. Thus $\mathcal{E}^{(m)}(\widehat{H}_n f, \widehat{H}_n f) = \mathcal{E}^{(n)}(f,f)$ for any m, so that $\widehat{H}_n f \in \mathcal{D}'$ and

$$\mathcal{E}(\widehat{H}_n f, \widehat{H}_n f) = \mathcal{E}^{(n)}(f,f), \qquad f \in C(V^{(n)}).$$

If $w \in \mathbb{W}_m$, and $x, y \in V^{(\infty)} \cap F_w$ then by Proposition 7.10

$$(7.11) \qquad |\widehat{H}_n f(x) - \widehat{H}_n f(y)|^2 \le c_2 r_w \rho^{-m} \mathcal{E}^{(n)}(f,f).$$

Since $r_w \rho^{-m} \le \gamma^m$, (7.11) implies that $\mathrm{Osc}(\widehat{H}_n f, V^{(\infty)} \cap F_w)$ converges to 0 as $|w| = m \to \infty$. Thus $\widehat{H}_n f$ has a continuous extension $H_n f$, and $H_n f \in \mathcal{D}$ since $\widehat{H}_n f \in \mathcal{D}'$.
(b) Note that, by polarization, we have

$$\mathcal{E}^{(n+1)}(\widetilde{H}_{n+1}f, \widetilde{H}_{n+1}g) = \mathcal{E}^{(n)}(f,g).$$

Since $\mathcal{E}^{(n+1)}(\widetilde{H}_{n+1}f, h) = 0$ for any h such that $h|_{V^{(n)}} = 0$, it follows that

$$\mathcal{E}^{(n+1)}(\widetilde{H}_{n+1}f, g) = \mathcal{E}^{(n)}(f,g).$$

Iterating, we obtain (7.10). $\qquad \square$

Theorem 7.14. $(\mathcal{E}, \mathcal{D})$ is an irreducible, regular, local Dirichlet form on $L^2(F,\mu)$.

Proof. Let $f \in C(F)$. Since for any $n \ge 1$, $w \in \mathbb{W}_n$ we have

$$\inf_{F_w} f \le H_n f(x) \le \sup_{F_w} f, \qquad x \in F_w$$

it follows that $H_n f \to f$ uniformly. As $H_n f \in \mathcal{D}$, we deduce that $\mathcal{D}$ is dense in $C(F)$ in the uniform norm. Hence also $\mathcal{D}$ is dense in $L^2(F,\mu)$. As (4.5) is immediate, we deduce that $\mathcal{D}$ is a regular Dirichlet form. If $\mathcal{E}(f,f) = 0$ then $\mathcal{E}^{(n)}(f,f) = 0$ for each n. Since $\mathcal{E}^{(n)}$ is irreducible, $f|_{V^{(n)}}$ is constant for each n. As f is continuous, f is therefore constant. Thus $\mathcal{E}$ is irreducible.

To prove that $\mathcal{E}$ is local, let f, g be functions in $\mathcal{D}$ with disjoint closed supports, S_f, S_g say. If $\mathcal{E}^{(n)}(f,g) \neq 0$ then one of the terms in the sum (7.1) must be non-zero, so there exists $w_n \in W_n$, and points $x_n \in S_f \cap V_{w_n}^{(0)}$, $y_n \in S_g \cap V_{w_n}^{(0)}$. Passing to a subsequence, there exists z such that $x_n \to z$, $y_n \to z$, and as therefore $z \in S_f \cap S_g$, this is a contradiction. $\qquad\square$

By Theorem 4.8 there exists a continuous μ-symmetric Hunt process $(X_t, t \geq 0, \mathbb{P}^x, x \in F)$ associated with $(\mathcal{E}, \mathcal{D})$ and $L^2(F, \mu)$.

Remark 7.15. Note that we have constructed a process $X = X^{(\mu)}$ for each Radon measure μ on F. So, at first sight, the construction given here has built much more than the probabilistic construction outlined in Section 2. But this added generality is to a large extent an illusion: Theorem 4.17 implies that these processes can all be obtained from each other by time-change.

On the other hand the regularity of $(\mathcal{E}, \mathcal{D})$ was established without much pain, and here the advantage of the Dirichlet form approach can be seen: all the probabilistic approaches to the Markov property are quite cumbersome.

The general probabilistic construction, such as given in [L1] for example, encounters another obstacle which the Dirichlet form construction avoids. As well as finding a decimation invariant set of transition probabilities, it also appears necessary (see e.g. [L1, Chapter VI]) to find associated transition times. It is not clear to me why these estimates appear essential in probabilistic approaches, while they do not seem to be needed at all in the construction above.

We collect together a number of properties of $(\mathcal{E}, \mathcal{D})$.

Proposition 7.16. (a) For each $n \geq 0$

$$(7.12) \qquad \mathcal{E}(f,g) = \sum_{w \in W_n} \rho^n r_w^{-1} \mathcal{E}(f \circ \psi_w, g \circ \psi_w).$$

(b) For $f \in \mathcal{D}$,

$$(7.13) \qquad |f(x) - f(y)|^2 \leq c_1 r_w \rho^{-n} \mathcal{E}(f,f) \quad \text{if} \quad x, y \in F_w, \, w \in W_n$$

$$(7.14) \qquad \int f^2 d\mu \leq c_2 \mathcal{E}(f,f) + \left(\int f d\mu \right)^2,$$

$$(7.15) \qquad f(x)^2 \leq 2 \int f^2 d\mu + 2c_1 \mathcal{E}(f,f), \qquad x \in F.$$

Proof. (a) is immediate from Lemma 7.5, while (b) follows from Proposition 7.10 and the continuity of f. Taking $n = 0$ in (7.13) we deduce that

$$\left(f(x) - f(y) \right)^2 \leq c_1 \mathcal{E}(f,f), \qquad f \in \mathcal{D}.$$

So as $\mu(F) = 1$,

$$\int\int c_1 \mathcal{E}(f,f)\mu(dx)\mu(dy) = c_1\mathcal{E}(f,f)$$

$$\leq \int\int (f(x) - f(y))^2 \mu(dx)\mu(dy)$$

$$= 2\int f^2\,d\mu - 2\left(\int f\,d\mu\right)^2,$$

proving (7.14).

Since $f(x)^2 \leq 2f(y)^2 + 2|f(x) - f(y)|^2$ we have from (7.13) that

$$f(x)^2 = \int f(x)^2 \mu(dy)$$

$$\leq 2\int f(y)^2 \mu(dy) + 2c_1\int \mathcal{E}(f,f)\mu(dy),$$

which proves (7.15). $\qquad\qquad\qquad\qquad\qquad\qquad\qquad\qquad\qquad\square$

We need to examine further the resistance metric introduced in Section 4.

Definition 7.17. Let $R(x,x) = 0$, and for $x \neq y$ set

$$R(x,y)^{-1} = \inf\{\mathcal{E}(f,f) : f(x) = 0, f(y) = 1, f \in \mathcal{D}\}.$$

Note that

$$(7.16) \qquad R(x,y) = \sup\left\{\frac{|f(x) - f(y)|^2}{\mathcal{E}(f,f)} : f \in \mathcal{D}, \quad f \text{ non constant}\right\}.$$

Proposition 7.18. (a) If $x \neq y$ then $0 < R(x,y) \leq c_1 < \infty$.
(b) If $w \in \mathbb{W}_n$ then

$$(7.17) \qquad\qquad\qquad R(x,y) \leq c_1 r_w \rho^{-n}, \qquad x,y \in F_w.$$

(c) For $f \in \mathcal{D}$

$$(7.18) \qquad\qquad\qquad |f(x) - f(y)|^2 \leq R(x,y)\mathcal{E}(f,f).$$

(d) R is a metric on F, and the topology induced by R is equal to the original topology on F.

Proof. Let x, y be distinct points in F. As $\mathcal{D}$ is dense in $C(F)$, there exists $f \in \mathcal{D}$ with $f(x) \geq 1$, $f(y) \leq 0$. Since $\mathcal{E}$ is irreducible, $\mathcal{E}(f,f) > 0$, and so by (7.16) $R(x,y) > 0$. (7.17) is immediate from Proposition 7.16, proving (b). Taking $n = 0$, and w to be the empty word in (7.17) we deduce $R(x,y) \leq c_1$ for any x, $y \in F$, completing the proof of (a).
(c) is immediate from (7.16).
(d) R is clearly symmetric. The triangle inequality for R is proved exactly as in Proposition 4.25, by considering the trace of $\mathcal{E}$ on the set $\{x,y,z\}$.

It remains to show that the topologies induced by R and d (the original metric on F) are the same. Let $R(x_n, x) \to 0$. If $\varepsilon > 0$, there exists $f \in \mathcal{D}$ with $f(x) = 1$ and $\mathrm{supp}(f) \subset B_d(x, \varepsilon)$. By (7.16) $R(x, y) \geq \mathcal{E}(f, f)^{-1} > 0$ for any $y \in B_d(x, \varepsilon)^c$. So $x_n \in B_d(x, \varepsilon)$ for all sufficiently large n, and hence $d(x_n, x) \to 0$.

If $d(x_n, x) \to 0$ then writing

$$N_m(x) = \bigcup \{F_w : w \in \mathbb{W}_m, \quad x \in F_w\}$$

we have by Lemma 5.12 that $x_n \in N_m(x)$ for all sufficiently large n. However if $\gamma = \max_i r_i / \rho < 1$ we have by, (7.17), $R(x, y) \leq c_1 \gamma^m$ for $y \in N_m(x)$. Thus $R(x_n, x) \to 0$. $\qquad\square$

Remark 7.19. The resistance metric R on F is quite well adapted to the study of the diffusion X on F. Note however that $R(x, y)$ is obtained by summing (in a certain sense) the resistance of all paths from x to y. So it is not surprising that R is not a geodesic metric. (Unless F is a tree).

Also, R is not a geometrically natural metric on F. For example, on the Sierpinski gasket, since $r_i = 1$, and $\rho = 5/3$, we have that if x, y are neighbours in $(V^{(n)}, \mathbf{E}_n)$ then

$$R(x, y) \asymp (3/5)^n.$$

However, for general p.c.f.s.s. sets it is not easy to define a metric which is well-adapted to the self-similar structure. (And, if one imposes strict conditions of exact self-similarity, it is not possible in general – see the examples in [Ki6]). So, for these general sets the resistance metric plays an extremely useful role. The next section contains some additional results on R.

It is also worth remarking that the balls $B_R(x, r) = \{y : R(x, y) < r\}$ need not in general be connected. For example, consider the wire network corresponding to the graph consisting of two points x, y, connected by n wires each of conductivity 1. Let z be the midpoint of one of the wires. Then $R(x, y) = 1/n$, while the conductivities in the network $\{x, y, z\}$ are given by $C(x, z) = C(z, y) = 2$, $C(x, y) = n - 1$. So, after some easy calculations,

$$R(x, z) = \frac{n+1}{4n-1} > \tfrac{1}{4}.$$

So if $n = 4$, $R(x, y) = \tfrac{1}{4}$ while $R(x, z) = \tfrac{1}{3}$. Hence if $\tfrac{1}{4} < r < \tfrac{1}{3}$ the ball $B_R(x, r)$ is not connected. (In fact, y is an isolated point of $\overline{B}_R(x, \tfrac{1}{4}) = \{x' : d(x, x') \leq \tfrac{1}{4}\}$). (Are the balls $B_R(x, r)$ in the Sierpinski gasket connected? I do not know).

Recall the notation $\mathcal{E}_\alpha(f, g) = \mathcal{E}(f, g) + \alpha(f, g)$. Let $(U_\alpha, \alpha > 0)$ be the resolvent of X. Since by (4.8) we have

$$\mathcal{E}_\alpha(U_\alpha f, g) = (f, g),$$

if U_α has a density $u_\alpha(x, y)$ with respect to μ, then a formal calculation suggests that

$$\mathcal{E}_\alpha\big(u_\alpha(x, \cdot), g\big) = \mathcal{E}_\alpha(U_\alpha \delta_x, g) = (\delta_x, g) = g(x).$$

We can use this to obtain the existence and continuity of the resolvent density u_α. (See [FOT, p. 73]).

Theorem 7.20. (a) *For each* $x \in F$ *there exists* $u_\alpha^x \in \mathcal{D}$ *such that*

(7.19)
$$\mathcal{E}_\alpha(u_\alpha^x, f) = f(x) \qquad \text{for all} \quad f \in \mathcal{D}.$$

(b) *Writing* $u_\alpha(x, y) = u_\alpha^x(y)$, *we have*

$$u_\alpha(x, y) = u_\alpha(y, x) \qquad \text{for all} \quad x, y \in F.$$

(c) $u_\alpha(\cdot, \cdot)$ *is continuous on* $F \times F$ *and in particular*

(7.20)
$$|u_\alpha(x, y) - u_\alpha(x, y')|^2 \le R(y, y') u_\alpha(x, x).$$

(d) $u_\alpha(x, y)$ *is the resolvent density for* X: *for* $f \in C(F)$,

$$E^x \int_0^\infty e^{-\alpha t} f(X_t) dt = U_\alpha f(x) = \int u_\alpha(x, y) f(y) \mu(dy).$$

(e) *There exists* $c_2(\alpha)$ *such that*

(7.21)
$$u_\alpha(x, y) \le c_2(\alpha), \qquad x, y \in F.$$

Proof. (a) The existence of u_α^x is given by a standard argument with reproducing kernel Hilbert spaces. Let $x \in F$, and for $f \in \mathcal{D}$ let $\phi(f) = f(x)$. Then by (7.15)

$$|\phi(f)|^2 = |f(x)|^2 \le 2\|f\|_2^2 + 2c_1 \mathcal{E}(f, f) \le c_\alpha \mathcal{E}_\alpha(f, f),$$

where $c_\alpha = 2\max(c_1, \alpha^{-1})$. Thus ϕ is a bounded linear functional on the Hilbert space $(\mathcal{D}, \| \ \|_{\mathcal{E}_\alpha})$, and so there exists a $u_\alpha^x \in \mathcal{D}$ such that

$$\phi(f) = \mathcal{E}_\alpha(u_\alpha^x, f) = f(x), \qquad f \in \mathcal{D}.$$

(b) This is immediate from (a) and the symmetry of $\mathcal{E}$:

$$u_\alpha^y(x) = \mathcal{E}_\alpha(u_\alpha^x, u_\alpha^y) = \mathcal{E}_\alpha(u_\alpha^y, u_\alpha^x) = u_\alpha^x(y).$$

(c) As $u_\alpha^x \in \mathcal{D}$, $u_\alpha(x, x) < \infty$. Since $\mathcal{E}(u_\alpha^x, u_\alpha^x) = u_\alpha(x, x) < \infty$, the estimate (7.20) follows from (7.18). It follows immediately that u is jointly continuous on $F \times F$.
(d) This follows from (7.19) and linearity. For a measure ν on F set

$$V_\nu f(x) = \int u_\alpha(x, y) f(y) \nu(dy), \qquad f \in C(F).$$

As u_α is uniformly continuous on $F \times F$, we can choose $\nu_n \xrightarrow{w} \mu$ so that $V_{\nu_n} f \to V f$ uniformly, and ν_n are atomic with a finite number of atoms. Write $V_n = V_{\nu_n}$, $V = V_\mu$. Since by (7.19)

$$\mathcal{E}_\alpha(V_n f, g) = \sum_x \nu_n(\{x\}) f(x) \mathcal{E}_\alpha(u_\alpha^x, g)$$

$$= \sum_x f(x) g(x) \nu_n(\{x\}) = \int fg \, d\nu_n,$$

we have

$$\mathcal{E}_\alpha(V_n f - V_m f, V_n f - V_m f) =$$
$$\int f(V_n f - V_m f) \, d\nu_n - \int f(V_n f - V_m f) \, d\nu_m.$$

Thus $\mathcal{E}_\alpha(V_n f - V_m f, V_n f - V_m f) \to 0$ as $m, n \to \infty$, and so, as $\mathcal{E}$ is closed, we deduce that $Vf \in \mathcal{D}$ and $\mathcal{E}_\alpha(Vf, g) = \lim_n \mathcal{E}_\alpha(V_n f, g) = \lim_n \int f g \, d\nu_n = \int f g \, d\mu$. So $\mathcal{E}_\alpha(Vf, g) = \mathcal{E}_\alpha(U_\alpha f, g)$ for all g, and hence $Vf = U_\alpha f$.

(e) As $R(y, y') \le c_1$ for $y, y' \in F$, we have from (7.20) that

$$(7.22) \qquad u_\alpha(x, y) \ge u_\alpha(x, x) - \left(c_1 u_\alpha(x, x) \right)^{1/2}.$$

Since $\int u_\alpha(x, y) \mu(dy) = \alpha^{-1}$, integrating (7.22) we obtain

$$u_\alpha(x, x) \le \left(c_1 u_\alpha(x, x) \right)^{1/2} + \alpha^{-1},$$

and this implies that $u_\alpha(x, x) \le c_2(\alpha)$, where $c(\alpha)$ depends only on α and c_1. Using (7.20) again we obtain (7.21). $\qquad\square$

Theorem 7.21. (a) For each $x \in F$, x is regular for $\{x\}$.

(b) X has a jointly continuous local time $(L_t^x, x \in F, t \ge 0)$ such that for all bounded measurable f

$$\int_0^t f(X_s) \, ds = \int f(a) L_t^a \mu(da), \quad \text{a.s.}$$

Proof. These follow from the estimates on the resolvent density u_α. As u_α is bounded and continuous, we have that x is regular for $\{x\}$. Thus X has jointly measurable local times $(L_t^x, x \in F, t \ge 0)$.

Since X is a symmetric Markov process, by Theorem 8.6 of [MR], L_t^x is jointly continuous in (x, t) if and only if the Gaussian process $Y_x, x \in F$ with covariance function given by

$$EY_a Y_b = u_1(a, b), \qquad a, b \in F$$

is continuous. Necessary and sufficient conditions for continuity of Gaussian processes are known (see [Tal]), but here a simple sufficient condition in terms of metric entropy is enough. We have

$$E(Y_a - Y_b)^2 = u_1(a, a) - 2u_1(a, b) + u_1(b, b) \le c_1 R(a, b)^{1/2}.$$

Set $r(a, b) = R(a, b)^{1/2}$: r is a metric on F. Write $N_r(\varepsilon)$ for the smallest number of sets of r-diameter ε needed to cover F. By (7.17) we have $R(a, b) \le c\gamma^n$ if $a, b \in F_w$ and $w \in \mathbb{W}_n$. So $N_r(c'\gamma^{n/2}) \le \#\mathbb{W}_n = M^n$, and it follows that

$$N_r(\varepsilon) \le c_2 \varepsilon^{-\beta},$$

where $\beta = 2 \log M / \log \theta^{-1}$. So

$$\int_{0+} \left(\log N_r(\varepsilon) \right)^{1/2} d\varepsilon < \infty,$$

and thus by [Du, Thm. 2.1] Y is continuous. □

We can use the continuity of the local time of X to give a simple proof that X is the limit of a natural sequence of approximating continuous time Markov chains. For simplicity we take μ to be a Bernouilli measure of the form $\mu = \mu_\theta$, where $\theta_i > 0$. Let μ_n be the measure on $V^{(n)}$ given in (5.21). Set

$$A_t^n = \int_F L_t^x \mu_n(dx),$$
$$\tau_t^n = \inf\{s : A_s^n > t\},$$
$$X_t^n = X_{\tau_t^n}.$$

Theorem 7.22. (a) $\left(X_t^n, t \geq 0, \mathbb{P}^x, x \in V^{(n)}\right)$ is the symmetric Markov process associated with $\mathcal{E}^{(n)}$ and $L^2(V^{(n)}, \mu_n)$.
(b) $X_t^n \to X_t$ a.s. and uniformly on compacts.

Proof. (a) By Theorem 7.21(a) points are non-polar for X. So by the trace theorem (Theorem 4.17) X^n is the Markov process associated with the trace of $\mathcal{E}$ on $L^2(V^{(n)}, \mu_n)$. But for $f \in \mathcal{D}$, by the definition of $\mathcal{E}$,

$$Tr\left(\mathcal{E}|V^{(n)}\right)(f, f) = \mathcal{E}^{(n)}\left(f|_{V^{(n)}}, f|_{V^{(n)}}\right).$$

(b) As F is compact, for each $T > 0$, $(L_t^x, 0 \leq t \leq T, x \in F)$ is uniformly continuous. So, using (5.22), if $T_2 < T_1 < T$ then $A_t^n \to t$ uniformly in $[0, T_1]$, and so $\tau_t^n \to t$ uniformly on $[0, T_2]$. As X is continuous, $X_t^n \to X$ uniformly in $[0, T_2]$. □

Remark 7.23. As in Example 4.21, it is easy to describe the generator L_n of X^n. Let $a^{(n)}(x, y)$, $x, y \in V^{(n)}$ be the conductivity matrix such that

$$\mathcal{E}^{(n)}(f, f) = \tfrac{1}{2} \sum_{x,y} a^{(n)}(x, y)\left(f(x) - f(y)\right)^2.$$

Then by (7.1) we have

(7.23) $$a^{(n)}(x, y) = \sum_{w \in W_n} 1_{(x,y \in V_w^{(0)})} \rho^n r_w^{-1} A\left(\psi_w^{-1}(x), \psi_w^{-1}(y)\right),$$

where A is such that $\mathcal{E}^{(0)} = \mathcal{E}_A$, and $A(x, y) = A_{xy}$. Then for $f \in L^2(V^{(n)}, \mu_n)$,

(7.24) $$L_n f(x) = \mu_n(\{x\})^{-1} \sum_{y \in V^{(n)}} a^{(n)}(x, y)\left(f(y) - f(x)\right).$$

Of course Theorem 7.22 implies that if (Y^n) is a sequence of continuous time Markov chains, with generators given by (7.24), then $Y^n \overset{w}{\longrightarrow} X$ in $D([0, \infty), F)$.

8. Transition Density Estimates.

In this section we fix a connected p.c.f.s.s. set $(F, (\psi_i))$, a resistance vector r_i, and a non-degenerate regular fixed point $\mathcal{E}_A$ of the renormalization map Λ. Let $\mu = \mu_\theta$ be a measure on F, and let $X = (X_t, t \geq 0, \mathbb{P}^x, x \in F)$ be the diffusion process constructed in Section 7. We investigate the transition densities of the process X: initially in fairly great generality, but as the section proceeds, I will restrict the class of fractals.

We begin by fixing the vector θ which assigns mass to the 1-complexes $\psi_i(F)$, in a fashion which relates $\mu_\theta(\psi_i(F))$ with r_i. Let $\beta_i = r_i \rho^{-1}$: by (7.8) we have

(8.1)
$$\beta_i < 1, \quad 1 \leq i \leq M.$$

Let $\alpha > 0$ be the unique positive real such that

(8.2)
$$\sum_{i=1}^{M} \beta_i^\alpha = 1.$$

Set

(8.3)
$$\theta_i = \beta_i^\alpha, \quad 1 \leq i \leq M,$$

and let $\mu = \mu_\theta$ be the associated Bernouilli type measure on F. Write $\beta_+ = \max_i \beta_i$, $\beta_- = \min_i \beta_i$: we have $0 < \beta_- \leq \beta_i \leq \beta_+ < 1$.

We wish to split the set F up into regions which are, "from the point of view of the process X", all roughly the same size. The approximation Theorem 7.22 suggests that if $w \in \mathbb{W}_n$ then the 'crossing time' of the region F_w is of the order of $\rho^{-n} r_w \theta_w^{-1} = \beta_w \theta_w^{-1} = \beta_w^{1-\alpha}$. (See Proposition 8.10 below for a more precise statement of this fact). So if $r.$ is non-constant the decomposition $F = \cup \{F_w, w \in \mathbb{W}_n\}$ of F into n complexes is unsuitable; instead we need to use words w of different lengths. (This idea is due to Hambly – see [Ham2]).

Let $\mathbb{W}_\infty = \cup_{n=0}^\infty \mathbb{W}_n$ be the space of all words of finite length. $\mathbb{W}_\infty$ has a natural tree structure: if $w \in \mathbb{W}_n$ then the parent of w is $w|n-1$, while the offspring of w are the words $w \cdot i$, $1 \leq i \leq M$. (We define the truncation operator τ on $\mathbb{W}_\infty$ by $\tau w = w|(|w|-1)$.) Write also for $w \in \mathbb{W}_\infty$

$$w \cdot \mathbb{W} = \{w \cdot v, v \in \mathbb{W}\} = \{v \in \mathbb{W} : v_i = w_i, \ 1 \leq i \leq |w|\}.$$

Lemma 8.1. *(a) For $\lambda > 0$ let*

$$\mathbb{W}_\lambda = \{w \in \mathbb{W}_\infty : \beta_w \leq \lambda, \ \beta_{\tau w} > \lambda\}.$$

Then the sets $\{w \cdot \mathbb{W}, w \in \mathbb{W}_\lambda\}$ are disjoint, and

$$\bigcup_{w \in \mathbb{W}_\lambda} w \cdot \mathbb{W} = \mathbb{W}.$$

(b) For $f \in L^1(F, \mu)$,

$$\int f \, d\mu = \sum_{w \in W_\lambda} \theta_w \int f_w \, d\mu$$

$$\mathcal{E}(f, f) = \sum_{w \in W_\lambda} \beta_w^{-1} \mathcal{E}(f_w, f_w).$$

Proof. (a) Suppose w, $w' \in W_\lambda$ and $v \in (w \cdot W) \cap (w' \cdot W)$. Then there exist u, $u' \in W$ such that $v = w \cdot u = w' \cdot u'$. So one of w, w' (say w) is an ancestor of the other. But if $\beta_w \leq \lambda$, $\beta_{\tau w} > \lambda$ then as $\beta_i < 1$ we can only have $\beta_{\tau w'} > \lambda$ if $w' = w$. So if $w \neq w'$, $w \cdot W$ and $w' \cdot W$ are disjoint.

Let $v \in W$. Then $\beta_{v|n} = \prod_{i=1}^n \beta_{v_i} \to 0$ as $n \to \infty$. So there exists m such that $v|m \in W_\lambda$, and then $v \in (v|m) \cdot W$, completing the proof of (a).
(b) This follows in a straightforward fashion from the decompositions given in (7.12) and Lemma 5.28. $\qquad \square$

Note that $\beta_- > 0$ and that

$$(8.4) \qquad \beta\lambda \leq \beta_w \leq \lambda, \qquad (\beta_-)^\alpha \lambda^\alpha \leq \theta_w \leq \lambda^\alpha, \quad w \in W_\lambda.$$

Definition 8.2. The *spectral dimension* of F is defined by

$$d_s = d_s(F, \mathcal{E}_A) = 2\alpha/(1 + \alpha).$$

Theorem 8.3. For $f \in \mathcal{D}$,

$$(8.5) \qquad \|f\|_2^{2+4/d_s} \leq c_1 \left(\mathcal{E}(f, f) + \|f\|_2^2 \right) \|f\|_1^{4/d_s}.$$

Proof. It is sufficient to consider the case f non-negative, so let $f \in \mathcal{D}$ with $f \geq 0$. Let $0 < \lambda < 1$: by Lemma 8.1, (7.14) and (8.4) we have

$$
\begin{aligned}
(8.6) \qquad \|f\|_2^2 &= \sum_{w \in W_\lambda} \theta_w \int f_w^2 \, d\mu \\
&\leq \sum_w \theta_w \left(c_1 \mathcal{E}(f_w, f_w) + \left(\int f_w \, d\mu \right)^2 \right) \\
&\leq c_2 \sum_w \lambda^\alpha \mathcal{E}(f_w, f_w) + c_2 \sum_w \lambda^\alpha \left(\int f_w \, d\mu \right)^2 \\
&\leq c_3 \lambda^{\alpha+1} \sum_w \beta_w^{-1} \mathcal{E}(f_w, f_w) + c_2 \lambda^\alpha \left(\sum_w \int f_w \, d\mu \right)^2 \\
&\leq c_3 \lambda^{\alpha+1} \mathcal{E}(f, f) + c_4 \lambda^{-\alpha} \left(\sum_w \theta_w \int f_w \, d\mu \right)^2 \\
&= c_3 \lambda^{\alpha+1} \mathcal{E}(f, f) + c_4 \lambda^{-\alpha} \|f\|_1^2.
\end{aligned}
$$

The final line of (8.6) is minimized if we take $\lambda^{2\alpha+1} = c_5\|f\|_1^2/\mathcal{E}(f,f)$. If $\mathcal{E}(f,f) \geq c_5\|f\|_1^2$ then $\lambda < 1$ and so we obtain from (8.6) that

$$(8.7) \qquad \|f\|_2^2 \leq c\mathcal{E}(f,f)^{\alpha/(2\alpha+1)}\left(\|f\|_1^2\right)^{(\alpha+1)/(2\alpha+1)},$$

which implies that that

$$(8.8) \qquad \|f\|_2^{2+4/d_s} \leq c\mathcal{E}(f,f)\|f\|_1^{4/d_s} \qquad \text{if} \quad \mathcal{E}(f,f) \geq c_5\|f\|_1^2.$$

If $\mathcal{E}(f,f) \leq c_5\|f\|_1^2$ then by (7.14)

$$\|f\|_2^2 \leq c_1\left(\mathcal{E}(f,f) + \|f\|_1^2\right) \leq c\|f\|_1^2,$$

and so

$$(8.9) \qquad \|f\|_2^{2+4/d_s} \leq c\|f\|_2^2\|f\|_1^{4/d_s} \qquad \text{if} \quad \mathcal{E}(f,f) \leq c_5\|f\|_1^2.$$

Combining (8.8) and (8.9) we obtain (8.5). $\qquad\square$

From the results in Section 4 we then deduce

Theorem 8.4. *X has a transition density $p(t,x,y)$ which satisfies*

$$(8.10) \qquad p(t,x,y) \leq c_1 t^{-d_s/2}, \qquad 0 < t \leq 1, \quad x,y \in F,$$

$$(8.11) \qquad \left|p(t,x,y) - p(t,x,y')\right|^2 \leq c_2 t^{-1-d_s/2}R(y,y'), \qquad 0 \leq t \leq 1, \quad x,y,y' \in F.$$

Proof. By Proposition 4.14 X has a jointly measurable transition density, and by Corollary 4.15 we have for $x, y \in F$, $0 < t \leq 1$,

$$p(t,x,y) \leq ct^{-d_s/2}e^{ct} \leq c't^{-d_s/2}.$$

By (4.17) the function $q_{t,x} = p(t,x,\cdot)$ satisfies $\mathcal{E}(q_{t,x},q_{t,x}) \leq ct^{-1-d_s/2}$, and so $q_{t,x} \in \mathcal{D}$ and is continuous. Further, by Proposition 7.18

$$\left|p(t,x,y) - p(t,x,y')\right|^2 \leq cR(y,y')t^{-d_s/2-1}, \qquad x,y,y' \in F.$$

Thus $p(t,\cdot,\cdot)$ is jointly Hölder continuous in the metric R on F. $\qquad\square$

Remarks 8.5. 1. As $\alpha > 0$, we have $0 < d_s = 2\alpha(1+\alpha)^{-1} < 2$.

2. The estimate (8.10) is good if $t \in (0,1]$ and x close to y. It is poor if t is small compared with $R(x,y)$, and in this case we can obtain a better estimate by chaining, as was done for fractional diffusions in Section 3. For this we need some additional properties of the resistance metric.

Lemma 8.6. *If $v, w \in \mathbb{W}_\lambda$ and $v \neq w$ then $F_v \cap F_w = V_v^{(0)} \cap V_w^{(0)}$.*

Proof. This follows easily from the corresponding property for $\mathbb{W}_n$. Let $v, w \in \mathbb{W}_\lambda$, with $|v| = m \leq |w| = n$, $v \neq w$. Let $x \in F_v \cap F_w$. Set $w' = w|m$; then as $F_w \subset F_{w'}$, $x \in F_v \cap F_{w'}$, and so by Lemma 5.17(a) $x \in V_v^{(0)} \cap V_{w'}^{(0)}$. Further, as $x \in F_v$ there exists $v' \in \mathbb{W}_n$ such that $v'|m = v$, and $x \in F_{v'}$. Then $x \in F_{v'} \cap F_w = V_{v'}^{(0)} \cap V_w^{(0)}$. So $x \in V_v^{(0)} \cap V_w^{(0)}$. $\qquad\square$

Definition 8.7. Set

$$V_\lambda^{(0)} = \bigcup_{w \in \mathbb{W}_\lambda} V_w^{(0)}.$$

Let $G_\lambda = \left(V_\lambda^{(0)}, \mathbf{E}_\lambda\right)$ be the graph with vertex set $V_\lambda^{(0)}$, and edge set $\mathbf{E}_\lambda$ such that $\{x, y\}$ is an edge if and only if x, $y \in V_w^{(0)}$ for some $w \in \mathbb{W}_\lambda$. For $A \subset F$ set

$$N_\lambda(A) = \bigcup \{F_w : w \in \mathbb{W}_\lambda, \, F_w \cap A \neq \emptyset\},$$
$$\tilde{N}_\lambda(x) = N_\lambda\left(N_\lambda(\{x\})\right).$$

As we will see, $\tilde{N}_\lambda(x)$ is a neighbourhood of x with a structure which is well adapted to the geometry of F in the metric R. We write $N_\lambda(y) = N_\lambda(\{y\})$.

Lemma 8.8. (a) If x, $y \in V_\lambda^{(0)}$ and $x \neq y$ then

$$R(x, y) \geq c_1 \lambda.$$

(b) If $\{x, y\} \in \mathbf{E}_\lambda$ then $R(x, y) \leq c_2 \lambda$.

Proof. (b) is immediate from the definition of $\mathbb{W}_\lambda$ and Proposition 7.18(b). For (a), note first that if $x \in F$ then by Proposition 5.21 x can belong to at most $M_1 = M \#(P)$ n–complexes, for any n. So there are at most M_1 distinct elements $w \in \mathbb{W}_\lambda$ such that $x \in F_w$.

As $V^{(0)}$ is a finite set, and $\mathcal{E}_A^{(0)}$ is non-degenerate, there exists c_3, $c_4 > 0$ such that,

(8.12) $$c_4 \geq R(x, V^{(0)} - \{x\}) \geq c_3, \qquad x \in V^{(0)}.$$

(Recall that this resistance is, by the construction of $\mathcal{E}$, the same in $(F, \mathcal{E})$ as in $(V^{(0)}, \mathcal{E}_A^{(0)})$). Now fix $x \in V_\lambda^{(0)}$. If $w \in \mathbb{W}_\lambda$, and $x \in V_w^{(0)}$, let $x' = \psi_w^{-1}(x)$, and g_w be the function on F such that $g_w(x') = 1$, $g_w(y) = 0$, $g \in V^{(0)} - \{x'\}$, and

$$\mathcal{E}(g_w, g_w)^{-1} = R\left(x', V^{(0)} - \{x'\}\right) \geq c_3.$$

Define g'_w on F_w by $g'_w = g_w \circ \psi_w^{-1}$, and extend g'_w to F by setting $g'_w = 0$ on $F - F_w$.
Now let $g'_v = 0$ if $x \notin V_v^{(0)}$, $V \in \mathbb{W}_\lambda$, and set

$$g = \sum_{v \in \mathbb{W}_\lambda} g'_v.$$

Then $g(x) = 1$, $g(y) = 0$ if $y \in V_\lambda^{(0)}$, $y \neq x$, and

$$\mathcal{E}(g, g) = \sum_{w \in \mathbb{W}_\lambda} \beta_w^{-1} \mathcal{E}(g \circ \psi_w, g \circ \psi_w)$$
$$= \sum_w \beta_w^{-1} 1_{(x \in F_w)} \mathcal{E}(g_w, g_w) \leq c_5 \lambda^{-1} M_1.$$

Hence if $y \neq x$, $y \in V_\lambda^{(0)}$, we have

$$R(x,y)^{-1} \leq \mathcal{E}(g,g) \leq \lambda^{-1} M_1 c_5^{-1},$$

so that $R(x,y) \geq c_6 \lambda$. $\qquad\square$

Remark. For $x \in V_\lambda^{(0)}$ the function g constructed above is zero outside $N_\lambda(\{x\})$. So we also have

(8.13) $\qquad\qquad R(x,y) \geq c_6 \lambda, \qquad x \in V_\lambda^{(0)}, \quad y \in N_\lambda(\{x\})^c.$

Proposition 8.9. *There exist constants c_i such that for $x \in F$, $\lambda > 0$,*

(8.14) $\qquad\qquad B_R(x, c_1 \lambda) \subset \tilde{N}_\lambda(x) \subset B_R(x, c_2 \lambda),$

(8.15) $\qquad\qquad c_3 \lambda^\alpha \leq \mu(B_R(x,\lambda)) \leq c_4 \lambda^\alpha$

(8.16) $\qquad\qquad c_5 \lambda \leq R(x, \tilde{N}_\lambda(x)^c) \leq c_6 \lambda,$

(8.17) $\qquad\qquad c_7 \lambda \leq R(x, B_R(x,\lambda)^c) \leq c_8 \lambda.$

Proof. Let $x \in F$. If $y \in N_\lambda(\{x\})$ then by (7.17), $R(x,y) \leq c\lambda$. So if $z \in \tilde{N}_\lambda(x)$, since there exists $y \in N_\lambda(\{x\})$ with $z \in N_\lambda(\{y\})$, $R(x,z) \leq c'\lambda$, proving the right hand inclusion in (8.14).

If $x \in V_\lambda^{(0)}$ then by (8.13), if $c_9 = c_{8.7.6}$,

$$B_R(x, c_9 \lambda) \subset N_\lambda(x).$$

Now let $x \notin V_\lambda^{(0)}$, so that there exists a unique $w \in \mathbb{W}_\lambda$ with $x \in F_w$. For each $y \in V_w^{(0)}$ let $f_y(\cdot)$ be the function constructed in Lemma 8.8, which satisfies $f_y(y) = 1$, $f_y = 0$ outside $N_\lambda(y)$, $f_y(z) = 0$ for each $z \in V_\lambda^{(0)} - \{y\}$, and $\mathcal{E}(f_y, f_y) \leq c_{10}\lambda^{-1}$. Let $f = \sum_y f_y$: then $f(y) = 1$ for each $y \in V_w^{(0)}$. So if

$$g = 1_{F_w} + 1_{F_w^c} f,$$

$\mathcal{E}(g,g) \leq \mathcal{E}(f,f) \leq \#(V_w^{(0)}) c_{10}\lambda^{-1} \leq c_{11}\lambda^{-1}$. As $g(x) = 1$, and $g(z) = 0$ for $z \notin \tilde{N}_\lambda(x)$, we have for $z \notin \tilde{N}_\lambda(x)$ that $R(x,z)^{-1} \leq \mathcal{E}(g,g) \leq c_{11}\lambda^{-1}$. So $B_R(x, c_{11}\lambda) \subset N_\lambda(x)$. This proves (8.14), and also that $R(x, \tilde{N}_\lambda(x)^c) \geq c_{11}^{-1}\lambda$.

The remaining assertions now follow fairly easily. For $w \in \mathbb{W}_\lambda$ we have $c_{12}\lambda^\alpha \leq \mu(F_w) \leq c_{13}\lambda^\alpha$. As $\tilde{N}_\lambda(x)$ contains at least one λ–complex, and at most $M^2 \#(P)^2$ λ–complexes, we have

$$\mu(\tilde{N}_\lambda(x)) \asymp \lambda^\alpha,$$

and using (8.14) this implies (8.15).

If $A \subset B$ then it is clear that $R(x,A) \geq R(x,B)$. So (provided λ is small enough) if $x \in F$ we can find a chain x, y_1, y_2, y_3 where $y_i \in V_\lambda^{(0)}$, $\{y_i, y_{i+1}\}$ is an edge in $\mathbf{E}_\lambda$, $y_3 \notin \tilde{N}_\lambda(x)$, and x and y are in the same λ–complex. Then $R(x,y_3) \leq c\lambda$ by (7.17), and so, using Lemma 8.8(b) we have $R(x,y_3) \leq c'\lambda$. Thus $R(x, \tilde{N}_\lambda(x)^c) \leq R(x,y_3) \leq c'\lambda$ proving the right hand side of (8.16): the left hand side was proved above.

(8.17) follows easily from (8.14) and (8.16). $\qquad\square$

Corollary 8.10. *In the metric R, the Hausdorff dimension of F is α, and further*

$$0 < \mathcal{H}_R^\alpha(F) < \infty.$$

Proof. This is immediate from Corollary 2.8 and (8.15). □

Proposition 8.11. *For $x \in F$, $r > 0$ set $\tau(x, r) = T_{B_R(x,r)^c}$. Then*

$$(8.18) \qquad c_1 r^{\alpha+1} \le \mathbb{E}^x \tau(x, r) \le c_2 r^{\alpha+1}, \qquad x \in F, \quad r > 0.$$

Proof. Let $B = B_R(x, r)$. Then by Theorem 4.25 and the estimates (8.15) and (8.17)

$$\mathbb{E}^x \tau(x, r) \le \mu(B) R(x, B^c) \le c_3 r^{\alpha+1},$$

which proves the upper bound in (8.18).

Let $(X_t^B, t \ge 0)$ be the process X killed at $\tau = T_{B^c}$, and let $g(x, y)$ be the Greens' function for X^B. In view of Theorem 7.19, we can write

$$g(x, y) = \mathbb{E}^x L_\tau^y, \qquad x, y \in F.$$

Then if $f(y) = g(x, y)/g(x, x)$, $f \in \mathcal{D}$ and by the reproducing kernel property of g we have

$$\mathcal{E}(f, f) = g(x, x)^{-2} \mathcal{E}\big(g(x, \cdot), g(x, \cdot)\big) = g(x, x)^{-1},$$

and as in Theorem 4.25 $g(x, x) = R(x, B^c) \ge c_4 r$. By (7.18)

$$|f(x) - f(y)|^2 \le R(x, y) \mathcal{E}(f, f) \le R(x, y)(c_4 r)^{-1} \le \tfrac{1}{4}$$

if $R(x, y) \le \tfrac{1}{4} c_4 r$. Thus $f(y) \ge \tfrac{1}{2}$ on $B_R(x, \tfrac{1}{4} c_4 r)$, and hence

$$\mathbb{E}^x \tau = \int_B g(x, y) \mu(dy)$$
$$\ge \tfrac{1}{2} g(x, x) \mu\big(B_R\big(x, \tfrac{1}{4} c_4 r\big)\big) \ge c_5 r^{1+\alpha},$$

proving (8.18). □

We have a spectral decomposition of $p(t, x, y)$. Write $(f, g) = \int_F f g \, d\mu$.

Theorem 8.12. *There exist functions $\varphi_i \in \mathcal{D}$, $\lambda_i \ge 0$, $i \ge 0$, such that $(\varphi_i, \varphi_i) = 1$, $0 = \lambda_0 < \lambda_1 \le \cdots$, and*

$$\mathcal{E}(\varphi_i, f) = \lambda_i(\varphi_i, f), \qquad f \in \mathcal{D}.$$

The transition density $p(t, x, y)$ of X satisfies

$$(8.19) \qquad p(t, x, y) = \sum_{i=0}^\infty e^{-\lambda_i t} \varphi_i(x) \varphi_i(y),$$

where the sum in (8.19) converges uniformly and absolutely. So p is jointly continuous in (t, x, y).

Proof. This follows from Mercer's Theorem, as in [DaS]. Note that $\varphi_0 = 1$ as $\mathcal{E}$ is irreducible and $\mu(F) = 1$. □

The following is an immediate consequence of (8.19)

Corollary 8.13. *(a) For x, $y \in F$, $t > 0$,*

$$p(t,x,y)^2 \le p(t,x,x)p(t,y,y).$$

(b) For each $x, y \in F$

$$\lim_{t \to \infty} p(t,x,y) = 1.$$

Lemma 8.14.

(8.20) $p(t,x,y) \ge c_0 t^{-d_s/2}, \qquad 0 \le t \le 1, \quad R(x,y) \le c_1 t^{1/(1+\alpha)}.$

Proof. We begin with the case $x = y$. From Proposition 8.11 and Lemma 3.16 we deduce that there exists $c_2 > 0$ such that

$$\mathbb{P}^x\big(\tau(x,r) \le t\big) \le (1 - 2c_2) + c_3 tr^{-\alpha-1}.$$

Choose $c_4 > 0$ such that $c_3 tr_0^{-\alpha-1} = c_2$ if $r_0 = c_4 t^{1/(1+\alpha)}$. Then

$$\mathbb{P}^x\big(X_t \in B_R(x,r_0)\big) \ge \mathbb{P}^x\big(\tau(x,r_0) \le t\big) \ge c_2.$$

So using Cauchy-Schwarz and the symmetry of p, and writing $B = B_R(x,r_0)$,

$$0 < c_2^2 \le \left(\int_B p(t,x,y)\mu(dy)^2 \right)$$

$$\le \int_{B(x,r_0)} \mu(dy) \int_B p(t,x,y)p(t,y,x)\mu(dy)$$

$$\le \mu(B))\, p(2t,x,x)$$

$$\le c_5 t^{\alpha/(1+\alpha)} p(2t,x,x).$$

Replacing t by $t/2$ we have

$$p(t,x,x) \ge c_0 t^{-d_s/2}.$$

Fix t,x, and write $q(y) = p(t,x,y)$. By (4.16) and (8.5) $\mathcal{E}(q,q) \le c_6 t^{-1-d_s/2}$ for $t \le 1$, so using (7.18), if $R(x,y) \le c_7 t^{1/(1+\alpha)}$ then, as $1 + d_s/2 = (1 + 2\alpha)/(1 + \alpha)$,

$$q(y) \ge q(x) - |q(x) - q(y)|$$

$$\ge c_0 t^{-\alpha/(1+\alpha)} - \big(R(x,y)\mathcal{E}(q,q)\big)^{1/2}$$

$$\ge c_0 t^{-\alpha/(1+\alpha)} - \left(c_7 c_6 t^{-2\alpha/(1+\alpha)}\right)^{1/2}$$

$$= t^{-\alpha/(1+\alpha)}(c_0 - (c_7 c_6)^{1/2}).$$

Choosing c_7 suitably gives (8.20). $\qquad\qquad\qquad\qquad\qquad\qquad\qquad\qquad\quad\square$

We can at this point employ the chaining arguments used in Theorem 3.11 to extend these bounds to give upper and lower bounds on $p(t,x,y)$. However, as R is not in general a geodesic metric, the bounds will not be of the form given in Theorem 3.11. The general case is given in a paper of Hambly and Kumagai [HK2], but since the proof of Theorem 3.11 does not use the geodesic property for the upper bound we do obtain:

Theorem 8.15. *The transition density* $p(t, x, y)$ *satisfies*

(8.21) $$p(t, x, y) \le c_1 t^{-\alpha/(1+\alpha)} \exp\left(-c_2 \left(R(x,y)^{1+\alpha}/t\right)^{1/\alpha}\right).$$

Note. The power $1/\alpha$ in the exponent is not in general best possible.

Theorem 8.16. *Suppose that there exists a metric* ρ *on* F *with the midpoint property such that for some* $\theta > 0$

(8.22) $$c_1 \rho(x, y)^\theta \le R(x, y) \le c_2 \rho(x, y)^\theta \quad x, y \in F.$$

Then if $d_w = \theta(1 + \alpha)$, $d_f = \alpha\theta$, (F, ρ, μ) *is a fractional metric space of dimension* d_f, *and* X *is a fractional diffusion with indices* d_f, d_w.

Proof. Since $B_\rho(x, (r/c_2)^\theta) \subset B_R(x, r) \subset B_\rho(x, (r/c_1)^\theta)$, it is immediate from (8.15) that (F, ρ) is a $FMS(d_f)$. Write $\tau_\rho(x, r) = \inf\{t : X_t \notin B_\rho(x, r)\}$. Then from (8.18) and (8.22)

$$c r^{\theta(1+\alpha)} \le \mathbb{E}^x \tau_\rho(x, r) \le c_2 r^{\theta(1+\alpha)}.$$

So, by (8.10) and (8.20), X satisfies the hypotheses of Theorem 3.11, and so X is a $FD(d_f, d_w)$. $\square$

Remark. Note that in this case the estimate (7.20) on the Hölder continuity of $u_\lambda(x, y)$ implies that

(8.23) $$|u_\lambda(x, y) - u_\lambda(x', y)| \le c R(x, x')^{\frac{1}{2}} \le c' \rho(x, x')^{\theta/2},$$

while by Theorem 3.40 we have

(8.24) $$|u_\lambda(x, y) - u_\lambda(x', y)| \le c \rho(x, x')^\theta.$$

The difference is that (8.23) used only the fact that $u_\lambda(., y) \in \mathcal{D}$, while the proof of (8.24) used the fact that it is the λ-potential density.

Diffusions on nested fractals.

We conclude by treating briefly the case of nested fractals. Most of the necessary work has already been done. Let $(F, (\psi_i))$ be a nested fractal, with length, mass, resistance and shortest path scaling factors L, M, ρ, γ. Recall that in this context we take $r_i = 1$, $\theta_i = 1/M$, $1 \le i \le M$, and $\mu = \mu_\theta$ for the measure associated with θ. Write $d = d_F$ for the geodesic metric on F defined in Section 5.

Lemma 8.17. *Set* $\theta = \log \rho / \log \gamma$. *Then*

(8.24) $$c_1 d(x, y)^\theta \le R(x, y) \le c_2 d(x, y)^\theta, \quad x, y \in F.$$

Proof. Let $\lambda \in (0, 1)$. Since all the r_i are equal, $\tilde{N}_\lambda(x)$ is a union of n-complexes, where $\rho^{-n} \le \lambda \le \rho^{-n+1}$. So by Theorem 5.43 and Proposition 8.8, since $\gamma^{-n} = (\rho^{-n})^\theta$,

(8.26) $$y \in \tilde{N}_\lambda(x) \text{ implies that } R(x, y) \le c_1 \lambda, \text{ and } d(x, y) \le c_2 \lambda^\theta,$$

(8.27) $y \notin \widetilde{N}_\lambda(x)$ implies that $R(x,y) \geq c_3\lambda$, and $d(x,y) \geq c_4\lambda^\theta$.

The result is immediate from (8.26) and (8.27).   □

Applying Lemma 8.17 and Theorem 8.15 we deduce:

Theorem 8.18. *Let F be a nested fractal, with scaling factors L, M, ρ, γ. Set*

$$d_f = \log M / \log \gamma, \quad d_w = \log M\rho / \log \gamma.$$

Then (F, d_F, μ) is a fractional metric space of dimension d_f, and X is a $FD(d_f, d_w)$. In particular, the transition density $p(t, x, y)$ of X is jointly continuous in (t, x, y) and satisfies

$$(8.28) \quad c_1 t^{-d_f/d_w} \exp\left(-c_2 \big(d(x,y)^{d_w}/t\big)^{1/(d_w-1)}\right)$$
$$\leq p(t, x, y) \leq c_3 t^{-d_f/d_w} \exp\left(-c_4 \big(d(x,y)^{d_w}/t\big)^{1/(d_w-1)}\right).$$

References.

[AO] S. Alexander and R. Orbach: Density of states on fractals: "fractons". *J. Physique (Paris) Lett.* **43**, L625–L631 (1982).

[AT] W.N. Anderson and G.E. Trapp: Shorted operators, II, *SIAM J. Appl. Math.* **28**, 60–71 (1975).

[Ar] D.G. Aronson, Bounds on the fundamental solution of a parabolic equation. *Bull. Amer. Math. Soc.* **73**, 890–896 (1967).

[BS] C. Bandt and J. Stahnke: Self-similar sets 6. Interior distance on deterministic fractals. Preprint 1990.

[Bar1] M.T. Barlow: Random walks, electrical resistance and nested fractals. In *Asymptotic Problems in Probability Theory*, ed. K.D. Elworthy, N. Ikeda, 131–157, Longman Scientific, Harlow UK, 1990.

[Bar2] M.T. Barlow: Random walks and diffusions on fractals. *Proc. Int. Congress Math. Kyoto 1990*, 1025–1035. Springer, Tokyo 1991.

[Bar3] M.T. Barlow: Harmonic analysis on fractal spaces. *Séminaire Bourbaki Volume 1991/1992, Astérisque* **206** (1992).

[BB1] M. T. Barlow and R.F. Bass: The construction of Brownian motion on the Sierpinski carpet. *Ann. Inst. H. Poincaré* **25**, 225–257 (1989).

[BB2] M. T. Barlow and R.F. Bass: Local times for Brownian motion on the Sierpinski carpet. *Probab. Th. Rel. Fields,* **85**, 91–104 (1990).

[BB3] M. T. Barlow and R.F. Bass: On the resistance of the Sierpinski carpet. *Proc. R. Soc. London A.* **431**, 345–360 (1990).

[BB4] M.T. Barlow and R.F. Bass: Transition densities for Brownian motion on the Sierpinski carpet. *Probab. Th. Rel. Fields* **91**, 307–330 (1992).

[BB5] M.T. Barlow and R.F. Bass: Coupling and Harnack inequalities for Sierpinski carpets. *Bull. A.M.S.* **29**, 208–212 (1993).

[BB6] M.T. Barlow and R.F. Bass: Brownian motion and harmonic analysis on Sierpinski carpets. Preprint 1997.

[BB7] M.T. Barlow and R.F. Bass: Random walks on graphical Sierpinski carpets. In preparation.

[BBS] M.T. Barlow, R.F. Bass, and J.D. Sherwood: Resistance and spectral dimension of Sierpinski carpets. *J. Phys. A*, **23**, L253–L258 (1990).

[BH] M.T. Barlow and B.M. Hambly: Transition density estimates for Brownian motion on scale irregular Sierpinski gaskets. To appear *Ann. IHP*.

[BHHW] M.T. Barlow, K. Hattori, T. Hattori and H. Watanabe: Weak homogenization of anisotropic diffusion on pre-Sierpinski carpets. To appear *Comm Math. Phys.*

[BK] M.T. Barlow and J. Kigami: Localized eigenfunctions of the Laplacian on p.c.f. self-similar sets. To appear *J. Lond. Math. Soc.*

[BP] M. T. Barlow and E. A. Perkins: Brownian motion on the Sierpinski gasket. *Probab. Th. Rel. Fields* **79**, 543–623 (1988).

[Bas] R.F. Bass: Diffusions on the Sierpinski carpet. *Trends in Probability and related Analysis: Proceedings of SAP '96.* World Scientific, Singapore, 1997. To appear.

[BAH] D. Ben-Avraham and S. Havlin: Exact fractals with adjustable fractal and fracton dimensionalities. *J. Phys. A.* **16**, L559–563 (1983).

[Blu] L.M. Blumenthal: *Theory and applications of distance geometry.* Oxford, 1953.

[BG] R.M. Blumenthal and R.K. Getoor: *Markov processes and potential theory.* Academic Press, New York, 1968.

[CKS] E.A. Carlen, S. Kusuoka and D.W. Stroock: Upper bounds for symmetric Markov transition functions. *Ann. Inst. H. Poincare Sup. no. 2*, 245–287 (1987).

[Ca1] D. Carlson: What are Schur complements, anyway? *Linear Alg. Appl.* **74**, 257–275 (1986).

[CRRST] A.K. Chandra, P. Raghavan, W.L. Ruzzo, R. Smolensky and P. Tiwari: The electrical resistance of a graph captures its commute and cover times. Proceedings of the 21st ACM Symposium on theory of computing, 1989.

[Co] T.C. Coulhon: Ultracontractivity and Nash type inequalities. *J. Funct. Anal.* **141**, 510-539 (1996).

[DSV] K. Dalrymple, R.S. Strichartz and J.P. Vinson: Fractal differential equations on the Sierpinski gasket. Preprint 1997.

[Da] E.B. Davies: *Heat kernels and spectral theory.* Cambridge University Press 1989.

[DaSi] E.B. Davies and B. Simon: Ultracontractivity and the heat kernel for Schrödinger operators and Dirichlet Laplacians. *J. Funct. Anal.* **59**, 335–395, (1984).

[DS] P.G. Doyle and J.L. Snell: *Random walks and electrical networks.* Washington, Math. Assoc. of America, 1984.

[Dud] R.M. Dudley: Sample functions of the Gaussian process. *Ann. Prob.* **1**, 66–103 (1971).

[FaS] E.B. Fabes and D.W. Stroock, A new proof of Moser's parabolic Harnack inequality via the old ideas of Nash. *Arch. Mech. Rat. Anal.* **96**, 327–338 (1986).

[Fa1] K.J. Falconer: *Geometry of fractal sets.* Cambridge Univ. Press, 1985.

[Fa2] K.J. Falconer: *Fractal Geometry.* Wiley, 1990

[Fe] H. Federer: *Geometric measure theory.* Springer, New York, 1969.

[FHK] P.J. Fitzsimmons, B.M. Hambly, T. Kumagai: Transition density estimates for Brownian motion on affine nested fractals. *Comm. Math. Phys.* **165**, 595–620 (1995).

[FOT] M. Fukushima, Y. Oshima, and M. Takeda, *Dirichlet Forms and Symmetric Markov Processes.* de Gruyter, Berlin, 1994.

[Fu1] M. Fukushima, Dirichlet forms, diffusion processes, and spectral dimensions for nested fractals. *Ideas and methods in stochastic analysis, stochastics and applications*, 151–161. Cambridge Univ. Press., Cambridge, 1992.

[FS1] M. Fukushima and T. Shima: On a spectral analysis for the Sierpinski gasket. *J. of Potential Analysis.* **1**, 1–35 (1992).

[FS2] M. Fukushima and T. Shima, On discontinuity and tail behaviours of the integrated density of states for nested pre-fractals. *Comm. Math. Phys.* **163**, 461–471 (1994).

[FST] M. Fukushima, T. Shima and M. Takeda: Large deviations and related LILs for Brownian motions on nested fractals. Preprint 1997.

[GAM1] Y. Gefen, A. Aharony and B.B. Mandelbrot: Phase transitions on fractals.I. Quasilinear lattices. *J. Phys. A* **16**, 1267–1278 (1983).

[GAM2] Y. Gefen, A. Aharony, Y. Shapir and B.B. Mandelbrot: Phase transitions on fractals. II. Sierpinski gaskets. *J. Phys. A* **17**, 435–444 (1984).

[GAM3] Y. Gefen, A. Aharony and B. Mandelbrot: Phase transitions on fractals. III. Infinitely ramified lattices. *J. Phys. A* **17**, 1277–1289 (1984).

[GK] R.K. Getoor and H. Kesten: Continuity of local times of Markov processes. *Comp. Math.* **24**, 277-303 (1972).

[Go] S. Goldstein: Random walks and diffusion on fractals. In: Kesten, H. (ed.) *Percolation theory and ergodic theory of infinite particle systems* (IMA Math. Appl., vol.8.) Springer, New York, 1987, pp.121–129.

[Gra] P.J. Grabner: Functional equations and stopping times for the Brownian motion on the Sierpinski gasket. To appear in *Mathematica.*

[GrT] P.J. Grabner and R.F. Tichy: Equidistribution and Brownian motion on the Sierpinski gasket. Preprint 1997.

[GrW] P.J. Grabner and W Woess: Functional iterations and periodic oscillations for simple random walk on the Sierpinski gasket. To appear in *Stoch. Proc. Appl.*

[Gre] R.F. Green: A simple model for the construction of Brownian motion on Sierpinski's hexagonal gasket. Technical report, Univ. of Minnesota, Duluth. 1989.

[HS] L. deHaan and U. Stadtmuller: Dominated variation and related concepts and Tauberian theorems for Laplace transforms. *J. Math. Anal. Appl.* **108**, 344-365 (1985).

[Ham1] B.M. Hambly: Brownian motion on a homogeneous random fractal. *Probab. Th. Rel. Fields* **94**, 1-38, (1992).

[Ham2] B.M. Hambly: Brownian motion on a random recursive Sierpinski gasket. *Ann. Prob.* **25**, 1059-1102 (1997).

[HK1] B.M. Hambly, T. Kumagai: Heat kernel estimates and homogenization for asymptotically lower dimensional processes on some nested fractals. Preprint 1996.

[HK2] B.M. Hambly, T. Kumagai: Transition density estimates for diffusion processes on p.c.f. self-similar fractals. Preprint 1997.

[Har] T.E. Harris: *The theory of branching processes.* Springer 1963.

[Hat1] T. Hattori: Asympotically one-dimensional diffusions on scale-irregular gaskets. Preprint 1994.

[HH] K. Hattori and T. Hattori: Self-avoiding process on the Sierpinski gasket. *Probab. Th. Rel. Fields* **88**, 405-428 (1991).

[HatK] T. Hattori, S. Kusuoka: The exponent for mean square displacement of self-avoiding random walk on Sierpinski gasket. *Probab. Th. Rel. Fields* **93**, 273-284 (1992).

[HHK] K. Hattori, T. Hattori, and S. Kusuoka, Self-avoiding paths on the three-dimensional Sierpinski gasket. *Publ. Res. Inst. Math. Sci.* **29**, 455-509 (1993).

[HHW] K. Hattori, T. Hattori and H. Watanabe: Reasoning out the empirical rule $\bar{d} < 2$, *Physics Letters A* **115**, 207-212 (1986).

[HHW1] K. Hattori, T. Hattori and H. Watanabe: Gaussian field theories and the spectral dimensions. *Prog. Th. Phys. Supp. No.* **92**, 108-143, (1987).

[HHW2] K. Hattori, T. Hattori and H. Watanabe: New approximate renormalisation method on fractals. *Phys Rev. A* **32**, 3730-3733 (1985).

[HHW3] K. Hattori, T. Hattori and S. Kusuoka: Self-avoiding paths on the pre-Sierpinski gasket. *Probab. Th. Rel. Fields* **84**, 1-26 (1990).

[HHW4] K. Hattori, T. Hattori and H. Watanabe: Asymptotically one-dimensional diffusions on the Sierpinski gasket and the *abc*-gaskets. *Prob. Th. Rel. Fields* **100**, 85-116 (1994).

[HBA] S. Havlin and D. Ben-Avraham: Diffusion in disordered media, *Adv. Phys.* **36**, 695-798 (1987).

[He] M.K. Heck: Homogeneous diffusions on the Sierpinski gasket. Preprint 1996.

[Hu] J.E. Hutchinson: Fractals and self-similarity. *Indiana J. Math.* **30**, 713-747 (1981).

[Jo] O.D. Jones: Transition probabilities for the simple random walk on the Sierpinski graph. *Stoch. Proc. Appl.* **61**, 45-69 (1996).

[J] A. Jonsson: Brownian motion on fractals and function spaces. *Math. Zeit.* **122**, 495-504 (1986).

[Ka] M. Kac: On some connections between probability theory and differential and integral equations. *Proc. 2nd Berkelely Symposium, 1950*, 189-215, (1951).

[Ke] H. Kesten: Subdiffusive behaviour of random walk on random cluster. *Ann. Inst. H. Poincare B* **22**, 425–487 (1986).

[Ki1] J. Kigami: A harmonic calculus on the Sierpinski space. *Japan J. Appl. Math.* **6**, 259–290 (1989).

[Ki2] J. Kigami: A harmonic calculus for p.c.f. self-similar sets. *Trans. A.M.S.* **335**, 721–755 (1993).

[Ki3] J. Kigami: Harmonic metric and Dirichlet form on the Sierpinski gasket. In *Asymptotic Problems in Probability Theory*, ed. K.D. Elworthy, N. Ikeda, 201–218, Longman Scientific, Harlow UK, 1990.

[Ki4] J. Kigami: A calculus on some self-similar sets. To appear in Proceedings on the 1st IFIP Conference "Fractal 90", Elsevier.

[Ki5] J. Kigami: Effective resistance for harmonic structures on P.C.F. self-similar sets. *Math. Proc. Camb. Phil. Soc.* **115**, 291–303 (1994).

[Ki6] J. Kigami: Hausdorff dimension of self-similar sets and shortest path metric. *J. Math. Soc. Japan* **47**, 381–404 (1995).

[Ki7] J. Kigami: Harmonic calculus on limits of networks and its application to dendrites. *J. Functional Anal.* **128**, 48–86 (1995).

[KL] J. Kigami, M. Lapidus: Weyl's spectral problem for the spectral distribution of Laplacians on P.C.F. self-similar fractals. *Comm. Math. Phys.* **158**, 93–125 (1993).

[Kn] F.B. Knight: *Essentials of Brownian motion and diffusion.* AMS Math. Surveys, Vol. 18. AMS 1981.

[Koz] S.M. Kozlov: Harmonization and homogenization on fractals. *Comm. Math. Phys.* **153**, 339-357 (1993).

[Kr1] W.B. Krebs: A diffusion defined on a fractal state space. *Stoch. Proc. Appl.* **37**, 199–212 (1991).

[Kr2] W.B. Krebs: Hitting time bounds for Brownian motion on a fractal. *Proc. A.M.S.* **118**, 223–232 (1993).

[Kr3] W.B. Krebs: Brownian motion on the continuum tree. *Prob. Th. Rel. Fields* **101**, 421-433 (1995). Preprint 1993.

[Kum1] T. Kumagai: Construction and some properties of a class of non- symmetric diffusion processes on the Sierpinski gasket. In *Asymptotic Problems in Probability Theory*, ed. K.D. Elworthy, N. Ikeda, 219–247, Longman Scientific, Harlow UK, 1990.

[Kum2] T. Kumagai: Estimates of the transition densities for Brownian motion on nested fractals. *Prob. Th. Rel. Fields* **96**, 205–224 (1993).

[Kum3] T. Kumagai: Regularity, closedness, and spectral dimension of the Dirichlet forms on p.c.f. self-similar sets. *J. Math. Kyoto Univ.* **33**, 765–786 (1993).

[Kum4] T. Kumagai: Rotation invariance and characterization of a class of self-similar diffusion processes on the Sierpinski gasket. *Algorithms, fractals and dynamics*, 131–142. Plenum, New York, 1995.

[Kum5] T. Kumagai: Short time asymptotic behaviour and large deviation for Brownian motion on some affine nested fractals. *Publ. R.I.M.S. Kyoto Univ.* **33**, 223-240 (1997).

[KK1] T. Kumagai and S. Kusuoka, Homogenization on nested fractals. *Probab. Th. Rel. Fields* **104**, 375–398 (1996).

[Kus1] S. Kusuoka: A diffusion process on a fractal. In: Ito, K., N. Ikeda, N. (ed.) Symposium on Probabilistic Methods in Mathematical Physics, Taniguchi, Katata. Academic Press, Amsterdam, 1987, pp.251–274

[Kus2] S. Kusuoka: Dirichlet forms on fractals and products of random matrices. *Publ. RIMS Kyoto Univ.*, **25**, 659–680 (1989).

[Kus3] S. Kusuoka: Diffusion processes on nested fractals. In: *Statistical mechanics and fractals*, Lect. Notes in Math. **1567**, Springer, 1993.

[KZ1] S. Kusuoka and X.Y. Zhou: Dirichlet form on fractals: Poincaré constant and resistance. *Prob. Th. Rel. Fields* **93**, 169–196 (1992).

[KZ2] S. Kusuoka and X.Y. Zhou: Waves on fractal-like manifolds and effective energy propagation. Preprint 1993.

[L1] T. Lindstrøm: Brownian motion on nested fractals. *Mem. A.M.S.* **420**, 1990.

[L2] T. Lindstrøm: Brownian motion penetrating the Sierpinski gasket. In *Asymptotic Problems in Probability Theory*, ed. K.D. Elworthy, N. Ikeda, 248–278, Longman Scientific, Harlow UK.

[Mae] F-Y Maeda: *Dirichlet Integrals on Harmonic Spaces*. Springer L.N.M. **803** 1980.

[Man] B.B. Mandelbrot: *The fractal geometry of nature*. W.H. Freeman, San Fransisco, 1982.

[MR] M.B. Marcus, J. Rosen: Sample path functions of the local times of strongly symmetric Markov processes via Gaussian processes. *Ann. Prob.* **20**, 1603–1684 (1992).

[MW] R.D. Mauldin, S.C. Williams: Hausdorff dimension in graph directed constructions. *Trans. A.M.S.* **309**, 811–829 (1988).

[Me1] V. Metz: Potentialtheorie auf dem Sierpinski gasket. *Math. Ann.* **289**, 207–237 (1991).

[Me2] V. Metz, Renormalization of finitely ramified fractals. *Proc. Roy. Soc. Edinburgh Ser A* **125**, 1085–1104 (1995).

[Me3] V. Metz: How many diffusions exist on the Vicsek snowflake? *Acta Appl. Math.* **32**, 224–241 (1993).

[Me4] V. Metz: Hilbert's projective metric on cones of Dirichlet forms. *J. Funct. Anal.* **127**, 438–455, (1995).

[Me5] V. Metz, Renormalization on fractals. *Proc. International Conf. Potential Theory 94*, 413–422. de Gruyter, Berlin, 1996.

[Me6] V. Metz: Maeda's energy norm explains effective resistance. Preprint.

[Me7] V. Metz, Renormalization contracts on nested fractals. *C.R. Acad. Sci. Paris* **332**, 1037–1042 (1996).

[MS] V. Metz and K.-T. Sturm, Gaussian and non-Gaussian estimates for heat kernels on the Sierpinski gasket. In: *Dirichlet forms and stochastic processes*, 283–289, de Gruyter, Berlin, 1995.

[Mor] P.A.P. Moran: Additive functions of intervals and Hausdorff measure. *Proc. Camb. Phil. Soc.* **42**, 15–23 (1946).

[Mos] U. Mosco: Composite media and asymptotic Dirichlet forms. *J. Funct. Anal.* **123**, 368–421 (1994).

[Nus] R.D. Nussbaum: Hilbert's projective metric and iterated non-linear maps. *Mem. A.M.S.* **75**, 1988.

[OSS] M. Okada, T. Sekiguchi and Y. Shiota: Heat kernels on infinte graph networks and deformed Sierpinski gaskets. *Japan J. App. Math.* **7**, 527–554 (1990).

[O1] H. Osada: Isoperimetric dimension and estimates of heat kernels of pre-Sierpinski carpets. *Probab. Th. Rel. Fields* **86**, 469–490 (1990).

[O2] H. Osada: Cell fractals and equations of hitting probabilities. Preprint 1992.

[OS] B. O'Shaughnessy, I. Procaccia: Analytical solutions for diffusion on fractal objects. *Phys. Rev. Lett.* **54**, 455–458, (1985).

[Pe] R. Peirone: Homogenization of functionals on fractals. Preprint, 1996.

[PP] K. Pietruska-Paluba: The Lifchitz singularity for the density of states on the Sierpinski gasket. *Probab. Th. Rel. Fields* **89**, 1–34 (1991).

[RT] R. Rammal and G. Toulouse: Random walks on fractal structures and percolation clusters, *J. Physique Lettres* **44**, L13–L22 (1983).

[R] R. Rammal: Spectrum of harmonic excitations on fractals. *J. de Physique* **45**, 191–206 (1984).

[Rog] L.C.G. Rogers, Multiple points of Markov processes in a complete metric space. *Sém. de Probabilités XXIII*, 186–197. Springer, Berlin, 1989.

[Sab1] C.Sabot: Existence and uniqueness of diffusions on finitely ramified self-similar fractals. Preprint 1996.

[Sab2] C.Sabot: Espaces de Dirichlet reliés par un nombre fini de points et application aux diffusions sur les fractales. Preprint 1996.

[Sha] M. Sharpe: *General Theory of Markov Processes*. Academic Press. New York, 1988.

[Sh1] T. Shima: On eigenvalue problems for the random walk on the Sierpinski pre-gaskets. *Japan J. Appl. Ind. Math.*, **8**, 127–141 (1991).

[Sh2] T. Shima: The eigenvalue problem for the Laplacian on the Sierpinski gasket. In *Asymptotic Problems in Probability Theory* , ed. K.D. Elworthy, N. Ikeda, 279–288, Longman Scientific, Harlow UK.

[Sh3] T. Shima: On Lifschitz tails for the density of states on nested fractals. *Osaka J. Math.* **29**, 749–770 (1992).

[Sh4] T. Shima: On eigenvalue problems for Laplacians on P.C.F. self- similar sets. *Japan J. Indust. Appl. Math.* **13**, 1–23 (1996).

[Sie1] W. Sierpinski: Sur une courbe dont tout point est un point de ramification. *C.R. Acad. Sci. Paris* **160**, 302–305 (1915).

[Sie2] W. Sierpinski: Sur une courbe cantorienne qui contient une image biunivoque et continue de toute courbe donnée. *C.R. Acad. Sci. Paris* **162**, 629–632 (1916).

[Stu1] K.T. Sturm: Diffusion processes and heat kernels on metric spaces. To appear *Ann. Probab.*

[Stu2] K.T. Sturm: Analysis on local Dirichlet spaces I. Recurrence, conservativeness and L^p-Liouville properties. *J. reine angew. Math.* **456**, 173–196 (1994).

[Stu3] K.T. Sturm: Analysis on local Dirichlet spaces II. Upper Gaussian estimates for the fundamental solutions of parabolic equations. *Osaka J. Math.* **32**, 275–312 (1995).

[Stu4] K.T. Sturm: Analysis on local Dirichlet spaces III. The parabolic Harnack inequality. *J. Math. Pures Appl.* **75**, 273–297 (1996).

[Tal] M. Talagrand: Regularity of Gaussian processes *Acta Math.* **159**, 99-149 (1987).

[Tel] A. Telcs: Random walks on graphs, electric networks and fractals. *Prob. Th. Rel. Fields* **82**, 435-451 (1989).

[Tet] P. Tetali: Random walks and the effective resistance of networks. *J. Theor. Prob.* **4**, 101–109 (1991).

[Tri] C. Tricot: Two definitions of fractional dimension. *Math. Proc. Camb. Phil. Soc.* **91**, 54–74, (1982).

[V1] N. Th. Varopoulos: Isoperimetric inequalities and Markov chains. *J. Funct. Anal.* **63**, 215–239 (1985).

[Wat1] H. Watanabe: Spectral dimension of a wire network. *J. Phys. A.* **18**, 2807–2823 (1985).

[Y1] J.-A. Yan: A formula for densities of transition functions. *Sem. Prob. XXII*, 92-100. Lect Notes Math. **1321**.

ANALYSIS ON WIENER SPACE
AND ANTICIPATING STOCHASTIC CALCULUS

David NUALART

Contents

Introduction

The first two chapters of these notes are devoted to present the basic elements of the stochastic calculus of variations or *Malliavin calculus* on a Gaussian space. That is, we develop an infinite dimensional differential calculus for functionals of an arbitrary Gaussian process. The stochastic calculus of variations on the Wiener space was introduced by Malliavin in [62] in order to provide a probabilistic proof of Hörmander's hypoellipticity theorem. One of the main results of this subject is the equivalence between the L^p norms of the iterations of the derivative and the Ornstein-Uhlenbeck operators, that was established by Meyer in [66].

Chapter 3 deals with the application of the techniques of the stochastic calculus of variations to study regularity properties of probability laws. First we discuss the existence and smoothness of a density for an m-dimensional random vector assuming that its Malliavin matrix is nondegenerate. The basic tool to obtain these results is an integration by parts formula. We deduce also L^p estimates of the density and apply them to the problem of the existence, uniqueness and convergence of approximation schemes for ordinary and partial stochastic differential equations. In Chapter 4 we discuss the properties of the support of the probability distribution of a random element in the underlying Gaussian space. For nondegenerate smooth random vectors, the points where the density is strictly positive are characterized by means of the notion of skeleton. We show this type of characterization and we apply it to deduce Varadhan-type estimates when the variance of the white noise tends to zero.

In chapters 5 and 6 we present some results on the anticipating stochastic calculus. The divergence operator coincides with an extension of the Itô stochastic integral, due to Skorohod, for processes which are not adapted to the Brownian motion. The Skorohod integral turns out to have properties which are similar to the classical Itô integral. For instance, one can deduce a change-of-variable formula which generalizes the Itô formula. On the other hand, one can also define an extended Stratonovich integral by means of suitable approximation by Riemann sums. We discuss the relationship between the Skorohod integral and the extended Stratonovich integral and we establish a change-of-variable formula for the Stratonovich integral which is similar to that in classical analysis. The anticipating stochastic integrals allow us to formulate and solve stochastic differential equations whose coefficients are random, or such that some boundary conditions are in force. We deduce some existence and uniqueness results for these type of equations in Chapter 6.

David Nualart

Barcelona, July, 1995

Chapter 1

Derivative and divergence operators on a Gaussian space

In this chapter we study the differential calculus on a Gaussian space. That is, we introduce the derivative operator and its adjoint the divergence operator. We show that these operators are local, and we study their behavior when they are applied to a Wiener chaos expansion.

1.1 Derivative operator

We work in the following general context. Suppose that H is a real separable Hilbert space whose norm and inner product are denoted by $\|\cdot\|_H$ and $\langle\cdot,\cdot\rangle_H$, respectively. We associate with H a Gaussian and centered family of random variables $\mathcal{H}_1 = \{W(h), h \in H\}$ such that

$$E(W(h)W(g)) = \langle h, g \rangle_H,$$

for all $h, g \in H$. The family $\mathcal{H}_1$ is defined in some complete probability space $(\Omega, \mathcal{F}, P)$. This means that $\mathcal{H}_1$ is a closed Gaussian subspace of $L^2(\Omega)$ isometric to H. We will assume that the σ-field $\mathcal{F}$ is generated by $\mathcal{H}_1$.

Examples:

1. Let $\{W_t^i, t \geq 0, i = 1, \ldots, d\}$ be a d-dimensional Brownian motion. In this case the Gaussian space is obtained by taking $H = L^2(\mathbb{R}_+; \mathbb{R}^d)$, and for any $h \in H$ the variable $W(h)$ is the Wiener stochastic integral $\sum_{i=1}^d \int_0^\infty h_t^i dW_t^i$.

2. Suppose that $(T, \mathcal{B}, \mu)$ is a measure space such that $\mathcal{B}$ is countably generated, and $\{W(A), A \in \mathcal{B}, \mu(A) < \infty\}$ is a family of random variables such that each $W(A)$ has the distribution $N(0, \mu(A))$, $W(A \cup B) = W(A) + W(B)$ if A and B are disjoint, and $\{W(A_1), \ldots, W(A_n)\}$ are independent whenever the sets $\{A_1, \ldots, A_n\}$ are pairwise-disjoint. In this case the Gaussian space is obtained by taking $H = L^2(T, \mathcal{B}, \mu)$, and $W(h)$ is given by the stochastic integral

$$W(h) = \int_T h(t)W(dt).$$

We will say that W is a *white noise* on $(T, \mathcal{B}, \mu)$. Note that Example 1 is a particular case of example 2 by setting $T = \mathbb{R}_+ \times \{1, \ldots, d\}$ and μ equals to the product of the Lebesgue measure times the uniform measure on $\{1, \ldots, d\}$.

3. Let ν be a zero mean Gaussian measure on a real separable Banach space $\mathbb{B}$ with full support. Consider the inclusion $j : \mathbb{B}^* \to L^2(\nu)$ which is linear, continuous and one-to-one. Set $\mathcal{H}_1 = \overline{j(\mathbb{B}^*)}$. Then $\mathcal{H}_1$ is a Gaussian subspace of $L^2(\nu)$.

Remarks:

1. Consider the case of a d-dimensional Brownian motion W defined on the canonical probability space $\Omega = C_0(\mathbb{R}_+, \mathbb{R}^d)$. The subspace H^1 of Ω formed by the functions of the form $\varphi(t) = \int_0^t \dot{\varphi}(s)ds$, $\dot{\varphi} \in H = L^2(\mathbb{R}_+; \mathbb{R}^d)$ is called the Cameron-Martin space. Equipped with the scalar product

$$\langle \varphi_1, \varphi_2 \rangle_{H^1} = \langle \dot{\varphi}_1, \dot{\varphi}_2 \rangle_H$$

H^1 is isometric to H, and the inclusion $i : H^1 \to \Omega$ is continuous.

2. In the context of Example 3, one can show (cf. Gross [34], Kuo [55]) that the dual map j^* of j is a compact operator from $\mathcal{H}_1$ into $\mathbb{B}$ with a dense image. Then the image $H^1 := j^*(\mathcal{H}_1)$ is a subspace of $\mathbb{B}$, and the triple $(\mathbb{B}, H^1, \nu)$ is called an abstract Wiener space. When $\mathbb{B} = C_0(\mathbb{R}_+, \mathbb{R}^d)$, and ν is the Wiener measure, then H^1 is the Cameron-Martin space. In fact, given a point measure $m = c\delta_{t_0} \in \mathbb{B}^*$, $c \in \mathbb{R}^d$, $t_0 \geq 0$, clearly $j(m) = c \cdot W(t_0)$ and $j^*j(m)(t) = c(t_0 \wedge t)$ because

$$\begin{aligned} j^*j(m)_i(t) &= \langle j^*j(m), e_i\delta_t \rangle_{\mathbb{B},\mathbb{B}^*} = \langle j(m), j(e_i\delta_t) \rangle_{L^2(\nu)} \\ &= E((c \cdot W(t_0))(e_i \cdot W(t)) = c_i(t_0 \wedge t), \end{aligned}$$

where e_i is the ith vector of the canonical basis of $\mathbb{R}^d$. Notice that the function $\varphi(t) = c(t_0 \wedge t)$ is absolutely continuous and its derivative has an L^2-norm equals to $\sqrt{t_0}\|c\|$ which is the L^2-norm of $j(m)$ in $L^2(\nu)$. As a consequence, $j^*(\mathcal{H}_1)$ consists of all absolutely continuous functions $\varphi : \mathbb{R}_+ \to \mathbb{R}^d$ which vanish at zero at have a square integrable density, that is, $j^*(\mathcal{H}_1)$ is the Cameron-Martin space.

Let us first introduce the derivative operator D. We will follow an approach analogous to the definition of the Sobolev spaces in finite dimensions. We denote by $C_p^\infty(\mathbb{R}^n)$ (resp. $C_b^\infty(\mathbb{R}^n)$) the set of all infinitely differentiable functions $f : \mathbb{R}^n \to \mathbb{R}$ such that f and all of its partial derivatives have polynomial growth (resp. are bounded).

Let $\mathcal{S}$ (resp. $\mathcal{S}_b$) denote the class of random variables of the form

$$F = f(W(h_1), \ldots, W(h_n)), \tag{1.1}$$

where f belongs to $C_p^\infty(\mathbb{R}^n)$ (resp. $C_b^\infty(\mathbb{R}^n)$), $h_1, \ldots, h_n$ are in H, and $n \geq 1$. These random variables are called *smooth*. We will denote by $\mathcal{P}$ the subset of $\mathcal{S}$ formed by random variables of the form (1.1) where f is a polynomial. If F has the form (1.1) we define its derivative DF as the H-valued random variable given by

$$DF = \sum_{i=1}^n \frac{\partial f}{\partial x_i}(W(h_1), \ldots, W(h_n))h_i. \tag{1.2}$$

Notice that for any element $h \in H$ the scalar product $\langle DF, h \rangle_H$ coincides with the directional derivative $\frac{d}{d\epsilon}F^{\epsilon h}|_{\epsilon=0}$, where $F^{\epsilon h}$ is the shifted random variable

$$F^{\epsilon h} = f(W(h_1) + \epsilon\langle h, h_1 \rangle_H, \ldots, W(h_n) + \epsilon\langle h, h_n \rangle_H).$$

In order to show that the operator D is closable we need the following integration-by-parts formula:

Lemma 1.1.1 *Suppose that F is a smooth random variable and $h \in H$. Then*

$$E(\langle DF, h \rangle_H) = E(FW(h)). \tag{1.3}$$

Proof: We can assume that the norm of h is one. There exist orthonormal elements of H, $e_1, \ldots, e_n$, such that $h = e_1$ and F is a smooth random variable of the form

$$F = f(W(e_1), \ldots, W(e_n)).$$

Let $\phi(x)$ denote the density of the standard normal distribution on $\mathbb{R}^n$. Then we have

$$E(\langle DF, h \rangle_H) = \int_{\mathbb{R}^n} \frac{\partial f}{\partial x_1}(x)\phi(x)dx = \int_{\mathbb{R}^n} f(x)\phi(x)x_1 dx = E(FW(e_1)),$$

which completes the proof of the lemma. □

Applying the previous result to a product FG, we obtain the following consequence.

Lemma 1.1.2 *Suppose that F and G are smooth functionals, and let $h \in H$. Then we have*

$$E(G\langle DF, h \rangle_H) = E(-F\langle DG, h \rangle_H + FGW(h)). \tag{1.4}$$

As a consequence of the above lemma, D is closable as an operator from $L^p(\Omega)$ to $L^p(\Omega; H)$ for any $p \geq 1$. In fact, let $\{F_N, N \geq 1\}$ be a sequence of smooth random variables such that F_N converges to zero in $L^p(\Omega)$ and the sequence of derivatives DF_N converges to η in $L^p(\Omega; H)$. Then, from Lemma 1.1.2 it follows that η is equal to zero. Indeed, for any $h \in H$ and for any smooth random variable $F \in S_b$ such that $FW(h)$ is bounded, we have

$$
\begin{aligned}
E(\langle \eta, h \rangle_H F) &= \lim_N E(\langle DF_N, h \rangle_H F) \\
&= \lim_N E(-F_N \langle DF, h \rangle_H + F_N FW(h)) = 0,
\end{aligned}
$$

because F_N converges to zero in L^p as N tends to infinity, and the random variables $\langle DF, h \rangle_H$ and $FW(h)$ are bounded. This implies $\eta = 0$.

We will denote the domain of D in $L^p(\Omega)$ by $\mathbb{D}^{1,p}$, meaning that $\mathbb{D}^{1,p}$ is the closure of the class of smooth random variables S with respect to the norm

$$\|F\|_{1,p} = [E(|F|^p) + E(\|DF\|_H^p)]^{\frac{1}{p}}.$$

For $p = 2$, the space $\mathbb{D}^{1,2}$ is a Hilbert space with the scalar product

$$\langle F, G \rangle_{1,2} = E(FG) + E(\langle DF, DG \rangle_H).$$

We can define the iteration of the operator D in such a way that for a smooth random variable F, the derivative $D^k F$ is a random variable with values on $H^{\otimes k}$.

Then for every $p \geq 1$ and any natural number k we introduce a seminorm on S defined by

$$\|F\|_{k,p}^p = E(|F|^p) + \sum_{j=1}^{k} E(\|D^j F\|_{H^{\otimes j}}^p). \qquad (1.5)$$

As in the case $k = 1$ one can show that the operator D^k is closable from $S \subset L^p(\Omega)$ into $L^p(\Omega; H^{\otimes k})$, $p \geq 1$. For any real $p \geq 1$ and any natural number $k \geq 0$, we will denote by $\mathbb{D}^{k,p}$ the completion of the family of smooth random variables S with respect to the norm $\| \cdot \|_{k,p}$. Note that $\mathbb{D}^{j,p} \subset \mathbb{D}^{k,q}$ if $j \geq k$ and $p \geq q$.

Let V be a real separable Hilbert space. We can also introduce the corresponding Sobolev spaces of V-valued random variables. More precisely, S_V will denote the family of V-valued smooth random variables of the form

$$F = \sum_{j=1}^{n} F_j v_j, \qquad v_j \in V, \qquad F_j \in S.$$

We define $D^k F = \sum_{j=1}^{n} D^k F_j \otimes v_j$, $k \geq 1$. Then D^k is a closable operator from $S_V \subset L^p(\Omega; V)$ into $L^p(\Omega; H^{\otimes k} \otimes V)$ for any $p \geq 1$. For any integer $k \geq 1$ and any real number $p \geq 1$ we can define the seminorm on S_V by

$$\|F\|_{k,p,V}^p = E(\|F\|_V^p) + \sum_{j=1}^{k} E(\|D^j F\|_{H^{\otimes j} \otimes V}^p).$$

We denote by $\mathbb{D}^{k,p}(V)$ the completion of S_V with respect to the norm $\| \cdot \|_{k,p,V}$. For $k = 0$ we put $\|F\|_{0,p,V} = [E(\|F\|_V^p)]^{\frac{1}{p}}$, and $\mathbb{D}^{0,p}(V) = L^p(\Omega; V)$.

Consider the intersection

$$\mathbb{D}^\infty(V) = \cap_{p \geq 1} \cap_{k \geq 1} \mathbb{D}^{k,p}(V).$$

Then $\mathbb{D}^\infty(V)$ is a complete, countably normed, metric space. We will write $\mathbb{D}^\infty(\mathbb{R}) = \mathbb{D}^\infty$. For every integer $k \geq 1$ and any real number $p \geq 1$ the operator D is continuous from $\mathbb{D}^{k,p}(V)$ into $\mathbb{D}^{k-1,p}(H \otimes V)$. Consequently, D is a continuous linear operator from $\mathbb{D}^\infty(V)$ into $\mathbb{D}^\infty(H \otimes V)$. Moreover, if F and G are random variables in $\mathbb{D}^\infty$, then the scalar product $\langle DF, DG \rangle_H$ is also in $\mathbb{D}^\infty$.

The following result is the chain rule, which can be easily proved by approximating the random variable F by smooth random variables, and the function φ by $(\varphi * \psi_\epsilon)$, where ψ_ϵ is an approximation of the identity.

Proposition 1.1.1 *Let* $\varphi : \mathbb{R}^m \to \mathbb{R}$ *be a continuously differentiable function with bounded partial derivatives, and fix* $p \geq 1$. *Suppose that* $F = (F^1, \ldots, F^m)$ *is a random vector whose components belong to the space* $\mathbb{D}^{1,p}$. *Then* $\varphi(F) \in \mathbb{D}^{1,p}$, *and*

$$D(\varphi(F)) = \sum_{i=1}^{m} \frac{\partial \varphi}{\partial x_i}(F) DF^i.$$

As a consequence of the chain rule, the space $\mathbb{D}^\infty$ is an algebra. We remark that in the infinite dimensional case, and assuming that Ω is a Banach space, the elements of $\mathbb{D}^\infty$ may not be differentiable or even continuous. For instance, the random variable $F = \int_0^1 W_t^1 dW_t^2$ is not continuous on $C([0,1]; \mathbb{R}^2)$ and it belongs to $\mathbb{D}^\infty$.

We will write $D_h F = \langle DF, h \rangle_H$.

1.2 Divergence operator

We will denote by δ the adjoint of the operator D as an unbounded operator from $L^2(\Omega)$ into $L^2(\Omega; H)$. That is, the domain of δ, denoted by $\mathrm{Dom}\,\delta$, is the set of H-valued square integrable random variables u such that

$$|E(\langle DF, u\rangle_H)| \leq c\|F\|_2, \tag{1.6}$$

for all $F \in \mathbb{D}^{1,2}$, where c is some constant depending on u. If u belongs to $\mathrm{Dom}\,\delta$, then $\delta(u)$ is the element of $L^2(\Omega)$ characterized by

$$E(F\delta(u)) = E(\langle DF, u\rangle_H) \tag{1.7}$$

for any $F \in \mathbb{D}^{1,2}$. The operator δ will be called the *divergence operator* and is closed as the adjoint of an unbounded and densely defined operator.

We denote by $\mathcal{S}_H$ the class of smooth elementary elements of the form

$$u = \sum_{j=1}^{n} F_j h_j, \tag{1.8}$$

where the F_j are smooth random variables, and the h_j are elements of H. From the integration-by-parts formula established in Lemma 1.1.2 we deduce that an element u of this type belongs to the domain of δ and moreover that

$$\delta(u) = \sum_{j=1}^{n} F_j W(h_j) - \sum_{j=1}^{n} \langle DF_j, h_j\rangle_H. \tag{1.9}$$

Notice that if $u \in \mathcal{S}_H$ then Du is an $H \otimes H$-valued random variable.

For smooth elements u, v and F we have the three following basic relations between the operators D and δ:

$$D_h(\delta(u)) = \langle u, h\rangle_H + \delta(D_h u), \tag{1.10}$$

$$E(\delta(u)\delta(v)) = E(\langle u, v\rangle_H) + E(\mathrm{Tr}\,(Du \circ Dv)) \tag{1.11}$$

$$\delta(Fu) = F\delta(u) - \langle DF, u\rangle_H. \tag{1.12}$$

Proof of (1.10): Suppose that u has the form (1.8). From (1.9) we deduce

$$D_h(\delta(u)) = \sum_{j=1}^{n} F_j\langle h, h_j\rangle_H + \sum_{j=1}^{n} \left(D_h F_j W(h_j) - \langle D(D_h F_j), h_j\rangle_H\right),$$

which implies (1.10).

Proof of (1.11): Let $\{e_i, i \geq 1\}$ be a complete orthonormal system on H. Using the duality relationship (1.7) and property (1.10) we obtain

$$
\begin{aligned}
E(\delta(u)\delta(v)) &= E\left(\langle v, D(\delta(u))\rangle_H\right) = E\left(\sum_{i=1}^{\infty} \langle v, e_i\rangle_H D_{e_i}(\delta(u))\right) \\
&= E\left(\sum_{i=1}^{\infty} \langle v, e_i\rangle_H \left(\langle e_i, u\rangle_H + \delta(D_{e_i}u)\right)\right) \\
&= E(\langle u, v\rangle_H) + E\left(\sum_{i,j=1}^{\infty} D_{e_i}\langle u, e_j\rangle_H D_{e_j}\langle v, e_i\rangle_H\right).
\end{aligned}
$$

Proof of (1.12):   For any smooth random variable G we have

$$
\begin{aligned}
E(G\delta(Fu)) &= E(\langle DG, Fu\rangle_H) = E(\langle u, (D(FG) - GDF)\rangle_H) \\
&= E(G(F\delta(u) - \langle u, DF\rangle_H),
\end{aligned}
$$

and the result follows.

Property (1.11) implies the following estimate

$$
E[\delta(u)^2] \le E(\|u\|_H^2) + E(\|Du\|_{H\otimes H}^2) = \|u\|_{1,2,H}^2.
$$

As a consequence, the space of weakly differentiable H-valued variables $\mathbb{D}^{1,2}(H)$ is included into $\mathrm{Dom}\,\delta$, and (1.11) holds for any $u \in \mathbb{D}^{1,2}(H)$. By an approximation argument property (1.10) holds for an element $u \in \mathbb{D}^{2,2}(H)$ (actually, (1.10) holds if $u \in \mathbb{D}^{1,2}(H)$ and $D_h u$ belongs to the domain of δ, as it can be proved using Lemma 1.4.1). By the definition of the operator δ, property (1.12) holds whenever $F \in \mathbb{D}^{1,2}$, u belongs to $\mathrm{Dom}\,\delta$, $Fu \in L^2(\Omega; H)$, and the right-hand side of (1.12) belongs to $L^2(\Omega)$.

1.3   Local properties

In this section we will show that the operators D and δ are local. The next result shows that the operator D is local in the space $\mathbb{D}^{1,1}$.

Proposition 1.3.1 *Let F be a random variable in the space $\mathbb{D}^{1,1}$ such that $F = 0$ a.s. on some set $A \in \mathcal{F}$. Then $DF = 0$ a.e. on A.*

Proof: We can assume that $F \in \mathbb{D}^{1,1} \cap L^\infty(\Omega)$, replacing F by $\arctan(F)$. We want to show that $1_{\{F=0\}} DF = 0$ a.s. Suppose that $\phi : \mathbb{R} \to \mathbb{R}$ is an infinitely differentiable function such that $\phi \ge 0$, $\phi(0) = 1$ and its support is included in the interval $[-1, 1]$. Define the function $\phi_\epsilon(x) = \phi(\frac{x}{\epsilon})$ for all $\epsilon > 0$. Set

$$
\psi_\epsilon(x) = \int_{-\infty}^x \phi_\epsilon(y)dy.
$$

By the chain rule $\psi_\epsilon(F)$ belongs to $\mathbb{D}^{1,1}$ and $D\psi_\epsilon(F) = \phi_\epsilon(F)DF$. Let u be a smooth H-valued random variable of the form

$$
u = \sum_{j=1}^n F_j h_j,
$$

where $F_j \in \mathcal{S}_b$ and $h_j \in H$. Observe that the duality relation (1.7) holds for F in $\mathbb{D}^{1,1} \cap L^\infty(\Omega)$ and for an element u of this type. Note that the class of such processes u is total in $L^1(\Omega; H)$ in the sense that if $v \in L^1(\Omega; H)$ satisfies $E(\langle v, u\rangle_H) = 0$ for all u in the class, then $v \equiv 0$. Then we have

$$
\begin{aligned}
|E(\phi_\epsilon(F)\langle DF, u\rangle_H)| &= |E(\langle D(\psi_\epsilon(F)), u\rangle_H)| \\
&= |E(\psi_\epsilon(F)\delta(u))| \le \epsilon\|\phi\|_\infty E(|\delta(u)|).
\end{aligned}
$$

Letting $\epsilon \downarrow 0$, we obtain

$$
E\left(1_{\{F=0\}}\langle DF, u\rangle_H\right) = 0,
$$

which implies the desired result.   □

Let us now state and prove the local property of the divergence.

Proposition 1.3.2 *Let* $u \in \mathbb{D}^{1,2}(H)$ *and* $A \in \mathcal{F}$, *such that* $u(\omega) = 0$, P *a.e. on* A. *Then* $\delta(u) = 0$ *a.e. on* A.

Proof: Let F be a smooth random variable of the form

$$F = f(W(h_1), \ldots, W(h_n)),$$

with $f \in C_0^\infty(\mathbb{R}^n)$ (f is an infinitely differentiable function with compact support). We want to show that

$$\delta(u) 1_{\{\|u\|_H = 0\}} = 0, \quad \text{a.s.}$$

Consider a function $\phi : \mathbb{R} \to \mathbb{R}$ such as that in the proof of Proposition 1.3.1. It is easy to show that the product $F\phi_\epsilon(\|u\|_H^2)$ belongs to $\mathbb{D}^{1,2}$. Then by the duality relation (1.7) we obtain

$$
\begin{aligned}
E\left(\delta(u)\phi_\epsilon(\|u\|_H^2)F\right) &= E\left(\langle u, D[F\phi_\epsilon(\|u\|_H^2)]\rangle_H\right) \\
&= E\left(\phi_\epsilon(\|u\|_H^2)\langle u, DF\rangle_H\right) + 2E\left(F\phi_\epsilon'(\|u\|_H^2)\langle u, D_u u\rangle_H\right).
\end{aligned}
$$

We claim that the above expression converges to zero as ϵ tends to zero. In fact, first observe that the random variables

$$V_\epsilon = \phi_\epsilon(\|u\|_H^2)\langle u, DF\rangle_H + 2F\phi_\epsilon'(\|u\|_H^2)\langle u, D_u u\rangle_H$$

converge a.s. to zero as $\epsilon \downarrow 0$, since $\|u\|_H = 0$ implies $V_\epsilon = 0$. Second, we can apply the Lebesgue dominated convergence theorem because we have

$$
\begin{aligned}
\left|\phi_\epsilon(\|u\|^2)\langle u, DF\rangle_H\right| &\leq \|\phi\|_\infty \|u\|_H \|DF\|_H, \\
\left|\phi_\epsilon'(\|u\|_H^2)\langle u, D_u u\rangle_H\right| &\leq \sup_x |x\phi_\epsilon'(x)| \|Du\|_{H \otimes H} \leq \|\phi'\|_\infty \|Du\|_{H \otimes H}.
\end{aligned}
$$

The proof is now complete. $\qquad\square$

Note that D is also local when acting on Hilbert-valued random variables. We can localize the domains of the operators D and δ as follows. We will denote by $\mathbb{D}_{loc}^{1,p}(V)$, $p \geq 1$, the set of V-valued random variables F such that there exists a sequence $\{(\Omega_n, F_n), n \geq 1\} \subset \mathcal{F} \times \mathbb{D}^{1,p}(V)$ with the following properties:

(i) $\Omega_n \uparrow \Omega$, a.s.

(ii) $F = F_n$ a.s. on Ω_n.

We then say that (Ω_n, F_n) localizes F in $\mathbb{D}^{1,p}(V)$, and DF is defined without ambiguity by $DF = DF_n$ on Ω_n, $n \geq 1$. The spaces $\mathbb{D}_{loc}^{k,p}(V)$ can be introduced analogously.

Then, if $u \in \mathbb{D}_{loc}^{1,2}(H)$, the divergence $\delta(u)$ is defined as a random variable determined by the conditions

$$\delta(u)|_{\Omega_n} = \delta(u_n)|_{\Omega_n}, \quad \text{for all} \quad n \geq 1,$$

where (Ω_n, u_n) is a localizing sequence for u.

1.4 Wiener chaos expansions

We will denote by $\{H_n, n \in \mathbb{N}\}$ the sequence of the Hermite polynomials defined from the series expansion

$$\exp(tx - \frac{t^2}{2}) = \sum_{n=0}^{\infty} t^n H_n(x). \tag{1.13}$$

Then $\{\sqrt{n!}\, H_n, n \in \mathbb{N}\}$ is a complete orthonormal system in $L^2(\mathbb{R}, \mu)$ where μ is the normal distribution $N(0,1)$. We will denote by Λ the set of all sequences $a = (a_1, a_2, \ldots)$, $a_i \in \mathbb{N}$, such that $|a| = a_1 + a_2 + \cdots < \infty$. Let $\{e_i, i \geq 1\}$ be a complete orthonormal system in H. For any $a \in \Lambda$ we set

$$\Phi_a = \sqrt{a!} \prod_{i=1}^{\infty} H_{a_i}(W(e_i)),$$

where $a! = \prod_{i=1}^{\infty} a_i!$. Then the family $\{\Phi_a, a \in \Lambda\}$ constitutes an orthonormal basis of $L^2(\Omega, \mathcal{F}, P)$ (we recall that we assume that the σ-field $\mathcal{F}$ is generated by W). For any $n \geq 0$ we will denote by $\mathcal{H}_n$ the closed subspace of $L^2(\Omega)$ spanned by $\{\Phi_a, a \in \Lambda, |a| = n\}$. Then $\mathcal{H}_n$ is called the Wiener chaos of order n and we have the orthogonal decomposition

$$L^2(\Omega) = \oplus_{n=0}^{\infty} \mathcal{H}_n.$$

We will denote by J_n the orthogonal projection on the nth Wiener chaos $\mathcal{H}_n$. The Wiener chaos expansion can be extended to the space $L^2(\Omega; V)$ of Hilbert valued random variables: $L^2(\Omega; V) = \oplus_{n=0}^{\infty} \mathcal{H}_n(V)$, where $\mathcal{H}_n(V) = \mathcal{H}_n \otimes V$.

The following result characterizes the domain of the operator D in L^2 in terms of the Wiener chaos expansion.

Proposition 1.4.1 *We have*

$$E(\|DF\|_H^2) = \sum_{n=1}^{\infty} n \|J_n(F)\|_2^2, \tag{1.14}$$

in the sense that this sum is finite if and only if $F \in \mathbb{D}^{1,2}$. Moreover, for all $n \geq 1$ we have $D(J_n(F)) = J_{n-1}(DF)$.

Proof: It suffices to compute the derivative of a random variable of the form Φ_a, using the relationship $H_n' = H_{n-1}$, which follows immediately from (1.13):

$$D(\Phi_a) = \sqrt{a!} \sum_{j=1}^{\infty} \prod_{i=1, i \neq j}^{\infty} H_{a_i}(W(e_i)) H_{a_j-1}(W(e_j)) e_j.$$

Then $D(\Phi_a) \in \mathcal{H}_{n-1}(H)$ if $|a| = n$, and

$$E(\|D(\Phi_a)\|_H^2) = \sum_{j=1}^{\infty} \frac{a!}{\prod_{i=1, i \neq j}^{\infty} a_i!(a_j - 1)!} = |a|.$$

Now the proposition follows easily. $\qquad\square$

By iteration we obtain

$$E(\|D^k F\|_{H^{\otimes k}}^2) = \sum_{n=k}^{\infty} n(n-1) \cdots (n-k+1) \|J_n(F)\|_2^2, \tag{1.15}$$

and $F \in \mathbb{D}^{k,2}$ if and only if $\sum_{n=1}^{\infty} n^k \|J_n(F)\|_2^2 < \infty$. Moreover, for all $n \geq k$ we have $D^k(J_n(F)) = J_{n-k}(D^k F)$.

The next lemma is a consequence of the expression of the operator D in terms of the Wiener chaos.

Lemma 1.4.1 *Let $G \in L^2(\Omega)$ and $\eta \in L^2(\Omega; H)$ be such that*

$$E(G\delta(v)) = E(\langle \eta, v \rangle_H),$$

for all $v \in \mathbb{D}^{1,2}(H)$. Then $G \in \mathbb{D}^{1,2}$ and $DG = \eta$.

Proof: We have

$$E(\langle \eta, v \rangle_H) = E(G\delta(v)) = \sum_{n=1}^{\infty} E(J_n(G)\delta(v)) = \sum_{n=1}^{\infty} E(\langle D(J_n(G)), v \rangle_H),$$

hence, $J_{n-1}\eta = D(J_n(G))$ for each $n \geq 1$ and this implies the result. $\qquad \square$

Remarks:

If F is a random variable in the space $\mathbb{D}^{1,1}$ such that $DF = 0$, then $F = E(F)$. This is obvious when $F \in \mathbb{D}^{1,2}$ if we use the Wiener chaos expansion of F and Proposition 1.4.1. In the general case this property can be proved by a duality argument. Indeed, let ψ_N be a function in $C_b^{\infty}(\mathbb{R})$ such that $\psi_N(x) = x$ if $|x| \leq N$, and $|\psi_N(x)| \leq N + 1$. Then one shows that

$$E(\psi_N(F - E(F))\delta(u)) = 0 \qquad (1.16)$$

for any bounded H-valued random variable u in the domain of δ. Approximating an arbitrary random variable $u \in \mathbb{D}^{1,2}(H)$ by the sequence of bounded random variables $\{u\phi_k(\|u\|_H^2), k \geq 1\}$, where $\phi_k(x) = 1$ if $|x| \leq k$, $\phi_k(x) = 0$ if $|x| \geq 2k$, and $|\phi_k'(x)| \leq 2/k$, we obtain that (1.16) holds for any $u \in \mathbb{D}^{1,2}(H)$. Finally, by Lemma 1.4.1 this implies that $\psi_N(F - E(F)) = E(\psi_N(F - E(F)))$, and letting N tend to infinity we deduce the result.

As an application of the chain rule and the above property we can show the following result (see Sekiguchi and Shiota [96]).

Lemma 1.4.2 *Let $A \in \mathcal{F}$. Then the indicator function of A belongs to $\mathbb{D}^{1,1}$ if and only if $P(A)$ is equal to zero or one.*

Proof: Applying the chain rule to a function $\varphi \in C_0^{\infty}(\mathbb{R})$ which is equal to x^2 on $[0, 1]$ yields

$$D1_A = D(1_A)^2 = 21_A D1_A.$$

Consequently, $D1_A = 0$ because from the above equality we get that this derivative is zero on A^c and equal to twice its value on A. So, by the previous remark we obtain $1_A = P(A)$. $\qquad \square$

1.5　The white noise case

Consider the particular case of a white noise $\{W(A), A \in \mathcal{B}, \mu(A) < \infty\}$ defined on a measure space $(T, \mathcal{B}, \mu)$. Suppose in addition that the measure μ is σ-finite and without atoms. In this case (cf. Itô [49]) the multiple stochastic integral of order n provides an isometry between the space $L_s^2(T^n, \mathcal{B}^n, \mu^n)$ of square integrable and symmetric functions of n variables and the nth Wiener chaos $\mathcal{H}_n$. Let us briefly describe how the multiple stochastic integral is constructed. Consider the set $\mathcal{E}_n$ of elementary functions of the form

$$f(t_1, \ldots, t_n) = \sum_{i_1, \ldots, i_n = 1}^{k} a_{i_1 \cdots i_n} 1_{A_{i_1} \times \cdots \times A_{i_n}}(t_1, \ldots, t_n), \qquad (1.17)$$

where $A_1, A_2, \ldots, A_k$ are pairwise-disjoint sets of finite measure, and the coefficients $a_{i_1 \cdots i_n}$ are zero if any two of the indices $i_1, \ldots, i_n$ are equal. Then we set

$$I_n(f) = \sum_{i_1, \ldots, i_n = 1}^{k} a_{i_1 \cdots i_n} W(A_{i_1}) \cdots W(A_{i_n}).$$

The set $\mathcal{E}_n$ is dense in $L^2(T^n)$, and I_n is a linear map from $\mathcal{E}_n$ into $L^2(\Omega)$ which verifies $I_n(f_n) = I_n(\tilde{f}_n)$ and

$$E(I_n(f)I_q(g)) = \begin{cases} 0 & \text{if } n \neq q, \\ n! \langle \tilde{f}, \tilde{g} \rangle_{L^2(T^n)} & \text{if } n = q, \end{cases}$$

where $\tilde{f}$ denotes the symmetrization of f. As a consequence, I_n can be extended as a linear and continuous operator from $L^2(T^n)$ into $L^2(\Omega)$. The image of $L^2(T^n)$ by I_n is the nth Wiener chaos $\mathcal{H}_n$. This is a consequence of the fact that multiple stochastic integrals of different order are orthogonal, and that $I_n(f)$ is a polynomial of degree n in $W(A_1), \ldots, W(A_k)$ if f has the form (1.17), and those polynomials are included in the sum of the first n chaos. As a consequence, any square integrable random variable F admits an orthogonal expansion in the form

$$F = E(F) + \sum_{n=1}^{\infty} I_n(f_n), \qquad (1.18)$$

where the functions $f_n \in L^2(T^n)$ are symmetric and uniquely determined by F.

The operators D and δ can be represented in terms of the Wiener chaos expansion.

Proposition 1.5.1 *Let $F \in \mathbb{D}^{1,2}$ be a square integrable random variable with an expansion of the form (1.18). Then we have*

$$D_t F = \sum_{n=1}^{\infty} n I_{n-1}(f_n(\cdot, t)). \qquad (1.19)$$

Proof:　Suppose first that $F = I_n(f_n)$, where f_n is a symmetric and elementary function of the form (1.17). Then

$$D_t F = \sum_{j=1}^{n} \sum_{i_1, \ldots, i_n = 1}^{k} a_{i_1 \cdots i_n} W(A_{i_1}) \cdots 1_{A_{i_j}}(t) \cdots W(A_{i_n}) = n I_{n-1}(f_n(\cdot, t)).$$

Then the results follows easily. □

Note that the $L^2(\Omega; H)$-norm of the right-hand side of (1.19) coincides with expression (1.14). We remark that the derivative $D_t F$ is a random field parametrized by T.

Suppose that F is a random variable in the space $\mathbb{D}^{N,2}$, with a Wiener chaos expansion of the form $F = \sum_{n=0}^\infty I_n(f_n)$. Then, applying Proposition 1.5.1 N times we obtain that $D^N F$ is a random field parametrized by T^N given by

$$D^N_{t_1,\dots,t_N} F = \sum_{n=N}^\infty n(n-1)\cdots(n-N+1)I_{n-N}(f_n(\cdot, t_1, \dots, t_N)).$$

Note that the L^2-norm of this expression is given by (1.15). As a consequence, if F belongs to $\mathbb{D}^{\infty,2} = \cap_N \mathbb{D}^{N,2}$ then $f_n = \frac{1}{n!}E(D^n F)$ for every $n \geq 0$ (cf. Stroock [100]).

We will now describe the divergence operator in terms of the Wiener chaos expansion. Any element $u \in L^2(\Omega; H) \cong L^2(T \times \Omega)$ (which can be regarded as a square integrable process parametrized by T) has an orthogonal expansion of the form

$$u(t) = \sum_{n=0}^\infty I_n(f_n(\cdot, t)), \tag{1.20}$$

where for each $n \geq 1$, $f_n \in L^2(T^{n+1})$ is a symmetric function in the first n variables.

Proposition 1.5.2 *Let $u \in L^2(T \times \Omega)$ with the expansion (1.20). Then u belongs to* Dom δ *if and only if the series*

$$\delta(u) = \sum_{n=0}^\infty I_{n+1}(\tilde{f}_n) \tag{1.21}$$

converges in $L^2(\Omega)$.

Equation (1.21) can also be written without symmetrization, because for each n, $I_{n+1}(f_n) = I_{n+1}(\tilde{f}_n)$. However, the symmetrization is needed in order to compute the L^2 norm of the stochastic integrals.

Proof: Suppose that $G = I_n(g)$ is a multiple stochastic integral of order $n \geq 1$ where g is symmetric. Then we have the following equalities:

$$
\begin{aligned}
E\left(\langle u, DG\rangle_H\right) &= \int_T E\left(I_{n-1}(f_{n-1}(\cdot, t))nI_{n-1}(g(\cdot, t))\right)\mu(dt) \\
&= n(n-1)! \int_T \langle f_{n-1}(\cdot, t), g(\cdot, t)\rangle_{L^2(T^{n-1})}\mu(dt) \\
&= n!\langle f_{n-1}, g\rangle_{L^2(T^n)} = n!\langle \tilde{f}_{n-1}, g\rangle_{L^2(T^n)} \\
&= E\left(I_n(\tilde{f}_{n-1})I_n(g)\right) = E\left(I_n(\tilde{f}_{n-1})G\right).
\end{aligned}
$$

Suppose first that $u \in$ Dom δ. Then from the above computations and from formula (1.7) we deduce that

$$E(\delta(u)G) = E(I_n(\tilde{f}_{n-1})G)$$

for every multiple stochastic integral $G = I_n(g)$. This implies that $I_n(\tilde{f}_{n-1})$ coincides with the projection of $\delta(u)$ on the nth Wiener chaos. Consequently, the series in

(1.21) converges in $L^2(\Omega)$ and its sum is equal to $\delta(u)$. The converse can be proved by a similar argument. $\qquad\square$

For any set $A \subset T$ we will denote by $\mathcal{F}_A$ the σ-field generated by the random variables $\{W(B), B \subset A, B \in \mathcal{B}_0\}$, where $\mathcal{B}_0 = \{B \in \mathcal{B}, \mu(B) < \infty\}$.

Lemma 1.5.1 *Let $A \in \mathcal{B}$ and let F be a random variable in $L^2(\Omega, \mathcal{F}_{A^c}, P) \cap \mathbb{D}^{1,2}$. Then $D_t F = 0$ for all $(t, \omega) \in A \times \Omega$ a.e. with respect to the measure $\mu \otimes P$.*

Proof: Suppose that $F = f(W(B_1), \ldots, W(B_n))$, where $f \in C_p^\infty(\mathbb{R}^n)$, and $B_i \subset A^c$, $B_i \in \mathcal{B}_0$ for all $i = 1, \ldots, n$. Then we have

$$D_t F = \frac{\partial f}{\partial x_i}(W(B_1), \ldots, W(B_n)) \mathbf{1}_{B_i}(t),$$

which implies that $D_t F = 0$ for all $(t, \omega) \in A \times \Omega$. Finally in the general case it suffices to consider a sequence F_n as above converging to F in L^2. $\qquad\square$

The next lemma allow us to interpret the operator δ as a stochastic integral with respect to the white noise W.

Lemma 1.5.2 *Let $A \in \mathcal{B}_0$ and let F be a random variable in $L^2(\Omega, \mathcal{F}_{A^c}, P)$. Then, the process $u = F \mathbf{1}_A$ belongs to Dom δ and*

$$\delta(F \mathbf{1}_A) = F W(A).$$

Proof: Suppose first that $F \in \mathbb{D}^{1,2}$. Then, using (1.9) and Lemma 1.5.2 we obtain

$$\delta(F \mathbf{1}_A) = F W(A) - \int_T D_t F \mathbf{1}_A(t) \mu(dt) = F W(A).$$

The general case follows by a limit argument. $\qquad\square$

The following lemma provides additional examples of processes in the domain of δ which are not smooth.

Lemma 1.5.3 *Let $A \in \mathcal{B}_0$. Let $\{u(x), x \in \mathbb{R}^m\}$ be an $\mathcal{F}_{A^c}$-measurable random field with continuously differentiable paths such that*

$$E\left(\sup_{|x| \leq K} |u'(x)|^4\right) < \infty \quad and \quad E\left(|u(0)|^2\right) < \infty,$$

for every $K > 0$. Let F be a bounded random variable in the space $\mathbb{D}^{1,4}$. Then $u(F) \mathbf{1}_A$ belongs to the domain of δ and

$$\delta(u(F) \mathbf{1}_A) = u(F) W(A) - \sum_{i=1}^m (\partial_i u)(F) \int_A D_t F^i \mu(dt).$$

Idea of the proof: Approximate $u(F)$ by the convolution of $u(x)$ with an approximation of the identity. $\qquad\square$

Consider the following space of elementary processes:

$$\mathcal{R}_0 = \left\{ \sum_{i=1}^{n} F_i 1_{A_i}, A_i \in \mathcal{B}_0, F_i \in L^2(\Omega, \mathcal{F}_{A_i^c}, P) \right\}.$$

By Lemma 1.5.2 we have that $\mathcal{R}_0 \subset \text{Dom}\,\delta$ and

$$\delta(\sum_{i=1}^{n} F_i 1_{A_i}) = \sum_{i=1}^{n} F_i W(A_i).$$

As a consequence we deduce the following result:

Proposition 1.5.3 *Let $S_0 \subset \mathcal{R}_0$ be a subspace such that for all $u \in S_0$ the isometry property $E[\delta(u)^2] = E(\|u\|_H^2)$ holds. Then $\overline{S_0}^{\|\cdot\|_2} \subset \text{Dom}\,\delta$, and the isometry property still holds in the closure of S_0.*

Notice that the isometry property holds for a process u of the form $u = F 1_A \in \mathcal{R}_0$, but, in general, it is not true for a general element u in $\mathcal{R}_0$. Particular examples of subspaces of $\mathcal{R}_0$ where the isometry holds are the following:

Example 1:

Suppose that $W = \{W(t), t \in [0, 1]\}$ is a d-dimensional Wiener process. Set

$$S_0 = \{u : u(t) = \sum_{i=0}^{n-1} F_i 1_{(t_i, t_{i+1}]}(t), F_i \in L^2(\Omega, \mathcal{F}_{[0, t_i]}, P; \mathbb{R}^d),$$
$$0 = t_0 < \cdots < t_n = 1\}.$$

Then $\overline{S_0}^{\|\cdot\|_2}$ is the space L_a^2 of d-dimensional square integrable and adapted processes. Therefore, $L_a^2 \subset \text{Dom}\,\delta$ and $\delta(u)$ coincides with the Itô stochastic integral on this space.

If we take

$$S_1 = \{u : u(t) = \sum_{i=0}^{n-1} F_i 1_{(t_i, t_{i+1}]}(t), F_i \in L^2(\Omega, \mathcal{F}_{[t_{i+1}, 1]}, P; \mathbb{R}^d),$$
$$0 = t_0 < \cdots < t_n = 1\},$$

then $\overline{S_1}^{\|\cdot\|_2}$ is the space L_b^2 of d-dimensional backward adapted processes and $\delta(u)$ coincides with the backward Itô integral on this space.

Example 2:

We will denote by W a white noise on $[0, 1]^2$ with intensity equal to the Lebesgue measure. Then $\{W(s, t) = W([0, s] \times [0, t]), s, t, \in [0, 1]\}$ is a two-parameter Wiener process, that is, a zero mean Gaussian process with covariance given by

$$E[W(s_1, t_1)W(s_2, t_2)] = (s_1 \wedge s_2)(t_1 \wedge t_2).$$

Set

$$S_2 = \{u : u(s, t) = \sum_{i,j=0}^{n-1} F_{i,j} 1_{(s_i, s_{i+1}] \times (t_j, t_{j+1}]}(s, t),$$
$$F_{i,j} \in L^2(\Omega, \mathcal{F}_{[0, s_i] \times [0, 1]}, P), 0 = t_0 < \cdots < t_n = 1, 0 = s_0 < \cdots < s_n = 1\}.$$

Then $\overline{S_2}^{\|\cdot\|_2}$ is the space $L_{1,a}^2$ of square integrable and 1-adapted processes. Therefore, $L_{1,a}^2 \subset \text{Dom}\,\delta$ and $\delta(u)$ coincides with the Itô stochastic integral on this space.

1.6 Stochastic integral representation of random variables

Suppose that $W = \{W(t), t \in [0,1]\}$ is a one-dimensional Brownian motion. We know that any square integrable random variable F, measurable with respect to W, can be written as

$$F = E(F) + \int_0^1 \phi(t) dW_t,$$

where the process ϕ belongs to L_a^2. When the variable F belongs to the space $\mathbb{D}^{1,2}$, it turns out that the process ϕ can be identified as the optional projection of the derivative of F. This is called the Clark-Ocone representation formula (see [24], [83]):

Proposition 1.6.1 *Let $F \in \mathbb{D}^{1,2}$, and suppose that W is a one-dimensional Brownian motion. Then*

$$F = E(F) + \int_0^1 E(D_t F|\mathcal{F}_t) dW_t. \tag{1.22}$$

In order to proof this stochastic integral representation, we will make use of the Wiener chaos expansions, and we need the following technical result:

Lemma 1.6.1 *Let W be a white noise on $(T, \mathcal{B}, \mu)$. Suppose that F is a square integrable random variable with the representation $F = \sum_{n=0}^{\infty} I_n(f_n)$. Let $A \in \mathcal{B}$. Then*

$$E(F|\mathcal{F}_A) = \sum_{n=0}^{\infty} I_n(f_n 1_A^{\otimes n}). \tag{1.23}$$

Proof: It suffices to assume that $F = I_n(f_n)$, where f_n is a square integrable and symmetric kernel. Also, by linearity and density we can assume that the kernel f_n is of the form $1_{B_1 \times \cdots \times B_n}$, where $B_1, \ldots, B_n$ are mutually disjoint sets of finite measure. In this case we have

$$
\begin{aligned}
E(F|\mathcal{F}_A) &= E(W(B_1) \cdots W(B_n)|\mathcal{F}_A) \\
&= E\Big(\prod_{i=1}^{n} (W(B_i \cap A) + W(B_i \cap A^c)) \,|\, \mathcal{F}_A \Big) \\
&= I_n(1_{(B_1 \cap A) \times \cdots \times (B_n \cap A)}).
\end{aligned}
$$

$\square$

Proof of Proposition 1.6.1: Suppose that $F = \sum_{n=0}^{\infty} I_n(f_n)$. Using (1.19) and Lemma 1.6.1 we deduce

$$
\begin{aligned}
E(D_t F|\mathcal{F}_t) &= \sum_{n=1}^{\infty} n E(I_{n-1}(f_n(\cdot, t))|\mathcal{F}_t) \\
&= \sum_{n=1}^{\infty} n I_{n-1}\left(f_n(t_1, \ldots, t_{n-1}, t) 1_{\{t_1 \vee \cdots \vee t_{n-1} \leq t\}} \right).
\end{aligned}
$$

Set $\phi_t = E(D_t F|\mathcal{F}_t)$. We can compute $\delta(\phi)$ using the above expression for ϕ and (1.21), and we obtain

$$\delta(\phi) = \sum_{n=1}^{\infty} I_n(f_n) = F - E(F),$$

which shows the desired result because $\delta(\phi)$ is equal to the Itô stochastic integral of ϕ. $\qquad\qquad\qquad\qquad\qquad\qquad\qquad\qquad\qquad\qquad\qquad\qquad\qquad\qquad\quad$ $\square$

Bibliographical notes: The integral representation formula (1.22) can been extended to random variables in $\mathbb{D}^{1,1}$ (cf. [50]) and to elements of $\mathbb{D}^{-\infty}$ (cf. [104]). The iterated divergence operator δ^k is an extension of the multiple stochastic integrals introduced by Hajek and Wong [40] (see [80]).

Chapter 2

Ornstein-Uhlenbeck semigroup and equivalence of norms

In this chapter we introduce the operator L, which is the infinitesimal generator of the Ornstein-Uhlenbeck semigroup, and we show the equivalence between the $\|\cdot\|_{k,p}$ norms defined using the derivative operator and the norms $\||F\||_{k,p} = \|(I-L)(F)^{k/2}\|_p$.

Let $W = \{W(h), h \in H\}$ be a centered Gaussian family associated with a real and separable Hilbert space H, defined on a probability space $(\Omega, \mathcal{F}, P)$. Suppose that the σ-field $\mathcal{F}$ is generated by W. Consider the one-parameter semigroup $\{T_t, t \geq 0\}$ of contraction operators on $L^2(\Omega)$ defined by

$$T_t F = \sum_{n=0}^{\infty} e^{-nt} J_n F, \tag{2.1}$$

where J_n denotes the projection on the nth Wiener chaos. This semigroup is called the Ornstein-Uhlenbeck semigroup.

2.1 Mehler's formula

The following result is known as Mehler's formula.

Proposition 2.1.1 *Let $W' = \{W'(h), h \in H\}$ be an independent copy of W. Then for any $t \geq 0$ and $F \in L^2(\Omega)$ we have*

$$T_t F = E'(F(e^{-t}W + \sqrt{1 - e^{-2t}}W')), \tag{2.2}$$

where E' denotes the mathematical expectation with respect to W'.

Note that the right-hand side of (2.2) is well defined because

$$\{e^{-t}W(h) + \sqrt{1 - e^{-2t}}W'(h), h \in H\}$$

is a centered Gaussian family with the same covariance as W.

Proof: Both T_t and the right-hand side of (2.2) give rise to linear contraction operators on $L^2(\Omega)$. Thus, it suffices to show (2.2) when $F = \exp(\lambda W(h) - \frac{1}{2}\lambda^2)$, where $h \in H$ is an element of norm one and $\lambda \in \mathbb{R}$, We have

$$E' \left(\exp\left(e^{-t}\lambda W(h) + \sqrt{1 - e^{-2t}}\lambda W'(h) - \frac{1}{2}\lambda^2 \right) \right)$$

$$= \exp\left(e^{-t}\lambda W(h) - \frac{1}{2}e^{-2t}\lambda^2 \right) = \sum_{n=0}^{\infty} e^{-nt}\lambda^n H_n(W(h)) = T_t F,$$

because $J_n F = \lambda^n H_n(W(h))$. □

Mehler's formula implies that the operator T_t is nonnegative and that it is a contraction on $L^p(\Omega)$ for any $p \geq 1$. Moreover the semigroup $\{T_t, t \geq 0\}$ is continuous in $L^p(\Omega)$ for any $p \geq 1$.

2.2 Hypercontractivity

The operators T_t verify a property (cf. Nelson [73], Neveu [74]) called hypercontractivity, which is stronger than the contractivity in L^p, and which says that

$$\|T_t F\|_{q(t)} \leq \|F\|_p, \tag{2.3}$$

if $q(t) = e^{2t}(p - 1) + 1 > p$, $t > 0$ and $F \in L^p(\Omega, \mathcal{F}, P)$.

As a consequence of the hypercontractivity property it can be shown that for any $1 < p < q < \infty$ the norms $\|\cdot\|_p$ and $\|\cdot\|_q$ are equivalent on any Wiener chaos $\mathcal{H}_n$. In fact, let $t > 0$ such that $q = 1 + e^{2t}(p - 1)$. Then for every $F \in \mathcal{H}_n$ we have

$$e^{-nt}\|F\|_q = \|T_t F\|_q \leq \|F\|_p. \tag{2.4}$$

In addition, for each $n \geq 1$ the operator J_n is bounded in L^p for any $1 < p < \infty$, and

$$\|J_n F\|_p \leq \begin{cases} (p-1)^{\frac{n}{2}}\|F\|_p & \text{if } p > 2 \\ (p-1)^{-\frac{n}{2}}\|F\|_p & \text{if } p < 2. \end{cases}$$

In fact, suppose first that $p > 2$, and let $t > 0$ be such that $p - 1 = e^{2t}$. Using the hypercontractivity property with the exponents p and 2, we obtain

$$\|J_n F\|_p = e^{nt}\|T_t J_n F\|_p \leq e^{nt}\|J_n F\|_2 \leq e^{nt}\|F\|_2 \leq e^{nt}\|F\|_p. \tag{2.5}$$

If $p < 2$, we use a duality argument.

Consider a sequence of real numbers $\{\phi(n), n \geq 0\}$. This sequence determines a linear operator $T_\phi : \mathcal{P} \to \mathcal{P}$ defined by

$$T_\phi F = \sum_{n=0}^{\infty} \phi(n) J_n F, \qquad F \in \mathcal{P}.$$

We remark that the operators T_t are of this type, the corresponding sequence being e^{-nt}. It would be useful to know whether such a multiplication operator is bounded in all L^p, $p > 1$. The following examples provide sufficient conditions for this property to hold.

Examples:

1. Suppose that $\phi(n) = \sum_{k=0}^{\infty} a_k n^{-k}$ for $n \geq N$ and for some $a_k \in \mathbb{R}$ such that $\sum_{k=0}^{\infty} |a_k| N^{-k} < \infty$. Then the operator T_ϕ is bounded in L^p for any $1 < p < \infty$.

Proof: By duality, and taking into account that T_ϕ is selfadjoint, we can assume that $p \geq 2$. Moreover it suffices to show that T_ϕ is bounded in L^p on $\bigoplus_{n=N}^{\infty} \mathcal{H}_n$. Fix $F \in \bigoplus_{n=N}^{M} \mathcal{H}_n$ with $M \geq N$. We have

$$\|T_\phi F\|_p = \left\| \sum_{n=N}^{\infty} \left(\sum_{k=0}^{\infty} a_k n^{-k} \right) J_n F \right\|_p \leq \sum_{k=0}^{\infty} |a_k| \left\| \sum_{n=N}^{\infty} n^{-k} J_n F \right\|_p.$$

Now, using the equality

$$n^{-k} = \left(\int_0^{\infty} e^{-nt} dt \right)^k = \int_{[0,\infty)^k} e^{-n(t_1 + \cdots + t_k)} dt_1 \cdots dt_k,$$

we obtain

$$\|T_\phi F\|_p \leq \sum_{k=0}^{\infty} |a_k| \int_{[0,\infty)^k} \|T_{t_1 + \cdots + t_k} F\|_p dt_1 \cdots dt_k. \tag{2.6}$$

Then it suffices to have an estimate of the L^p-norm of the Ornstein-Uhlenbeck semigroup of the form

$$\|T_t F\|_p \leq K_{N,p} e^{-Nt} \|F\|_p. \tag{2.7}$$

Indeed, from (2.6) and (2.7) we deduce

$$\|T_\phi F\|_p \leq K_{N,p} \sum_{k=0}^{\infty} |a_k| N^{-k} \|F\|_p,$$

which allows us to complete the proof. The estimate (2.7) follows from the hypercontractivity property (2.3). In fact, suppose $p > 2$ (the case $p = 2$ is immediate) and choose t_0 such that $p = e^{2t_0} + 1$. For all $t \geq t_0$ we have

$$
\begin{aligned}
\|T_t F\|_p^2 &= \|T_{t_0} T_{t-t_0} F\|_p^2 \leq \|T_{t-t_0} F\|_2^2 = \sum_{n=N}^{\infty} e^{-2n(t-t_0)} \|J_n F\|_2^2 \\
&\leq e^{-2N(t-t_0)} \|F\|_2^2 \leq e^{-2N(t-t_0)} \|F\|_p^2,
\end{aligned}
$$

which implies (2.7) with $K_{N,p} = e^{Nt_0}$. If $t < t_0$ Eq. (2.7) is obvious again with $K_{N,p} = e^{Nt_0}$. $\qquad \square$

2. For any $\alpha > 0$ let us consider the sequence $\phi(n) = (1+n)^{-\alpha}$, $n \geq 0$. Then the operator T_ϕ is a contraction in L^p for any $p \geq 1$. This follows immediately from the equation

$$(1+n)^{-\alpha} = \Gamma(\alpha)^{-1} \int_0^{\infty} e^{-(n+1)t} t^{\alpha-1} dt,$$

which implies

$$T_\phi = \Gamma(\alpha)^{-1} \int_0^{\infty} e^{-t} t^{\alpha-1} T_t dt. \tag{2.8}$$

With the notations of the next section we have $T_\phi = (I - L)^{-\alpha}$.

Given a sequence of real numbers $\phi = \{\phi(n), n \geq 0\}$, define $T_{\phi+} = \sum_{n=0}^{\infty} \phi(n+1) J_n$. The following commutativity relationship holds for a multiplication operator T_ϕ:

$$D(T_\phi F) = T_{\phi+}(DF), \tag{2.9}$$

for any $F \in \mathcal{P}$. In fact, if F belongs to the nth Wiener chaos, we have

$$D(T_\phi F) = D(\phi(n)F) = \phi(n)DF = T_{\phi+}(DF).$$

In particular we have $D(T_t F) = e^{-t} T_t(DF)$, and by iteration we obtain

$$D^k(T_t F) = e^{-kt} T_t(D^k F) \tag{2.10}$$

for any $F \in \mathcal{P}$, and $k \geq 1$.

The following two properties hold:

(A) Let $F \in \mathbb{D}^{k,p}$. Then $T_t F$ also belongs to $\mathbb{D}^{k,p}$, and

$$\lim_{t \downarrow 0} \|T_t F - F\|_{k,p} = 0. \tag{2.11}$$

Indeed, from (2.10) it follows that T_t is a contraction operator with respect to any seminorm $\| \cdot \|_{k,p}$. This implies that $T_t F \in \mathbb{D}^{k,p}$. The continuity at the origin with respect to the norm of $\mathbb{D}^{k,p}$ is immediate.

(B) Let $F \in L^p(\Omega)$. Then $T_t F \in \cap_{k \geq 1} \mathbb{D}^{k,p}$, and

$$\|D^k(T_t F)\|_p \leq c_{k,p} t^{-k} \|F\|_p, \tag{2.12}$$

for any $k \geq 1$, $t > 0$, and $p > 1$. In particular $T_t F \in \mathbb{D}^\infty$ for any random variable F which has moments of all orders.

Proof: It suffices to show (2.12) for any polynomial random variable. Suppose first that $k = 1$. We have from Mehler's formula

$$
\begin{aligned}
D(T_t F) &= D(E'(F(e^{-t}W + \sqrt{1 - e^{-2t}}W'))) \\
&= \frac{e^{-t}}{\sqrt{1 - e^{-2t}}} E'(D'(F(e^{-t}W + \sqrt{1 - e^{-2t}}W'))).
\end{aligned}
$$

We recall that for any random variable $G \in \mathbb{D}^{1,2}$ we have

$$\|E(DG)\|_H = \|J_1 G\|_2 = c_p \|J_1 G\|_p \leq c_p' \|G\|_p.$$

Hence,

$$E\left(\|D(T_t F)\|_H^p\right) \leq c_p \left(\frac{e^{-t}}{\sqrt{1 - e^{-2t}}}\right)^p E(|F|^p),$$

and, by iteration, we deduce (2.12). $\qquad \square$

From properties (A) and (B) it follows that the family $\mathcal{P}$ of polynomial random variables is dense in $\mathbb{D}^{k,p}$ for any $k \geq 1$ and $p > 1$. Indeed, we approximate a given random variable F in $\mathbb{D}^{k,p}$ by a polynomial $G \in \mathcal{P}$ in L^p, and, then, we use the inequalities

$$\|F - T_t G\|_{k,p} \leq \|F - T_t F\|_{k,p} + \sum_{j=0}^{k} c_{j,p} t^{-j} \|F - G\|_p.$$

2.3   Generator of the Ornstein-Uhlenbeck semigroup

The infinitesimal generator of the semigroup T_t in $L^2(\Omega)$ is given by

$$LF = \sum_{n=1}^{\infty} -nJ_nF, \qquad (2.13)$$

and its domain is

$$\text{Dom } L = \{F \in L^2(\Omega) : \sum_{n=1}^{\infty} n^2\|J_nF\|_2^2 < \infty\}.$$

From Proposition 1.4.1 it follows that $\text{Dom } L \subset \mathbb{D}^{1,2}$. Actually, $\text{Dom } L = \mathbb{D}^{2,2}$.

The next proposition explains the relationship between the operators D, δ, and L.

Proposition 2.3.1 *Let $F \in L^2(\Omega)$. Then $F \in \text{Dom } L$ if and only if $F \in \mathbb{D}^{1,2}$ and $DF \in \text{Dom } \delta$, and in this case, we have $\delta(DF) = -LF$.*

Proof: For any polynomial random variable G we have, using the duality relationship (1.7) and Proposition 1.4.1,

$$
\begin{aligned}
E(G\delta(DF)) &= E(\langle DG, DF\rangle_H) = \sum_{n=0}^{\infty} nE(J_nGJ_nF) \\
&= E\left(G\sum_{n=0}^{\infty} nJ_nF\right) = -E(G\,LF),
\end{aligned}
$$

and the result follows easily. □

We are going to show that the operator L behaves as a second-order differential operator.

Proposition 2.3.2 *Suppose that $F = (F^1, \ldots, F^m)$ is a random vector whose components belong to $\mathbb{D}^{2,4}$. Let φ be a function in $C^2(\mathbb{R}^m)$ with bounded first and second partial derivatives. Then $\varphi(F) \in \text{Dom } L$, and*

$$L(\varphi(F)) = \sum_{i,j=1}^{m} \frac{\partial^2\varphi}{\partial x_i \partial x_j}(F)\langle DF^i, DF^j\rangle_H + \sum_{i=1}^{m} \frac{\partial\varphi}{\partial x_i}(F)LF^i. \qquad (2.14)$$

Proof: Suppose first that F is a smooth random variable of the form

$$F = f(W(h_1), \ldots, W(h_n)),$$

with $f \in C_p^\infty(\mathbb{R}^n)$. In this case using Proposition 2.3.1 we obtain

$$
\begin{aligned}
LF &= -\delta(DF) = -\delta\left(\sum_{i=1}^{n} \frac{\partial f}{\partial x_i}(W(h_1), \ldots, W(h_n))h_i\right) = \\
&= \sum_{i,j=1}^{n} \frac{\partial^2 f}{\partial x_i \partial x_j}(W(h_1), \ldots, W(h_n))\langle h_i, h_j\rangle_H \\
&\quad - \sum_{i=1}^{n} \frac{\partial f}{\partial x_i}(W(h_1), \ldots, W(h_n))W(h_i).
\end{aligned}
$$

In the general case we approximate F by smooth random variables in the norm $\|\cdot\|_{2,4}$ and use the continuity of the operator L in the norm $\|\cdot\|_{2,2}$. $\qquad\square$

For any natural $k \geq 0$ we can define the norm

$$\|F\|_{k,2}^2 = E(|(I - L)^{k/2}(F)|^2) = \sum_{n=0}^{\infty}(n + 1)^k E(|J_n F|^2).$$

Notice that this seminorm is equivalent to the norm $\|F\|_{k,2}$, taking into account (1.15). In the next section we generalize this equivalence of norms to the case of $p \neq 2$. Set $C = -\sqrt{-L}$, then $\mathbb{D}^{1,2}$ coincides with the domain of C.

2.4 Meyer's inequalities

We are going to establish Meyer inequalities following Pisier's approach. In order to do this we need some preliminaries. Consider the function $\varphi : [-\frac{\pi}{2}, 0) \cup (0, \frac{\pi}{2}] \to \mathbb{R}_+$ defined by

$$\varphi(\theta) = \frac{1}{\sqrt{2\pi|\log\cos^2\theta|}}\text{sign}\,\theta. \tag{2.15}$$

Notice that when θ is close to zero this function tends to infinity as $\frac{1}{\sqrt{2\pi|\theta|}}$. Moreover, we have, for all $n \geq 0$ (making the changes of variable $\cos\theta = y$ and $y = e^{-\frac{x^2}{2}}$)

$$\int_0^{\frac{\pi}{2}} \frac{\sin\theta\cos^n\theta}{\sqrt{\pi|\log\cos^2\theta|}}d\theta = \frac{1}{\sqrt{2(n + 1)}}. \tag{2.16}$$

Suppose that $\{W'(h), h \in H\}$ is an independent copy of the Gaussian family $\{W(h), h \in H\}$. For any $\theta \in \mathbb{R}$, $F \in L^0(\Omega, \mathcal{F}, P)$ we set

$$R_\theta F = F(W\cos\theta + W'\sin\theta).$$

With these notations we can write the following expression for the operator $D(-C)^{-1}$.

Lemma 2.4.1 *For every $F \in \mathcal{P}$ such that $E(F) = 0$ we have*

$$D(-C)^{-1}F = \int_{-\frac{\pi}{2}}^{\frac{\pi}{2}} E'(D'(R_\theta F))\varphi(\theta)d\theta. \tag{2.17}$$

Proof: Suppose that $F = p(W(h_1), \ldots, W(h_n))$, where $h_1, \ldots, h_n \in H$ and p is a polynomial in n variables. For any $\theta \in (-\frac{\pi}{2}, \frac{\pi}{2})$ we have

$$R_\theta F = p(W(h_1)\cos\theta + W'(h_1)\sin\theta, \ldots, W(h_n)\cos\theta + W'(h_n)\sin\theta),$$

and therefore

$$D'(R_\theta F) = \sum_{i=1}^{n}\frac{\partial p}{\partial x_i}(W(h_1)\cos\theta + W'(h_1)\sin\theta,$$
$$\ldots, W(h_n)\cos\theta + W'(h_n)\sin\theta)\sin\theta h_i = \sin\theta R_\theta(DF).$$

Consequently, using (2.2) we obtain

$$E'(D'(R_\theta F)) = \sin\theta E'(R_\theta(DF)) = \sin\theta T_t(DF),$$

where $t \geq 0$ is such that $\cos\theta = e^{-t}$. This implies

$$E'(D'(R_\theta F)) = \sum_{n=0}^{\infty} \sin\theta(\cos\theta)^n J_n(DF).$$

Note that since F is a polynomial random variable the above series is actually the sum of a finite number of terms. By (2.16) the right-hand side of (2.17) can be written as

$$\sum_{n=0}^{\infty} \left(\int_{-\frac{\pi}{2}}^{\frac{\pi}{2}} \sin\theta(\cos\theta)^n \varphi(\theta)d\theta \right) J_n(DF) = \sum_{n=0}^{\infty} \frac{1}{\sqrt{n+1}} J_n(DF).$$

Finally, applying the commutativity relationship (2.9) to the multiplication operator defined by the sequence $\phi(n) = \frac{1}{\sqrt{n}}$, $n \geq 1$, $\phi(0) = 0$, we get

$$T_{\phi^+} DF = DT_\phi F = D(-C)^{-1}F,$$

and the proof of the lemma is complete. $\qquad\square$

Now with the help of the preceding equation we can show that the operator DC^{-1} is bounded from $L^p(\Omega)$ into $L^p(\Omega; H)$ for any $p > 1$. This property will be proved using the boundedness in L^p of the Hilbert transform. We recall that the Hilbert transform of a function $f \in C_0^\infty(\mathbb{R})$ is defined by

$$Hf(x) = \int_{\mathbb{R}} \frac{f(x+t) - f(x-t)}{t} dt.$$

The transformation H is bounded in $L^p(\mathbb{R})$ for any $p > 1$ (see Dunford and Schwarz [27], Theorem XI.7.8). Henceforth c_p and C_p denote generic constants depending only on p, which can be different from one formula to another.

Proposition 2.4.1 *Let $p > 1$. There exists a finite constant $c_p > 0$ such that for any $F \in \mathcal{P}$ with $E(F) = 0$ we have*

$$\|DC^{-1}F\|_p \leq c_p \|F\|_p.$$

Proof: Using (2.17) we can write

$$E\left(\|DC^{-1}F\|_H^p\right) = E\left(\left\| \int_{-\frac{\pi}{2}}^{\frac{\pi}{2}} E'(D'(R_\theta F))\varphi(\theta)d\theta \right\|_H^p\right)$$

$$= \alpha_p^{-1} EE'\left(\left| W'\left(\int_{-\frac{\pi}{2}}^{\frac{\pi}{2}} E'(D'(R_\theta F))\varphi(\theta)d\theta \right) \right|^p\right),$$

where $\alpha_p = E(|\xi|^p)$ with ξ an $N(0,1)$ random variable. We recall that for any $G \in L^2(\Omega', \mathcal{F}', P')$ which belongs to the domain of D' the Gaussian random variable $W'(E'(D'G))$ is equal to the projection $J_1'G$ of G on the first Wiener chaos. Therefore, we obtain that

$$E\left(\|DC^{-1}F\|_H^p\right) = \alpha_p^{-1} EE'\left(\left| \int_{-\frac{\pi}{2}}^{\frac{\pi}{2}} J_1' R_\theta F \varphi(\theta)d\theta \right|^p\right).$$

If $g : [-\frac{\pi}{2}, \frac{\pi}{2}] \to \mathbb{B}$ is a Lipschitz function with values in some separable Banach space $\mathbb{B}$ the product $\varphi(\theta)g(\theta)$ does not belong to $L^1([-\frac{\pi}{2}, \frac{\pi}{2}]; \mathbb{B})$ unless $g(0) = 0$. Nevertheless we can define the integral of the product φg in the following way

$$\int_{-\frac{\pi}{2}}^{\frac{\pi}{2}} \varphi(\theta)g(\theta)d\theta = \int_0^{\frac{\pi}{2}} \varphi(\theta)[g(\theta) - g(-\theta)]d\theta.$$

Notice that $J_1' R_\theta F$ vanishes at $\theta = 0$ but this is no longer true for $R_\theta F$. Taking into account this remark we can write

$$E\left(\|DC^{-1}F\|_H^p\right) = \alpha_p^{-1} EE'\left(\left|J_1'\left(\int_{-\frac{\pi}{2}}^{\frac{\pi}{2}} R_\theta F\varphi(\theta)d\theta\right)\right|^p\right)$$

$$\leq c_p EE'\left(\left|\int_{-\frac{\pi}{2}}^{\frac{\pi}{2}} R_\theta F\varphi(\theta)d\theta\right|^p\right),$$

for some constant $c_p > 0$. For any $\xi \in \mathbb{R}$ we define the process

$$\overline{W}_\xi(h) = (W(h)\cos\xi + W'(h)\sin\xi, -W(h)\sin\xi + W'(h)\cos\xi).$$

The law of this process is the same as that of $\{(W(h), W'(h)), h \in H\}$. On the other hand, $\overline{R}_\xi R_\theta F = R_{\xi+\theta}F$, where we set

$$\overline{R}_\xi G((W(h_1), W'(h_1)), \ldots, (W(h_n), W'(h_n))) = G(\overline{W}_\xi(h_1), \ldots, \overline{W}_\xi(h_n)).$$

Therefore, we get

$$\left\|\int_{-\frac{\pi}{2}}^{\frac{\pi}{2}} R_\theta F\varphi(\theta)d\theta\right\|_p = \left\|\overline{R}_\xi\left[\int_{-\frac{\pi}{2}}^{\frac{\pi}{2}} R_\theta F\varphi(\theta)d\theta\right]\right\|_p$$

$$= \left\|\int_{-\frac{\pi}{2}}^{\frac{\pi}{2}} R_{\xi+\theta}F\varphi(\theta)d\theta\right\|_p,$$

where $\|\cdot\|_p$ denotes the L^p norm with respect to $P \times P'$. Integration with respect to ξ yields

$$E\left(\|DC^{-1}F\|_H^p\right) \leq c_p EE'\left(\int_{-\frac{\pi}{2}}^{\frac{\pi}{2}} \left|\int_{-\frac{\pi}{2}}^{\frac{\pi}{2}} R_{\xi+\theta}F\varphi(\theta)d\theta\right|^p d\xi\right). \tag{2.18}$$

Furthermore, there exists a bounded continuous function $\tilde{\varphi}$ and a constant $c > 0$ such that

$$\varphi(\theta) = \tilde{\varphi}(\theta) + \frac{c}{\theta},$$

on $[-\frac{\pi}{2}, \frac{\pi}{2}]$. Consequently, using the L^p boundedness of the Hilbert transform, we see that the right-hand side of (2.18) is dominated up to a constant by

$$EE'\left(\int_{-\frac{\pi}{2}}^{\frac{\pi}{2}} |R_\theta F|^p d\theta\right) = \pi\|F\|_p^p.$$

In fact, the term $\tilde{\varphi}(\theta)$ is easy to treat. On the other hand, to handle the term $\frac{1}{\theta}$ we write

$$\int_{-\frac{\pi}{2}}^{\frac{\pi}{2}} \left|\int_{-\frac{\pi}{2}}^{\frac{\pi}{2}} \frac{R_{\xi+\theta}F}{\theta}d\theta\right|^p d\xi = \int_{-\frac{\pi}{2}}^{\frac{\pi}{2}} \left|\int_{-\frac{\pi}{2}}^{\frac{\pi}{2}} \frac{\tilde{R}_{\xi+\theta}F}{\theta}d\theta\right|^p d\xi$$

$$\leq c_p \left(\int_{-\frac{\pi}{2}}^{\frac{\pi}{2}} \left| \int_{\mathbb{R}} \frac{\tilde{R}_{\xi+\theta}F}{\theta} d\theta \right|^p d\xi + \int_{-\frac{\pi}{2}}^{\frac{\pi}{2}} \left| \int_{[-\frac{\pi}{2},\frac{\pi}{2}]^c} \frac{\tilde{R}_{\xi+\theta}F}{\theta} d\theta \right|^p d\xi \right)$$

$$\leq c_p' \left(\int_{\mathbb{R}} |H(\tilde{R}.F)(\xi)|^p d\xi + \left(\frac{2}{\pi} \right)^p \int_{-\frac{\pi}{2}}^{\frac{\pi}{2}} \left| \int_{\mathbb{R}} \tilde{R}_{\xi+\theta}F d\theta \right|^p d\xi \right)$$

$$\leq c_p'' \left(\int_{\mathbb{R}} |\tilde{R}_\xi F|^p d\xi \right) \leq c_p''' \left(\int_{-\frac{3\pi}{2}}^{\frac{3\pi}{2}} |R_\xi F|^p d\xi \right),$$

where $\tilde{R}_\theta F = \psi(\theta) R_\theta F$, and ψ is a smooth function with support included in $[-\frac{3\pi}{2}, \frac{3\pi}{2}]$ such that $0 \leq \psi \leq 1$ and $\psi(\theta) = 1$ if $\theta \in [-\pi, \pi]$. Hence,

$$EE'\left(\int_{-\frac{\pi}{2}}^{\frac{\pi}{2}} \left| \int_{-\frac{\pi}{2}}^{\frac{\pi}{2}} \frac{R_{\xi+\theta}F}{\theta} d\theta \right|^p d\xi \right) \leq c_p E(|F|^p).$$

$\square$

Proposition 2.4.2 *Let $p > 1$. Then there exist positive and finite constants c_p and C_p such that for any $F \in \mathcal{P}$ we have*

$$c_p \|DF\|_{L^p(\Omega;H)} \leq \|CF\|_p \leq C_p \|DF\|_{L^p(\Omega;H)}. \tag{2.19}$$

Proof: Set $G = CF$. Applying Proposition 2.4.1 to the random variable $G = CF$ we have

$$\|DF\|_{L^p(\Omega;H)} = \|DC^{-1}G\|_{L^p(\Omega;H)} \leq c_p \|G\|_p = c_p \|CF\|_p,$$

which shows the left inequality. The right inequality is proved by means of a duality argument. Let $F, G \in \mathcal{P}$. Set $\tilde{G} = C^{-1}(I - J_0)(G)$. Let q be such that $\frac{1}{p} + \frac{1}{q} = 1$. Then we have

$$
\begin{aligned}
|E(GCF)| &= |E((I - J_0)(G)CF)| = |E(CFC\tilde{G})| = |E(\langle DF, D\tilde{G} \rangle_H)| \\
&\leq \|DF\|_{L^p(\Omega;H)} \|D\tilde{G}\|_{L^q(\Omega;H)} \leq c_q \|DF\|_{L^p(\Omega;H)} \|C\tilde{G}\|_q \\
&= c_q \|DF\|_{L^p(\Omega;H)} \|(I - J_0)(G)\|_q \leq c_q' \|DF\|_{L^p(\Omega;H)} \|G\|_q.
\end{aligned}
$$

Taking the supremum with respect to $G \in \mathcal{P}$ with $\|G\|_q \leq 1$, we obtain

$$\|CF\|_p \leq c_q' \|DF\|_{L^p(\Omega;H)}.$$

$\square$

Now we can state Meyer's inequalities in the general case.

Theorem 2.4.1 *For any $p > 1$ and any integer $k \geq 1$ there exist positive and finite constants $c_{p,k}$ and $C_{p,k}$ such that for any $F \in \mathcal{P}$,*

$$c_{p,k} E \left(\|D^k F\|_{H^{\otimes k}}^p \right) \leq E \left(|C^k F|^p \right) \leq C_{p,k} \left[E \left(\|D^k F\|_{H^{\otimes k}}^p \right) + E(|F|^p) \right]. \tag{2.20}$$

Proof: The proof can be done by induction on k. The case $k = 1$ corresponds to Proposition 2.4.2. In order to illustrate the method let us describe the proof of the left-hand side of (2.20) for $k = 2$.

Notice that if $\{\xi_n, n \geq 1\}$ is a family of independent random variables defined on the probability space $([0,1], \mathcal{B}([0,1]), \lambda)$ (λ is the Lebesgue measure), with distribution N(0,1), and $\{a_n, 1 \leq n \leq N\}$ is a sequence of real numbers, we have, for each $p > 1$,

$$\int_{[0,1]} \left| \sum_{n=1}^{N} a_n \xi_n(t) \right|^p dt = A_p \left(\sum_{n=1}^{N} a_n^2 \right)^{\frac{p}{2}}, \tag{2.21}$$

where $A_p = \int_{\mathbb{R}} |x|^p \frac{1}{\sqrt{2\pi}} e^{-\frac{x^2}{2}} dx$.

Suppose that $F = p(W(h_1), \ldots, W(h_n))$, where the h_i's are orthonormal elements of H. We fix a complete orthonormal system $\{e_i, i \geq 1\}$ in H which contains the h_i's. With these notations, using Eq. (2.21) and Proposition 2.4.2, we can write

$$
\begin{aligned}
E(\|D^2 F\|_{H \otimes H}^p) &= E\left(\left| \sum_{i,j=1}^{\infty} \langle D^2 F, e_i \otimes e_j \rangle_H^2 \right|^{\frac{p}{2}} \right) \\
&\leq c_p \int_0^1 \int_0^1 E\left(\left| \sum_{i,j=1}^{\infty} \langle D^2 F, e_i \otimes e_j \rangle_H \xi_j(t) \xi_i(s) \right|^p \right) dt\, ds \\
&\leq c_p' \int_0^1 E\left(\left| \sum_{i=1}^{\infty} \left\langle D\left(\sum_{j=1}^{\infty} \langle DF, e_j \rangle_H \xi_j(t) \right), e_i \right\rangle_H^2 \right|^{\frac{p}{2}} \right) dt \\
&= c_p' \int_0^1 E\left(\left\| D\left(\sum_{j=1}^{\infty} \langle DF, e_j \rangle_H \xi_j(t) \right) \right\|_H^p \right) dt \\
&\leq c_p'' \int_0^1 E\left(\left| C\left(\sum_{j=1}^{\infty} \langle DF, e_j \rangle_H \xi_j(t) \right) \right|^p \right) dt \\
&\leq c_p''' E\left(\left| \sum_{j=1}^{\infty} \langle CDF, e_j \rangle_H^2 \right|^{\frac{p}{2}} \right) = c_p''' E(\|CDF\|_H^p).
\end{aligned}
$$

Consider the operator $R = \sum_{n=1}^{\infty} \sqrt{1 - \frac{1}{n}} J_n$. This operator verifies $CDF = DCRF$ and it is bounded in L^p for all $p > 1$ (see example 1). Hence, we obtain

$$E(\|D^2 F\|_{H \otimes H}^p) \leq c_p E(\|DCRF\|_H^p) \leq c_p' E(|C^2 RF|^p) \leq c_p'' E(|C^2 F|^p).$$

$\square$

Khintchine's inequalities also allows us to establish the inequalities (2.20) for polynomial random variable taking values in a separable Hilbert space V. Let us now introduce a continuous family of Sobolev spaces.

For any $p > 1$ and $s \in \mathbb{R}$ we will denote by $\|\cdot\|_{s,p}$ the seminorm

$$\|F\|_{s,p} = \|(I - L)^{\frac{s}{2}} F\|_p,$$

where $F \in \mathcal{P}$ is a polynomial random variable. Note that $(I - L)^{\frac{s}{2}} = \sum_{n=0}^{\infty} (1+n)^{\frac{s}{2}} J_n$. These seminorms verify the following properties:

(i) $\|F\|_{s,p}$ is increasing in both coordinates s and p. The monotonicity in p is clear and in s follows from the fact that the operators $(I - L)^{\alpha}$ are contractions in L^p, for all $\alpha < 0$, $p > 1$ (see (2.8)).

(ii) The seminorms $\|\cdot\|_{s,p}$ are compatible, in the sense that for any sequence F_n in $\mathcal{P}$ converging to zero in the norm $\|\cdot\|_{s,p}$, and being a Cauchy sequence in any other norm $\|\cdot\|_{s',p'}$ they also converge to zero in the norm $\|\cdot\|_{s',p'}$.

We define $\mathbb{D}^{s,p}$ as the completion of $\mathcal{P}$ with respect to this norm.

Remarks:

1. $\|F\|_{0,p} = \|F\|_{0,p} = \|F\|_p$, and $\mathbb{D}^{0,p} = L^p(\Omega)$. For $k = 1, 2, \ldots$ the seminorms $\|\cdot\|_{k,p}$ and $\|\cdot\|_{k,p}$ are equivalent due to Meyer's inequalities. In fact, we have

$$\|F\|_{k,p} = \|(I - L)^{\frac{k}{2}} F\|_p \le |E(F)| + \|R(-L)^{\frac{k}{2}} F\|_p,$$

where $R = \sum_{n=1}^{\infty} \left(\frac{n+1}{n}\right)^{\frac{k}{2}} J_n$. Notice that $R = T_\phi$ with

$$\phi(n) = \begin{cases} 0, & n = 0 \\ h(\frac{1}{n}), & n \ge 1 \end{cases}$$

where $h(x) = (1 + x)^{\frac{k}{2}}$ in analytic in a neighbourhood of 0. Therefore, the operator R is bounded in L^p for all $p > 1$. Hence,

$$\|F\|_{k,p} \le c_p(\|F\|_p + \|(-L)^{\frac{k}{2}} F\|_p) \le c_p'(\|F\|_p + \|D^k F\|_{L^p(\Omega; H^{\otimes k})})$$
$$\le c_p'' \|F\|_{k,p}.$$

In a similar way one can show the converse inequality.

Thus, the Sobolev spaces $\mathbb{D}^{k,p}$ coincide with those defined by means of the derivative operator.

2. From property (ii) we have $\mathbb{D}^{s,p} \subset \mathbb{D}^{s',p'}$ if $p' \le p$ and $s' \le s$.

3. For $s > 0$ the operator $(I - L)^{-\frac{s}{2}}$ is an isometric isomorphism (in the norm $\|\cdot\|_{s,p}$) between L^p and $\mathbb{D}^{s,p}$ and between $\mathbb{D}^{-s,p}$ and L^p for all $p > 1$. As a consequence, the dual of $\mathbb{D}^{s,p}$ is $\mathbb{D}^{-s,q}$ where $\frac{1}{p} + \frac{1}{q} = 1$. If $s < 0$, the elements of $\mathbb{D}^{s,p}$ may not be ordinary random variables and they are interpreted as distributions on the Gaussian space. Set $\mathbb{D}^{-\infty} = \bigcup_{s,p} \mathbb{D}^{s,p}$. The space $\mathbb{D}^{-\infty}$ is the dual of the space $\mathbb{D}^{\infty}$ which is a countably normed space.

4. Suppose that V is a real separable Hilbert space. We can define the Sobolev spaces $\mathbb{D}^{s,p}(V)$ of V-valued functionals as the completion of the class $\mathcal{P}_V$ of V-valued polynomial random variable with respect to the seminorm $\|F\|_{s,p,V}$ defined in the same way as before. The above properties are still true for V-valued functionals.

The main application of Meyer's inequalities is the following continuity theorem.

Proposition 2.4.3 *Let V be a real separable Hilbert space. For every $p > 1$ and $s \in \mathbb{R}$, the operator D is continuous from $\mathbb{D}^{s,p}(V)$ to $\mathbb{D}^{s-1,p}(V \otimes H)$ and the operator δ (defined as the adjoint of D) is continuous from $\mathbb{D}^{s,p}(V \otimes H)$ to $\mathbb{D}^{s-1,p}(V)$.*

Proof: We will only proof the continuity of D for $V = \mathbb{R}$. The continuity of the adjoint operator follows by duality. For any $F \in \mathcal{P}$ we have

$$(I - L)^{\frac{s}{2}} DF = DR(I - L)^{\frac{s}{2}} F,$$

where $R = \sum_{n=1}^{\infty} \left(\frac{n}{n+1} \right)^{\frac{s}{2}} J_n$. Notice that $R = T_\phi$ with

$$\phi(n) = \begin{cases} 0, & n = 0 \\ h(\frac{1}{n}), & n \geq 1 \end{cases}$$

where $h(x) = (\frac{1}{1+x})^{\frac{k}{2}}$ in analytic near 0. Therefore, the operator R is bounded in L^p for all $p > 1$, and we obtain

$$
\begin{aligned}
\|(I - L)^{\frac{s}{2}} DF\|_p &= \|DR(I - L)^{\frac{s}{2}} F\|_p \leq c_p \|(I - L)^{\frac{1}{2}} R(I - L)^{\frac{s}{2}} F\|_p \\
&= c_p \|R(I - L)^{\frac{s+1}{2}} F\|_p \leq c_p' \|(I - L)^{\frac{s+1}{2}} F\|_p = c_p' \|F\|_{s+1,p},
\end{aligned}
$$

which implies the desired continuity. □

The following inequality is also an immediate consequence of Meyer's inequalities.

Proposition 2.4.4 *Let u be an element of $\mathbb{D}^{1,p}(H)$, $p > 1$. Then we have*

$$\|\delta(u)\|_p \leq c_p \left(\|J_0 u\|_H + \|Du\|_{L^p(\Omega; H \otimes H)} \right). \tag{2.22}$$

Proof: From Proposition 2.4.3 we know that the operator δ is continuous from $\mathbb{D}^{1,p}(H)$ into $L^p(\Omega)$. This implies that

$$\|\delta(u)\|_p \leq c_p \left(\|u\|_{L^p(\Omega; H)} + \|Du\|_{L^p(\Omega; H \otimes H)} \right).$$

On the other hand, we have

$$\|u\|_{L^p(\Omega; H)} \leq \|J_0 u\|_H + \|u - J_0 u\|_{L^p(\Omega; H)},$$

and

$$
\begin{aligned}
\|u - J_0 u\|_{L^p(\Omega; H)} &= \|(I - L)^{-\frac{1}{2}} RCu\|_{L^p(\Omega; H)} \leq c_p \|Cu\|_{L^p(\Omega; H)} \\
&\leq c_p' \|Du\|_{L^p(\Omega; H \otimes H)},
\end{aligned}
$$

where $R = \sum_{n=1}^{\infty} \left(1 + \frac{1}{n} \right)^{\frac{1}{2}} J_n$. □

The next lemmas are often useful.

Lemma 2.4.2 *Let $\{F_n, n \geq 1\}$ be a sequence of random variables converging to F in L^p, for some $p > 1$. Suppose that $\sup_n \|F_n\|_{s,p} < \infty$ for some $s > 0$. Then $F \in \mathbb{D}^{s,p}$.*

Proof: We know that

$$\sup_n \|(I - L)^{\frac{s}{2}} F_n\|_p < \infty.$$

Therefore, there exists a subsequence $F_{n(i)}$ such that $(I - L)^{\frac{s}{2}} F_{n(i)}$ converges weakly in $\sigma(L^p, L^q)$ (where q is the conjugate of p) to some element G. Then, for any polynomial random variable R we have

$$
\begin{aligned}
E[F(I - L)^{\frac{s}{2}} R] &= \lim_i E[F_{n(i)}(I - L)^{\frac{s}{2}} R] \\
&= \lim_i E[(I - L)^{\frac{s}{2}} F_{n(i)} R] = E[GR].
\end{aligned}
$$

Thus, $F = (I - L)^{-\frac{s}{2}} G$, and this implies that $F \in \mathbb{D}^{s,p}$. □

Lemma 2.4.3 *Let* $\eta \in \mathbb{D}^{-\infty}$ *be such that* $\langle \eta, F \rangle \geq 0$ *for any nonnegative smooth random variable* $F \in \mathcal{S}$. *Fix a complete orthonormal system* $\{e_i, i \geq 1\}$ *in* H, *and let* $\mathcal{G}$ *be the* σ-*field generated by the random variables* $\{W(e_i), i \geq 1\}$. *Then there exists a finite measure* μ_η *on the* σ-*field* $\mathcal{G}$ *such that*

$$\int_\Omega F d\mu_\eta = \langle \eta, F \rangle, \tag{2.23}$$

for any random variable F *of the form* $F = f(W(e_1), \ldots, W(e_n))$, $n \geq 1$, $f \in C_p^\infty(\mathbb{R}^n)$. *In particular* $\mu_\eta(\Omega) = \langle \eta, 1 \rangle$.

Proof: The proof will be done in two steps:

Step 1: Consider the increasing sequence of σ-fields $\mathcal{F}_n = \sigma\{W(e_1), \ldots, W(e_n)\}$. We claim that there exists a measure ν_n on the σ-field $\mathcal{F}_n$ such that (2.23) holds for any random variable F of the form $F = f(W(e_1), \ldots, W(e_n))$, where $f \in C_p^\infty(\mathbb{R}^n)$. Indeed, consider the function $\varphi : \mathbb{R}^n \to \mathbb{C}$ defined by

$$\varphi(t) = \left\langle \eta, \exp(i \sum_{j=1}^n t^j W(e_j)) \right\rangle.$$

This mapping is nonnegative definite and continuous, because for all $t_1, \ldots, t_N \in \mathbb{R}^n$, $c_1, \ldots, c_N \in \mathbb{C}$ we have

$$\sum_{l,k=1}^N \varphi(t_l - t_k) c_l \bar{c}_k = \left\langle \eta, \left| \sum_{k=1}^N c_k \exp\left(i \sum_{j=1}^n t_k^j W(e_j)\right) \right|^2 \right\rangle \geq 0.$$

Therefore, there exists a finite measure $\bar{\nu}_n$ on $\mathbb{R}^n$ which has moments of all orders and such that

$$\int_{\mathbb{R}^n} f(x) \bar{\nu}_n(dx) = \langle \eta, f(W(e_1), \ldots, W(e_n)) \rangle,$$

for any function f in $C_p^\infty(\mathbb{R}^n)$. This implies the existence of a finite measure μ_η on $\mathcal{F}_n$ such that (2.23) holds for any F of the form $f(W(e_1), \ldots, W(e_n))$ with $f \in C_p^\infty(\mathbb{R}^n)$.

Step 2: It suffices to show that μ_η can be extended to a measure on $\mathcal{G}$. Consider the random variable $G = \sum_{n=1}^\infty n^{-2} W(e_n)^2$. This variable belongs to $\mathbb{D}^\infty$. Let ψ be a smooth function such that $\psi(x) = 0$ is $x \leq 0$, $\psi(x) = 1$ is $x \geq 1$, and $0 \leq \psi \leq 1$. For any $a > 1$, set $\varphi_a(x) = \psi((|x| - (a-1))^+)$. For any random variable F of the form $f(W(e_1), \ldots, W(e_n))$ with $f \in C_p^\infty(\mathbb{R}^n)$ and such that $0 \leq F \leq 1$, we can write

$$\int_\Omega F d\mu_\eta = \langle \eta, F \rangle = \langle \eta, F(1 - \varphi_a(G)) \rangle + \langle \eta, F \varphi_a(G) \rangle$$
$$\leq \langle \eta, 1 \rangle \| F(1 - \varphi_a(G)) \|_\infty + \langle \eta, \varphi_a(G) \rangle.$$

Hence, the inequality

$$\int_\Omega F d\mu_\eta \leq \langle \eta, 1 \rangle \| F(1 - \varphi_a(G)) \|_\infty + \langle \eta, \varphi_a(G) \rangle \tag{2.24}$$

holds for any F in the class $\mathcal{L} := \{f(W(e_1), \ldots, W(e_n)), n \geq 1, 0 \leq f \leq 1, f \text{ Borel}\}$.

In order to complete the proof of the lemma and taking into account Daniell's theorem, it suffices to show that given a sequence of random variables $\{F_n, n \geq$

$1\} \subset \mathcal{L}$, uniformly bounded by one, such that $0 \leq F_n(\omega) \downarrow 0$ for all $\omega \in \Omega$, then $\int_\Omega F_n d\mu_\eta \downarrow 0$. We can write

$$\int_\Omega F_n d\mu_\eta \leq \langle \eta, 1 \rangle \| F_n(1 - \varphi_a(G)) \|_\infty + \langle \eta, \varphi_a(G) \rangle \qquad (2.25)$$

Suppose that $\eta \in \mathbb{D}^{-k,p}$. The second summand in the right-hand side of (2.25) can be estimated as follows:

$$\langle \eta, \varphi_a(G) \rangle = \langle (I - L)^{-\frac{k}{2}} \eta, (I - L)^{\frac{k}{2}} \varphi_a(G) \rangle \leq \| \eta \|_{-k,p} \| (I - L)^{\frac{k}{2}} \varphi_a(G) \|_q,$$

where $\frac{1}{p} + \frac{1}{q} = 1$, and this tends to zero as a tends to infinity. The first summand converges to zero as n tends to infinity due to Dini's theorem and the fact that the set $\{\{x_n, n \geq 1\} : \sum_{n=1}^\infty n^{-2} x_n^2 < a\}$ is compact in ℓ^2. □

Bibliographical notes: The fact that the positive distributions on the Wiener space are measures has been proved in the context of an abstract Wiener space by Sugita in [103]. Other contributions to this problem are [79], [110], [92] and [82]. Spaces of Banach-valued Wiener functionals have been studied in [29], [64] and [42].

Chapter 3

Application of Malliavin calculus to study probability laws

In this chapter we will discuss the application of the stochastic calculus of variations on a Gaussian space to the study of different properties of probability distributions of functionals of the underlying Gaussian process. The type of properties that we can analyze with the help of the Malliavin calculus include the absolute continuity, smoothness of the density, estimates (uniform and in L^p) of the density, and application of these estimates to deduce Krylov-type inequalities.

3.1 Computation of probability densities

Suppose that $W = \{W(h), h \in H\}$ is a centered Gaussian family associated with a real and separable Hilbert space, defined on a probability space $(\Omega, \mathcal{F}, P)$. We assume that that the σ-field $\mathcal{F}$ is generated by W.

Suppose that $F = (F^1, \ldots, F^m)$ is a random vector whose components belong to the space $\mathbb{D}^{1,1}_{\text{loc}}$. We associate with F the following random symmetric nonnegative definite matrix:

$$\gamma_F = (\langle DF^i, DF^j \rangle_H)_{1 \leq i,j \leq m}.$$

This matrix is called the *Malliavin matrix* of the random vector F. The following theorem was established by Bouleau and Hirsch [16] and provides a criterion for the absolute continuity of the law of F.

Theorem 3.1.1 *Let $F = (F^1, \ldots, F^m)$ be a random vector whose components belong to the space $\mathbb{D}^{1,p}_{\text{loc}}$, $p > 1$, and suppose that the random matrix $\gamma_F = (\langle DF^i, DF^j \rangle)_{1 \leq i,j \leq m}$ is invertible a.s. Then the law of F is absolutely continuous with respect to the Lebesgue measure on $\mathbb{R}^m$.*

In dimension one the nondegeneracy condition required in the previous theorem reduces to $\|DF\|_H > 0$ a.s. The proof of Theorem 3.1.1 is based on the coarea formula. A simple proof in dimension one can be given as follows (see [81]).

Proof of Theorem 3.1.1 for $m = 1$: We can assume that F belongs to the space $\mathbb{D}^{1,p}$ and that $|F| < k$ for some constant k. We have to show that for any measurable function $g : (-k, k) \to [0, 1]$ such that $\int_{-k}^{k} g(y)dy = 0$ we have $E(g(F)) = 0$. We can find a sequence of continuously differentiable functions with bounded derivatives

$g^n : (-k, k) \to [0, 1]$ such that as n tends to infinity $g^n(y)$ converges to $g(y)$ for almost all y with respect to the measure $P \circ F^{-1} + \lambda$, where λ is the Lebesgue measure. Set

$$\psi^n(y) = \int_{-k}^{y} g^n(x)dx, \quad y \in (-k, k)$$

and

$$\psi(y) = \int_{-k}^{y} g(x)dx, \quad y \in (-k, k).$$

By the chain rule, $\psi^n(F)$ belongs to the space $\mathbb{D}^{1,p}$ and we have $D[\psi^n(F)] = g^n(F)DF$. We have that $\psi^n(F)$ converges to $\psi(F)$ a.s. as n tends to infinity, because g^n converges to g a.e. with respect to the Lebesgue measure. This convergence also holds in $L^p(\Omega)$ by dominated convergence. On the other hand, $D[\psi^n(F)]$ converges a.s. to $g(F)DF$ because g^n converges to g a.e. with respect to the law of F. Again by dominated convergence, this convergence holds in $L^p(\Omega; H)$. Observe that $\psi(F) = 0$ a.s. Now we use the property that the operator D is closed to deduce that $g(F)DF = 0$ a.s. Consequently, $g(F) = 0$ a.s., which completes the proof of the theorem. $\qquad\square$

We remark that the above proof also works with $p = 1$. The following proposition provides an explicit expression for the density of a one-dimensional random variable.

Proposition 3.1.1 *Let F be a random variable in the space $\mathbb{D}^{1,2}$. Suppose that $\frac{DF}{\|DF\|^2}$ belongs to the domain of the operator δ in L^2. Then the law of F has a continuous and bounded density given by*

$$p(x) = E\left[\mathbf{1}_{\{F>x\}}\delta\left(\frac{DF}{\|DF\|^2}\right)\right]. \tag{3.1}$$

Proof: Let ψ be a nonnegative smooth function with compact support, and set $\varphi(y) = \int_{-\infty}^{y} \psi(z)dz$, $y \in \mathbb{R}$. We know that $\varphi(F)$ belongs to $\mathbb{D}^{1,2}$, and making the scalar product of its derivative with DF obtains

$$\langle D(\varphi(F)), DF \rangle_H = \psi(F)\|DF\|_H^2.$$

Using the duality relationship between the operators D and δ (see (1.7)), we obtain

$$E[\psi(F)] = E\left[\left\langle D(\varphi(F)), \frac{DF}{\|DF\|_H^2}\right\rangle_H\right] = E\left[\varphi(F)\delta\left(\frac{DF}{\|DF\|_H^2}\right)\right]. \tag{3.2}$$

By an approximation argument Eq. (3.2) holds for $\psi(y) = \mathbf{1}_{[a,b]}(y)$. As a consequence, we apply Fubini's theorem to get

$$
\begin{aligned}
P(a \le F \le b) &= E\left[\left(\int_{-\infty}^{F} \mathbf{1}_{[a,b]}(x)dx\right)\delta\left(\frac{DF}{\|DF\|_H^2}\right)\right] \\
&= \int_{a}^{b} E\left[\mathbf{1}_{\{F>x\}}\delta\left(\frac{DF}{\|DF\|_H^2}\right)\right]dx,
\end{aligned}
$$

which implies the desired result. $\qquad\square$

Proposition 3.1.1 also holds under the hypotheses $F \in \mathbb{D}^{1,1}$, and $\frac{DF}{\|DF\|^2} \in \mathbb{D}^{1,p}(H)$ for some $p, p' > 1$. From expression (3.1) we can deduce estimates for the density. Fix p and q such that $\frac{1}{p} + \frac{1}{q} = 1$. By Hölder's inequality we obtain

$$p(x) \le (P(F > x))^{1/q} \|\delta\left(\frac{DF}{\|DF\|_H^2}\right)\|_p.$$

In the same way and taking into account the relation $E[\delta(DF/\|DF\|_H^2)] = 0$ we can deduce the inequality

$$p(x) \le (P(F < x))^{1/q} \|\delta\left(\frac{DF}{\|DF\|_H^2}\right)\|_p.$$

As a consequence, we get

$$p(x) \le (P(|F| > |x|))^{1/q} \|\delta\left(\frac{DF}{\|DF\|_H^2}\right)\|_p, \tag{3.3}$$

for all $x \in \mathbb{R}$. Now using the L^p-estimate of the operator δ established in Proposition 2.4.4 we obtain

$$\|\delta\left(\frac{DF}{\|DF\|_H^2}\right)\|_p \le c_p \left(\left\|E\left(\frac{DF}{\|DF\|_H^2}\right)\right\|_H + \left\|D\left(\frac{DF}{\|DF\|_H^2}\right)\right\|_{L^p(\Omega;H\otimes H)}\right). \tag{3.4}$$

We have

$$D\left(\frac{DF}{\|DF\|_H^2}\right) = \frac{D^2 F}{\|DF\|_H^2} - 2\frac{\langle D^2 F, DF \otimes DF\rangle_{H\otimes H}}{\|DF\|_H^4},$$

and, hence,

$$\left\|D\left(\frac{DF}{\|DF\|_H^2}\right)\right\|_{H\otimes H} \le 3\frac{\|D^2 F\|_{H\otimes H}}{\|DF\|_H^2}. \tag{3.5}$$

Finally, from the inequalities (3.3), (3.4) and (3.5) we deduce the following result.

Lemma 3.1.1 *Let q, α, β be three positive real numbers such that $\frac{1}{q} + \frac{1}{\alpha} + \frac{1}{\beta} = 1$. Let F be a random variable in the space $\mathbb{D}^{2,\alpha}$, such that $E(\|DF\|_H^{-2\beta}) < \infty$. Then the density $p(x)$ can be estimated as follows:*

$$p(x) \le c_{q,\alpha,\beta}(P(|F| > |x|))^{1/q} \left(E(\|DF\|_H^{-1}) + \|D^2 F\|_{L^\alpha(\Omega;H\otimes H)}\|\|DF\|_H^{-2}\|_\beta\right). \tag{3.6}$$

Let us apply the preceding lemma to a Brownian martingale.

Example: We will assume that $W = \{W_t, t \in [0, T]\}$ is a Brownian motion, and $H = \{H_t, t \in [0, T]\}$ is an adapted process verifying the following hypotheses:

(i) $E \int_0^T H_s^2 ds < \infty$, H_t belongs to the space $\mathbb{D}^{2,2}$ for each $t \in [0, T]$, and

$$\lambda := \sup_{s,t\in[0,T]} E(|D_s H_t|^p) + \sup_{r,s\in[0,T]} E((\int_0^T |D_{r,s}^2 H_t|^2 dt)^{p/2}) < \infty,$$

for some $p > 3$.

(ii) $|H_t| \geq \rho > 0$ for some constant ρ.

Consider the martingale $M_t = \int_0^t H_s dW_s$, and denote by $p_t(x)$ the probability density of M_t. Then the following estimate holds

$$p_t(x) \leq \frac{c}{\sqrt{t}} P(|M_t| > |x|)^{\frac{1}{q}}, \tag{3.7}$$

where $q > \frac{p}{p-3}$. The constant c depends on λ, ρ, and p.

Proof of (3.7): We will apply Lemma 3.1.1 to the random variable M_t. We claim that $M_t \in \mathbb{D}^{2,2}$ for each $t \in [0,T]$. In fact, note first that $M_t \in \mathbb{D}^{1,2}$ because the process H belongs to $\mathbb{D}^{2,2}(H)$, due to condition (i), and the operator δ is continuous from $\mathbb{D}^{2,2}(H)$ into $\mathbb{D}^{1,2}$. Furthermore, using (1.10) we get

$$D_s M_t = H_s 1_{\{s \leq t\}} + \delta(D_s H.).$$

For each $s \in [0,T]$, the process $\{D_s H_t, t \in [0,T]\}$ is adapted, square integrable, and it vanishes for $t \leq s$ due to Lemma 1.5.1. Hence it belongs to the domain of δ and $\delta(D_s H.)$ coincides with the Itô stochastic integral $\int_s^t D_s H_r dW_r$. So, we have

$$\begin{aligned} D_s M_t &= H_s + \int_s^t D_s H_r dW_r, \quad s \leq t, \\ D_s M_t &= 0, \quad s > t. \end{aligned} \tag{3.8}$$

Notice that $D_{s_1} H. \in \mathbb{D}^{1,2}(H)$ for all s_1 a.e., the process $D_{s_2} D_{s_1} H.$ belongs to the domain of δ for all s_1, s_2 a.e. (because it is adapted and square integrable), and

$$E \int_0^T |\delta(D_{s_2} D_{s_1} H.)|^2 ds_2 = E \int_0^T \int_0^T |D_{s_2} D_{s_1} H_t|^2 ds_2 dt < \infty.$$

As a consequence, from property (1.10) deduce that for all s_1 a.e. $\delta(D_{s_1} H.)$ belongs to $\mathbb{D}^{1,2}$ and

$$D_{s_2}(\delta(D_{s_1} H.)) = D_{s_1} H_{s_2} + \delta(D_{s_2} D_{s_1} H.).$$

Hence, $M_t \in \mathbb{D}^{2,2}$ and from (3.8) we deduce

$$D^2_{s_1,s_2} M_t = D_{s_1} H_{s_2} + D_{s_2} H_{s_1} + \int_{s_1 \vee s_2}^t D^2_{s_1,s_2} H_r dW_r, \quad s_1, s_2 \leq t.$$

We will take $\alpha = p$ in Lemma 3.1.1. Using Hölder's and Burkholder's inequalities we obtain

$$\begin{aligned} E(\|D^2 M_t\|^p_{H \otimes H}) &= E\left(\left(\int_{[0,t]^2} (D^2_{s_1,s_2} M_t)^2 ds_1 ds_2\right)^{p/2}\right) \\ &\leq c_p \left\{ E\left(\left(\int_{[0,t]^2} (D_{s_1} H_{s_2})^2 ds_1 ds_2\right)^{p/2}\right) \right. \\ &\quad + E\left(\left(\int_{[0,t]^2} \left(\int_{s_1 \vee s_2}^t D^2_{s_1,s_2} H_r dW_r\right)^2 ds_1 ds_2\right)^{p/2}\right) \right\} \\ &\leq c_p' \lambda t^p. \end{aligned}$$

Set
$$\sigma(t) := \|DM_t\|_H^2 = \int_0^t (H_s + \int_s^t D_s H_r dW_r)^2 ds.$$

We have the following lower estimates for σ for any $h \leq 1$:

$$\sigma(t) \geq \int_{t(1-h)}^t (H_s + \int_s^t D_s H_r dW_r)^2 ds$$

$$\geq \int_{t(1-h)}^t \frac{1}{2} H_s^2 ds - \int_{t(1-h)}^t \left(\int_s^t D_s H_r dW_r \right)^2 ds \geq \frac{th\rho^2}{2} - I_h(t),$$

where

$$I_h(t) = \int_{t(1-h)}^t \left(\int_s^t D_s H_r dW_r \right)^2 ds.$$

Choose h of the form $h = \frac{4}{t\rho^2 y}$, and notice that $h \leq 1$ if $y \geq a := \frac{4}{t\rho^2}$. We have

$$P \left(\sigma(t) \leq \frac{1}{y} \right) \leq P \left(I_h(t) \geq \frac{1}{y} \right) \leq y^{p/2} E(|I_h(t)|^{p/2}). \tag{3.9}$$

Using Burkholder's inequality for square integrable martingales we get the following estimate

$$E(|I_h(t)|^{p/2}) \leq c_p (th)^{\frac{p}{2}-1} \int_{t(1-h)}^t E\left(\left(\int_s^t (D_s H_r)^2 dr \right)^{p/2} \right) ds$$

$$\leq c_p' \sup_{s,r \in [0,t]} E(|D_s H_r|^p)(th)^p. \tag{3.10}$$

Consequently, for $0 < \gamma < \frac{p}{2}$ we obtain, using (3.9) and (3.10),

$$E[\sigma(t)^{-\gamma}] = \int_0^\infty \gamma y^{\gamma-1} P(\sigma(t)^{-1} > y) dy$$

$$\leq a^\gamma + \gamma \int_a^\infty y^{\gamma-1} P(\sigma(t) < \frac{1}{y}) dy$$

$$\leq \left(\frac{4}{t\rho^2} \right)^\gamma + \gamma \int_{\frac{4}{t\rho^2}}^\infty E(|I_h(t)|^{p/2}) y^{\gamma-1+\frac{p}{2}} dy$$

$$\leq c \left(t^{-\gamma} + \int_{\frac{4}{t\rho^2}}^\infty y^{\gamma-1-\frac{p}{2}} dy \right) \leq c' \left(t^{-\gamma} + t^{\frac{p}{2}-\gamma} \right). \tag{3.11}$$

Substituting (3.11) in Eq. (3.6) with $\alpha = p$, $\beta < \frac{p}{2}$, and with $\gamma = \frac{1}{2}$ and $\gamma = \beta$, we get the desired estimation. $\qquad \square$

The inequality (3.7) implies

$$p_t(x) \leq c \frac{|x|^{-\frac{p}{q}}}{\sqrt{t}} \left(E \left(\left| \int_0^t |H_s|^2 ds \right|^{\frac{p}{2}} \right) \right)^{\frac{1}{q}}.$$

If we assume, in addition that the process H is bounded by a constant M, that is, instead of (ii) we impose:

(ii)' $M \geq |H_t| \geq \rho > 0$ for some constants ρ and M,

then using the martingale exponential inequality we obtain

$$p_t(x) \leq c \frac{1}{\sqrt{t}} \exp(-\frac{|x|^2}{qM^2 t}). \tag{3.12}$$

3.2 Regularity of densities and composition of tempered distributions with elements of $\mathbb{D}^{-\infty}$

The results obtained in the last section for onedimensional random variables can be extended to the multidimensional case. Furthermore, under additional smoothness and integrability conditions one can show that the probability density of a random vector F is infinitely differentiable. The basic assumptions are introduced in the following definition of nondegenerate random vector.

Definition 3.2.1 *We will say that the random vector $F = (F^1, \ldots, F^m) \in (\mathbb{D}^\infty)^m$ is nondegenerate if the matrix γ_F is invertible a.s. and*

$$(\det \gamma_F)^{-1} \in \bigcap_{p \geq 1} L^p(\Omega). \tag{3.13}$$

For a nondegenerate random vector the following integration by parts formula plays a basic role.

Proposition 3.2.1 *Let $F = (F^1, \ldots, F^m) \in (\mathbb{D}^\infty)^m$ be a nondegenerate random vector in the sense of Definition 3.2.1, let $G \in \mathbb{D}^\infty$ and let $g \in C_p^\infty(\mathbb{R}^m)$. Then $(\det \gamma_F)^{-1} \in \mathbb{D}^\infty$ and for any multi-index $\alpha \in \{1, \ldots, m\}^k$, $k \geq 1$, there exists an element $H_\alpha(F, G) \in \mathbb{D}^\infty$ such that:*

$$E[(\partial_\alpha g)(F)G] = E[g(F)H_\alpha(F, G)]. \tag{3.14}$$

Moreover the elements $H_\alpha(F, G)$ are recursively given by:

$$H_{(i)}(F, G) = \sum_{j=1}^m \delta\left(G(\gamma_F^{-1})^{ij} DF^j\right),$$

$$H_\alpha(F, G) = H_{\alpha_k}(F, H_{(\alpha_1, \ldots, \alpha_{k-1})}(F, G)).$$

Proof: Let us first show that $(\det \gamma_F)^{-1} \in \mathbb{D}^\infty$. For any $N \geq 1$ we have $(\det \gamma_F + \frac{1}{N})^{-1} \in \mathbb{D}^\infty$, because the random variable $(\det \gamma_F + \frac{1}{N})^{-1}$ can be written as the composition of $\det \gamma_F$ with a function in $C_p^\infty(\mathbb{R})$. It is not difficult to show using (3.13) that $\{(\det \gamma_F + \frac{1}{N})^{-1}, N \geq 1\}$ is a Cauchy sequence in the norms $\| \cdot \|_{k,p}$ for all k, p, and we obtain the desired result. Also $\gamma_F^{-1} \in \mathbb{D}^\infty(\mathbb{R}^{m \times m})$. By the chain rule we have

$$D[g(F)] = \sum_{i=1}^m (\partial_i g)(F) DF^i,$$

hence,

$$\langle D[g(F)], DF^j \rangle_H = \sum_{i=1}^m (\partial_i g)(F) \gamma_F^{ij},$$

and as a consequence,

$$(\partial_i g)(F) = \sum_{j=1}^m \langle D[g(F)], DF^j \rangle_H (\gamma_F^{-1})^{ji}.$$

Finally, taking expectations in the above equality and introducing the adjoint of the operator D we get

$$E[G(\partial_i g)(F)] = E[g(F)H_{(i)}(F, G)],$$

where $H_{(i)}(F,G) = \sum_{j=1}^{m} \delta\left(G(\gamma_F^{-1})^{ij} DF^j\right)$. We complete the proof by a recurrence argument. $\qquad\square$

Proposition 3.2.2 *For any $p > 1$, and for any multi-index α there exist a constant $C(p,\alpha)$, natural numbers n_1, n_2, and indices k, d, d', b, b', depending also on p and α such that*

$$\|H_\alpha(F,G)\|_p \leq C(p,\alpha)\left(\|\gamma_F^{-1}\|_k^{n_1}\|F\|_{d,b}^{n_2}\|G\|_{d',b'}\right).$$

Proof: This estimate is an immediate consequence of the continuity of the operator δ from $\mathbb{D}^{k+1,p}$ into $\mathbb{D}^{k,p}$, Hölder's inequality for the $\|\cdot\|_{k,p}$-norms, and the equality:

$$D[(\gamma_F^{-1})^{ij}] = -\sum_{k,l=1}^{m}(\gamma_F^{-1})^{ik}(\gamma_F^{-1})^{jl}D[\gamma_F^{kl}].$$

$\qquad\square$

From Proposition 3.2.1 it follows that the density of a smooth and nondegenerated random vector is infinitely differentiable. We recall that $\mathcal{S}(\mathbb{R}^m)$ is the space of infinitely differentiable functions $f : \mathbb{R}^m \to \mathbb{R}$ such that for any $k \geq 1$, and for all multi-index $\beta \in \{1,\ldots,m\}^j$ one has $\sup_{x\in\mathbb{R}^m}|x|^k|\partial_\beta f(x)| < \infty$.

Corollary 3.2.1 *Let $F = (F^1,\ldots,F^m) \in (\mathbb{D}^\infty)^m$ be a nondegenerate random vector in the sense of Definition 3.2.1. Then the density of F belongs to the Schwartz space $\mathcal{S}(\mathbb{R}^m)$, and*

$$p(x) = E[1_{\{F>x\}}H_{(1,2,\ldots,m)}(F,1)], \tag{3.15}$$

where

$$H_{(1,2,\ldots,m)}(F,1) = \delta((\gamma_F^{-1}DF)^m\delta((\gamma_F^{-1}DF)^{m-1}\cdots\delta((\gamma_F^{-1}DF)^1)\cdots)).$$

Proof: Consider the multi-index $\alpha = (1,2,\ldots,m)$. From (3.14) we obtain, for any function $\psi \in C_0^\infty(\mathbb{R}^m)$,

$$E[(\partial_\alpha\psi)(F)] = E[\psi(F)H_\alpha(F,1)].$$

By Fubini's theorem we can write

$$E[(\partial_\alpha\psi)(F)] = \int_{\mathbb{R}^m}(\partial_\alpha\psi)(x)E[1_{\{x<F\}}H_\alpha(F,1)]dx.$$

We can take $\partial_\alpha\psi = 1_B$, where B is a bounded Borel subset of $\mathbb{R}^m$, and in this way we deduce the absolute continuity of the law of F and the expression (3.15) for its density. Moreover, for any multi-index β we obtain

$$\partial_\beta p(x) = E[\partial_\beta(1_{\{x<F\}})H_\alpha(F,1)] = E[1_{\{x<F\}}H_\beta(F,H_\alpha(F,1))],$$

and the density of F is infinitely differentiable. Finally in order to show that the density belongs to $\mathcal{S}(\mathbb{R}^m)$ we need to prove that

$$\sup_{x\in\mathbb{R}^m} x_j^{2k}|E[1_{\{x<F\}}H_\beta(F,H_\alpha(F,1))]| < \infty,$$

for all $j = 1, \ldots, m$. If $x_j > 0$ we have

$$x_j^{2k}|E[1_{\{x<F\}}H_\beta(F, H_\alpha(F, 1))]| \leq E[|F^j|^{2k}|H_\beta(F, H_\alpha(F, 1))|] < \infty.$$

If $x_j < 0$, then we use the alternative expression for the density

$$p(x) = E[\prod_{i \neq j} 1_{\{x_i < F^i\}} 1_{\{x_j > F^j\}} H_\alpha(F, 1)],$$

and we deduce a similar estimate. $\qquad\square$

Let F be an m-dimensional random vector. The probability density of F at a point x can be formally defined as the generalized expectation $E[\delta_x(F)]$, where δ_x denotes the Dirac function at x. The expression $E[\delta_x(F)]$ can be defined as the coupling $\langle \delta_x(F), 1 \rangle$, provided we show that $\delta_x(F)$ is an element of $\mathbb{D}^{-\infty}$. The Dirac function δ_x is a measure, and more generally we can define the composition $T(F)$ of a Schwartz distribution $T \in S'(\mathbb{R}^m)$ with a random variable in $\mathbb{D}^\infty$ the result being a distribution in $\mathbb{D}^{-\infty}$. This approach was introduced by Watanabe in [107]. Furthermore, the differentiability of the mapping $x \to \delta_x(F)$ from $\mathbb{R}^m$ into some Sobolev space $\mathbb{D}^{-k,p}$ will provide an alternative proof of the smoothness of the density.

Let us describe the main steps of this approach. We introduce the following sequence of seminorms on the Schwartz space $S(\mathbb{R}^m)$:

$$\|\phi\|_{2k} = \|(1 + |x|^2 - \Delta)^k \phi\|_\infty, \quad \phi \in S(\mathbb{R}^m),$$

for $k \in \mathbf{Z}$. Let S_{2k}, $k \in \mathbf{Z}$ be the completion of $S(\mathbb{R}^m)$ by the seminorm $\| \cdot \|_{2k}$. Then we have

$$S_{2k+2} \subset S_{2k} \subset \cdots S_2 \subset S_0 \subset S_{-2} \subset \cdots \subset S_{-2k} \subset S_{-2k-2},$$

and $S_0 = \hat{C}(\mathbb{R}^m)$ is the space of continuous functions on $\mathbb{R}^m$ which vanish at infinity. Moreover, $\cap_{k \geq 1} S_{2k} = S(\mathbb{R}^m)$, and $\cup_{k \geq 1} S_{-2k} = S'(\mathbb{R}^m)$.

Proposition 3.2.3 *Let $F = (F^1, \ldots, F^m) \in (\mathbb{D}^\infty)^m$ be a nondegenerate random vector in the sense of Definition 3.2.1. For any $k \geq 1$, $p > 1$, there exists a constant $C(p, k, F)$ such that for any $\phi \in S(\mathbb{R}^m)$ we have*

$$\|\phi(F)\|_{-2k,p} \leq C(p, k, F)\|\phi\|_{-2k}.$$

Proof: By a duality principle we have

$$\|\phi(F)\|_{-2k,p} = \sup\{E(\phi(F)G), \|G\|_{2k,q} \leq 1\},$$

where q is the conjugate of p. Now using the expression (3.14) we can write

$$\begin{aligned} E(\phi(F)G) &= E[((1 + |x|^2 - \Delta)^k (1 + |x|^2 - \Delta)^{-k}\phi)(F)G] \\ &= E[(1 + |x|^2 - \Delta)^{-k}\phi(F)\eta_{2k}], \end{aligned}$$

where $\eta_{2k} \in \mathbb{D}^\infty$ verifies

$$C(p, k, F) := \sup\{E(|\eta_{2k}|), \|G\|_{2k,q} \leq 1\} < \infty,$$

and this completes the proof. □

As a consequence, the mapping $\phi \to \phi(F)$ can be extended uniquely to a continuous linear mapping $T \to T \circ F$ from $\mathcal{S}_{-2k}$ to $\mathbb{D}^{-2k,p}$ for every $p > 1$ and $k = 0, 1, 2, \ldots$. In particular for every Schwartz distribution $T \in \mathcal{S}'(\mathbb{R}^m)$, we can define a Wiener distribution $T \circ F \in \mathbb{D}^{-\infty}$. Actually

$$T \circ F \in \bigcup_{k=1}^{\infty} \bigcap_{p>1} \mathbb{D}^{-2k,p}.$$

We have that for any fixed point $x \in \mathbb{R}^m$, the Dirac function belongs to $\mathcal{S}_{-2k}$ if $k > \frac{m}{2}$, and the mapping $x \to \delta_x$ is $2j$ times continuously differentiable from $\mathbb{R}^m$ to $\mathcal{S}_{-2k-2j}$. This implies that we can define the "composition" $\delta_x(F)$ as an element of $\mathbb{D}^{-2k,p}$ for any $p > 1$ and any integer k such that $k > \frac{m}{2}$. The density of the random vector F is then given by $\langle \delta_x(F), 1 \rangle$.

Moreover, for any random variable $G \in \mathbb{D}^{\infty}$ and for any x such that $p(x) \neq 0$, we have

$$\langle \delta_x(F), G \rangle = p(x) E[G|F = x].$$

3.3 The case of diffusion processes

We can apply the previous results to derive the smoothness of the density for solutions to stochastic differential equations. This provides probabilistic arguments to study heat kernels. Suppose that $\{W(t), t \geq 0\}$ is a d-dimensional Brownian motion defined on the canonical probability space $\Omega = C_0(\mathbb{R}_+; \mathbb{R}^d)$. Let $A_j : \mathbb{R}^m \to \mathbb{R}^m$, $j = 0, \ldots, d$ a system of C^∞ functions with bounded derivatives of all orders, and consider the following stochastic Stratonovich differential equation on $\mathbb{R}^m$:

$$\begin{cases} dX_t = \sum_{j=1}^{d} A_j(X_t) \circ dW_t^j + A_0(X_t)dt, \\ X_0 = x_0 \end{cases} \qquad (3.16)$$

One can show that $X_t \in (\mathbb{D}^{\infty})^m$ for all $t \geq 0$. The following nondegeneracy condition assures that for each $t > 0$ the random vector X_t is nondegenerate:

(H) There exists an integer $k_0 \geq 0$ such that the vector space spanned by the vector fields

$$[A_{j_k}, [A_{j_{k-1}}, [\ldots [A_{j_1}, A_{j_0}]] \ldots]], \quad 0 \leq k \leq k_0,$$

where $j_0 \in \{1, 2, \ldots, d\}$, $j_i \in \{0, 1, 2, \ldots, d\}$ if $1 \leq i \leq k$, at point x_0 has dimension m.

In the above hypothesis $[A, B]$ denotes the Lie bracket of the differentiable functions $A, B : \mathbb{R}^m \to \mathbb{R}^m$, defined as

$$[A, B] = \sum_{i=1}^{m} \left(A^i \frac{\partial B}{\partial x_i} - B^i \frac{\partial A}{\partial x_i} \right).$$

Theorem 3.3.1 *Suppose that the above condition (H) is satisfied. Then there exists a positive integer ν depending only on k_0, and for each $p > 1$ a positive constant $c(p, x_0)$ such that*

$$\left\| (\det \gamma_{X_t})^{-1} \right\|_p \leq ct^{-\nu},$$

for all $t > 0$.

This theorem is a consequence of the precise estimates obtained by Kusuoka and Stroock in [57].

3.4 L^p estimates of the density and applications

The stochastic calculus of variations can be used to establish Krylov-type estimates which are a useful tool in deriving existence and uniqueness of a solution to stochastic differential equations and to partial stochastic differential equations whose diffusion coefficient is nondegenerate and the drift is not smooth. Let us first establish a preliminary result similar to Lemma 3.1.1.

Lemma 3.4.1 *Let α, β be two positive real numbers such that $\frac{1}{\alpha} + \frac{1}{\beta} < 1$. Let F be an m-dimensional random variable ($m > 1$) whose components belong to the space $\mathbb{D}^{2,\alpha}$ and $E(|(\gamma_F^{-1})^{ii}|^\beta) < \infty$, $i = 1, \ldots, m$. Then for any nonnegative and measurable function $f : \mathbb{R}^m \to \mathbb{R}$ we have:*

$$E[f(F)] \leq c_{\alpha,\beta,m} \|f\|_m \sup_i \left(E(\sqrt{(\gamma_F^{-1})^{ii}}) + \|D^2 F\|_{L^\alpha(\Omega; H \otimes H \otimes \mathbb{R}^m)} \|(\gamma_F^{-1})^{ii}\|_\beta \right).$$

Proof: We can assume that the function f is continuous and with compact support. Let ψ_ϵ be an approximation of the identity in $\mathbb{R}^m$. We can write

$$\begin{aligned}
E[f(F)] &= \lim_{\epsilon \downarrow 0} \int_{\mathbb{R}^m} E[\psi_\epsilon(x - F)] f(x) dx \\
&\leq \|f\|_m \limsup_{\epsilon \downarrow 0} \left(\int_{\mathbb{R}^m} |E[\psi_\epsilon(x - F)]|^{\frac{m}{m-1}} dx \right)^{\frac{m-1}{m}} \\
&\leq \|f\|_m \limsup_{\epsilon \downarrow 0} \prod_{i=1}^m \left(\int_{\mathbb{R}^m} |E[\partial_i \psi_\epsilon(x - F)]| dx \right)^{\frac{1}{m}}.
\end{aligned}$$

Here we have used the Gagliardo-Nirenberg inequality which says that for any function f in the space $C_0^\infty(\mathbb{R}^m)$ one has

$$\|f\|_{L^{\frac{m}{m-1}}} \leq \prod_{i=1}^m \|\partial_i f\|_{L^1}^{\frac{1}{m}}.$$

Applying Proposition 3.2.1 we can write

$$E[\partial_i \psi_\epsilon(x - F)] = E[\psi_\epsilon(x - F) H_i(F, 1)], \tag{3.17}$$

where

$$H_i(F, 1) = \delta \left(\sum_{j=1}^m DF^j (\gamma_F^{-1})^{ij} \right).$$

Now, using (3.17) and (2.22) we obtain, for any $p > 1$

$$E[f(F)] \leq \|f\|_m \prod_{i=1}^{m} (E(|H_i(F, 1)|))^{\frac{1}{m}}$$

$$\leq c_p \|f\|_m \prod_{i=1}^{m} \left(\|E(\sum_{j=1}^{m} DF^j (\gamma_F^{-1})^{ij})\|_H + \right.$$

$$\left. + \|D(\sum_{j=1}^{m} DF^j (\gamma_F^{-1})^{ij})\|_{L^p(\Omega; H \otimes H)} \right)^{\frac{1}{m}}.$$

We conclude the proof choosing p in such a way that $\frac{1}{p} = \frac{1}{\alpha} + \frac{1}{\beta}$, applying Hölder's inequality, and using the relations

$$\| \sum_{j=1}^{m} DF^j (\gamma_F^{-1})^{ij} \|_H^2 = (\gamma_F^{-1})^{ii},$$

and

$$\|D(\sum_{j=1}^{m} DF^j (\gamma_F^{-1})^{ij})\|_{H \otimes H} \leq \| \sum_{j=1}^{m} D^2 F^j (\gamma_F^{-1})^{ij} \|_{H \otimes H}$$

$$+ \| \sum_{j,k,s=1}^{m} DF^j \otimes (\gamma_F^{-1})^{ik} D(\langle DF^k, DF^s \rangle_H)(\gamma_F^{-1})^{sj} \|_{H \otimes H}.$$

$\square$

Lemma 3.4.1 allows us to deduce Krylov-type estimates for solutions to stochastic differential equations. Let $\{W_t, t \in [0,1]\}$ be an m-dimensional Brownian motion. Consider the m-dimensional stochastic differential equation

$$\begin{cases} dX_t = f(X,t)dt + \sum_{i=1}^{m} g_i(X,t)dW_t^i, & t \in [0,1], \\ X_0 = x_0, \end{cases} \tag{3.18}$$

where $f, g_i : C([0,1]; \mathbb{R}^m) \times [0,1] \to \mathbb{R}^m$, $i = 1, \ldots, m$, are progressively measurable functions verifying the following conditions

(a) $|f(x,t)| \leq K$ for some constant K.

(b) g is twice Fréchet differentiable in the first variable, with uniformly bounded first and second order derivatives, and $g(0,t)$ is bounded.

(c) There exist positive constants c_1 and c_2 such that

$$c_1 |\theta|^2 \leq |g(x,t)\theta|^2 \leq c_2 |\theta|^2,$$

for all $\theta \in \mathbb{R}^m$, $x \in C([0,1]; \mathbb{R}^m)$, and $t \in [0,1]$.

We will denote this equation by Eq(f,g). By a solution to this equation we will mean a continuous and adapted process that satisfies the corresponding stochastic integral equation.

Proposition 3.4.1 *Let $g_i : C([0,1]; \mathbb{R}^m) \times [0,1] \to \mathbb{R}^m$, $i = 1, \ldots, m$, be measurable functions verifying conditions (b) and (c), and let $\{F_t, t \in [0,1]\}$ be an m-dimensional progressively measurable process bounded by K. We denote by $\{X_t, t \in [0,1]\}$ the solution of the following equation (denoted in the sequel by Eq(F, g)):*

$$X_t = x_0 + \int_0^t F_s ds + \sum_{i=1}^m \int_0^t g_i(X, s) dW_s^i, \quad t \in [0,1]. \tag{3.19}$$

Then for any measurable function $h : \mathbb{R}^m \times [0,1] \to \mathbb{R}$, and for all $p > m$, $\gamma > 1$, we have

$$E\left(\int_0^1 |h(X_t, t)| dt\right) \leq C \left(\int_0^1 \left(\int_{\mathbb{R}} |h(x,t)|^p dx\right)^\gamma dt\right)^{\frac{1}{p\gamma}}, \tag{3.20}$$

where the constant C depends on p, γ, K, and the coefficient g.

Proof: Define a new probability $\tilde{P}$ by $\frac{d\tilde{P}}{dP} = Z$, where

$$Z = \exp\left(-\int_0^1 F_t^T (g^{-1})^T(X, t) dW_t - \frac{1}{2} \int_0^1 |(g^{-1})(X, t) F_t|^2 dt\right).$$

By Girsanov's theorem

$$\tilde{W}_t = W_t + \int_0^t (g^{-1})(X, s) F_s ds$$

is a Brownian motion under $\tilde{P}$. Thus, under $\tilde{P}$ the process X has the same law as the process Y solution of

$$\begin{cases} dY_t = g(Y, t) dW_t, & t \in [0,1], \\ Y_0 = x_0, \end{cases} \tag{3.21}$$

By Hölder's inequality

$$\begin{aligned} E\left(\int_0^1 |h(X_t, t)| dt\right) &\leq (\tilde{E} Z^{-\alpha})^{\frac{1}{\alpha}} \left(\tilde{E} \int_0^1 |h(X_t, t)|^\beta dt\right)^{\frac{1}{\beta}} \\ &= C(\alpha, K, g) \left(E \int_0^1 |h(Y_t, t)|^\beta dt\right)^{\frac{1}{\beta}} \end{aligned}$$

for every $\alpha, \beta > 1$ such that $\frac{1}{\alpha} + \frac{1}{\beta} = 1$. The random vector Y_t belongs to the space $\mathbb{D}^{2,p}$ for all $p > 1$ (see Lemma 3.4.2 below). When $m = 1$ one can use the estimate on the density of Y_t obtained in (3.12). Suppose $m > 1$. Applying Lemma 3.4.1 to the random variable Y_t we obtain if $\frac{1}{\alpha'} + \frac{1}{\beta'} = 1$

$$\begin{aligned} E\left(\int_0^1 |h(Y_t, t)|^\beta dt\right) &\leq c_{m,\alpha',\beta'} \left(\int_0^1 \left(\int_{\mathbb{R}^m} |h(y,t)|^{m\beta} dy\right)^{\frac{1}{m}}\right. \\ &\quad \left. \times \sup_i \left(E\left(\sqrt{(\gamma_{Y_t}^{-1})^{ii}}\right) + \|D^2 Y_t\|_{\alpha'} \|(\gamma_{Y_t}^{-1})^{ii}\|_{\beta'}\right) dt\right). \end{aligned}$$

The inequality (3.25) allows to control the term $\|D^2 Y_t\|_{\alpha'}$ by a constant times t. On the other hand, using hypothesis (c) one can show (a sketch of the proof is given below) that

$$E([(\gamma_{Y_t}^{-1})^{ii}]^p) < a_3 t^{-p} \tag{3.22}$$

for all $p \geq \frac{1}{2}$. Hence,

$$E\left(\int_0^1 |h(Y_t, t)|^\beta dt\right) \leq c_{m,\alpha',\beta'} \int_0^1 \left(\int_{\mathbb{R}^m} |h(y,t)|^{m\beta} dy\right)^{\frac{1}{m}} (\frac{1}{\sqrt{t}} + c) dt$$

$$\leq c'_{m,\alpha',\beta'} \int_0^1 \left(\int_{\mathbb{R}^m} |h(y,t)|^{m\beta} dy\right)^{\frac{2+\epsilon}{m}} dt.$$

This allows to complete the proof of the Proposition.

Sketch of the proof of (3.22): First we write

$$E([(\gamma_{Y_t}^{-1})^{ii}]^p) \leq c_{m,p} \left[E(\|DY_t\|_H^{4p(m-1)})\right]^{\frac{1}{2}} \left[E((\det \gamma_{Y_t})^{-2p})\right]^{\frac{1}{2}}.$$

From (3.24) we get that the first factor in the right-hand side of the above equality is bounded by a constant times $t^{p(m-1)}$. In order to estimate the second factor we write

$$\det \gamma_{Y_t} \geq \inf_{|v|=1} (v^T \gamma_{Y_t} v)^m = \inf_{|v|=1} \|\langle v, DY_t\rangle\|_H^{2m}.$$

From (3.28) we get

$$\langle v, D_s^j Y_t\rangle = \langle v, g_j(Y,s)\rangle + \langle v, \int_s^t \langle g'(Y,r), (D_s^j Y.)\rangle dW_r\rangle, \quad j = 1, \ldots, m. \tag{3.23}$$

Consequently, we deduce, with the convention of summation over repeated indices, for any $h < 1$,

$$\begin{aligned}
\|\langle v, DY_t\rangle\|_H^2 &= \int_0^t \sum_{k=1}^m \left|D_s^k(v_j Y_t^j)\right|^2 ds \\
&\geq \frac{1}{2} \int_{t(1-h)}^t \sum_{k=1}^m \left|v_j g_k^j(Y,s)\right|^2 ds \\
&\quad - \int_{t(1-h)}^t \sum_{k=1}^m \left|v_j \int_s^t \langle (g')_i^j(Y,r), D_s^k Y.\rangle dW_r^i\right|^2 ds \\
&\geq \frac{c_1 th}{2} - \int_{t(1-h)}^t |J(s)|^2 ds,
\end{aligned}$$

where

$$J^j(s) = \int_s^t \langle (g')_i^j(Y,r), D_s^k Y.\rangle dW_r^i.$$

Now, using the same technique as in the proof of (3.7) we can show that

$$E(\inf_{|v|=1} \|\langle v, DY_t\rangle\|_H^{-4pm}) \leq E\left(\left|\frac{c_1 th}{2} - \int_{t(1-h)}^t |J(s)|^2 ds\right|^{-2pm}\right) \leq a_4 t^{-2pm},$$

and this allows us to conclude the proof. □

Remark: In the preceding proposition we can take $\gamma = 1$ when $m = 2$.

Lemma 3.4.2 *Let $\{Y_t, t \in [0,1]\}$ be the solution to equation $Eq(0,g)$ where the function g satisfies hypothesis (b). Then $Y_t \in \cap_{p>1} \mathbb{D}^{2,p}$, and*

$$E(\|DY_t\|_H^p) < a_1 t^{\frac{p}{2}}, \tag{3.24}$$

$$E(\|D^2 Y_t\|_{H \otimes H}^p) < a_2 t^p, \tag{3.25}$$

for some constants a_1 and a_2.

Proof: We will denote by g' and by g'' the Fréchet derivatives of the function g. We introduce the sequence of Picard approximations defined recursively by $Y^0 = x_0$, and

$$Y_t^{n+1} = x_0 + \int_0^t g(Y^n, s) dW_s, \quad n \geq 0.$$

Then for each n the random variable Y_t^{n+1} belongs to $\cap_{p>1} \mathbb{D}^{1,p}$, and

$$D_s^j Y_t^{n+1} = g_j(Y^n, s) + \int_s^t \langle g'(Y^n, r), (D_s^j Y^n) \rangle dW_r, \quad j = 1, \ldots, m,$$

where D^j denotes the derivative with respect to the jth component of the Brownian motion. That is, $D_s^j W_t^k = 1_{\{s \leq t\}} \delta_{ik}$. We have

$$\lim_n E[\|Y^n - Y\|_\infty^p] = 0,$$

for all $p > 1$, and using Grownall's lemma we deduce

$$\sup_n \sup_{s \in [0,1]} E[\|D_s Y^{n+1}\|_\infty^p] < \infty, \tag{3.26}$$

and

$$\sup_n \sup_{s,r \in [0,1]} E[\|D_{s,r}^2 Y^{n+1}\|_\infty^p] < \infty, \tag{3.27}$$

for all $p > 1$. This implies that $Y_t \in \cap_{p>1} \mathbb{D}^{2,p}$. Finally we use the chain rule (properly extended to $C([0,1])$-valued functions) in order to compute the derivative of $g(Y,t)$, and we obtain:

$$D_s^j Y_t = g_j(Y, s) + \int_s^t \langle g'(Y, r), (D_s^j Y) \rangle dW_r, \quad j = 1, \ldots, m. \tag{3.28}$$

Using Burkholder's inequality we can show (3.24) from the expression (3.28). Similar computations can be done for the second derivative. □

Notice that the inequality (3.20) also holds for a process X such for each $t \in [0,1]$, X_t is the almost sure limit of a sequence of random variables X_t^n such that X^n solves an equation of the form (3.19) with coefficients g and F_n, where F_n is bounded by K.

We can now to derive a convergence criterion for solutions to equation (3.19):

Corollary 3.4.1 *Let $h_n : \mathbb{R}^m \times [0,1] \to \mathbb{R}$ be a sequence of measurable functions uniformly bounded by a constant K_1 and converging a.e. to h. Consider a sequence of processes X^n solutions of equations $Eq(F_n, g)$, where g satisfies conditions (b) and (c) and F_n is a progressively measurable process bounded by K. Then*

$$\lim_n E\left(\int_0^1 |h_n(X_t^n, t) - h(X_t, t)| dt\right) = 0.$$

Proof: Set

$$J_n := E\left(\int_0^1 |h_n(X_t^n, t) - h(X_t, t)| dt\right).$$

Fix $\eta > 0$. Let ψ be a nonnegative smooth function with support included in $[-1, 1]$, $\psi(0) = 1$ and bounded by 1. Choose $R > 0$ such that

$$E\left(\int_0^1 \left(1 - \psi\left(\frac{X_t}{R}\right)\right) dt\right) < \eta.$$

Consider a continuous function $g_{R,\eta}$ on $\mathbb{R} \times [0,1]$ bounded by $2K_1$ such that $g(x,t) = 0$ for all $|x| > R$ and for a fixed $p > m$

$$\int_{[0,1]} \left(\int_{-R}^{R} |h(x,t) - g_{R,\eta}(x,t)|^p dx \right)^\gamma dt \leq \eta^{p\gamma},$$

where $p > m$ and $\gamma > 1$ are the exponents appearing in Proposition 3.4.1. Consider the following decomposition

$$J_n = J_n^1 + J_n^2 + J_n^3 + J_n^4,$$

where

$$
\begin{aligned}
J_n^1 &= E\left(\int_0^1 |h_n(X_t^n, t) - h(X_t^n, t)| dt \right) \\
J_n^2 &= E\left(\int_0^1 |h(X_t^n, t) - g_{R,\eta}(X_t^n, t)| dt \right) \\
J_n^3 &= E\left(\int_0^1 |h(X_t, t) - g_{R,\eta}(X_t, t)| dt \right) \\
J_n^4 &= E\left(\int_0^1 |g_{R,\eta}(X_t^n, t) - g_{R,\eta}(X_t, t)| dt \right).
\end{aligned}
$$

We write

$$
\begin{aligned}
J_n^1 &= E\left(\int_0^1 \psi\left(\frac{X_t^n}{R}\right) |h_n(X_t^n, t) - h(X_t^n, t)| dt \right) \\
&+ E\left(\int_0^1 \left(1 - \psi\left(\frac{X_t^n}{R}\right)\right) |h_n(X_t^n, t) - h(X_t^n, t)| dt \right),
\end{aligned}
$$

and we make similar decompositions for the terms J_n^2 and J_n^3. In this way we obtain, using Proposition 3.4.1,

$$
\begin{aligned}
\limsup_n J_n &\leq C\left(\limsup_n \left(\int_0^1 \left(\int_{\mathbb{R}} \psi(\tfrac{x}{R})|h_n(x,t) - h(x,t)|^p dx \right)^\gamma dt \right)^{\frac{1}{p\gamma}} \right. \\
&+ 2\left(\int_0^1 \left(\int_{-R}^R |h(x,t) - g_{R,\eta}(x,t)|^p dx \right)^\gamma dt \right)^{\frac{1}{p\gamma}} \\
&+ 6K_1 E\left(\int_0^1 \left(1 - \psi\left(\frac{X_t}{R}\right)\right) dt \right) \\
&\left. + \limsup_n E\left(\int_0^1 |g_{R,\eta}(X_t^n, t) - g_{R,\eta}(X_t, t)| dt \right) \right).
\end{aligned}
$$

Hence,

$$\limsup_n J_n \leq C\eta.$$

$\square$

Corollary 3.4.2 *Let $g_i : C([0,1]; \mathbb{R}^m) \times [0,1] \to \mathbb{R}$, $1 \leq i \leq m$, be measurable functions satisfying conditions (b) and (c) of Proposition 3.4.1. Consider sequences of* *progressively measurable processes* $F, F_n : [0,1] \times \Omega \to \mathbb{R}^m$, *and measurable functions* $f, f_n : [0,1] \times \mathbb{R}^m \to \mathbb{R}^m$ *such that*

$$\lim_n F_n(t) = F(t), \quad dt \otimes dP - a.e.,$$
$$\lim_n f_n(t,x) = f(t,x), \quad dt \otimes dx - a.e..$$

Suppose that for each $n \geq 1$ *equation* $Eq(F_n, g)$ *admits a solution* X^n, *and for each* $t \in [0,1]$, X_t^n *converges a.s. to a process* X_t. *Then* X *solves equation* $Eq(F,g)$.

Proof: It suffices to pass to the limit each term in the equation

$$X_t^n = x_0 + \int_0^t F_n(s)ds + \sum_{i=1}^m \int_0^t g_i(X^n, s)dW_s^i.$$

$\square$

Applying the preceding results we can derive existence and uniqueness results for equations of the form $Eq(f,g)$ where f is a measurable and bounded function of X_t and g is a smooth (up to the second order) nondegenerate function. To do this one usually makes use of comparison theorems for stochastic differential equations, and for this reason, one has to consider particular type of equations. In order to illustrate this method we will describe the onedimensional case.

Proposition 3.4.2 *Suppose that* $f : \mathbb{R} \times [0,1] \to \mathbb{R}$ *is a measurable function bounded by* K *and* $g : \mathbb{R} \times [0,1] \to \mathbb{R}$ *is twice continuously differentiable, with bounded derivatives and such that* $0 < c_1 \leq |g(x,t)| \leq c_2 < \infty$. *Then equation (3.18) has a unique solution.*

Proof of the existence: Let ρ be a smooth nonnegative function with compact support in $\mathbb{R}$ such that $\int_{\mathbb{R}} \rho(x)dx = 1$. Define

$$\tilde{f}_j(x,t) = j \int_{\mathbb{R}} f(z,t)\rho(j(x-z))dz.$$

Moreover, let

$$f_{n,k} = \bigwedge_{j=n}^k \tilde{f}_j, \quad n \leq k$$

and

$$f_n = \bigwedge_{j=n}^\infty \tilde{f}_j.$$

Cleraly, $f_{n,k}(x,t)$ is Lipschitz in x uniformly with respect to t, $f_{n,k}(x,t) \downarrow f_n(x,t)$ as $k \uparrow \infty$, and $f_n(x,t) \uparrow f(x,t)$ as $n \uparrow \infty$, dx-a.e. for each t. For each $n \geq k$ equation $Eq(f_{n,k}, g)$ has a unique solution $X_{n,k}$. From the comparison theorem for stochastic differential equations the sequence $\{X_{n,k}, k = n, n+1, \ldots\}$ is decreasing. Hence it has a limit

$$X_n = \lim_{k \to \infty} X_{n,k}.$$

Since $X_{n,k}$ is bounded above by the solution of $Eq(K,g)$ and below by the solution of $Eq(-K,g)$, we can apply Corollary 3.4.1 and deduce that

$$\lim_k \int_0^t f_{n,k}(X_{n,k}(t), t)dt = \int_0^t f_n(X_n(t), t)dt, \quad a.s..$$

On the other hand, by the continuity of the function g we get

$$\lim_k \int_0^t g(X_{n,k}(t), t) dW_t = \int_0^t g(X_n(t), t) dW_t, \quad \text{a.s..}$$

Hence the process X_n is a solution to Eq(f_n, g).

Proof of the uniqueness: The proof of the uniqueness has been inspired by the work [36]. Let X denote the solution constructed as the limit of the sequence X_n, and let Z be another solution. We can write with the above notations

$$dZ_t = \tilde{F}_j(t) dt + \tilde{f}_j(Z_t, t) dt + g(Z_t, t) dW_t,$$

where

$$\tilde{F}_j(t) = f(Z_t, t) - \tilde{f}_j(Z_t, t), \quad , j \geq 1.$$

Let us denote by $Z_{n,k,-}$ and by $Z_{n,k,+}$ the solutions of the following equations

$$dZ_{n,k,-}(t) = \left(\bigwedge_{j=n}^k \tilde{F}_j(t) \right) \wedge 0 \, dt + f_{n,k}(Z_{n,k,-}(t), t) dt + g(Z_{n,k,-}(t), t) dW_t,$$

and

$$dZ_{n,k,+}(t) = \left(\bigvee_{j=n}^k \tilde{F}_j(t) \right) \vee 0 \, dt + f_{n,k}(Z_{n,k,+}(t), t) dt + g(Z_{n,k,+}(t), t) dW_t.$$

By the comparison theorem we have

$$Z_{n,k,-} \leq X_{n,k} \leq Z_{n,k,+},$$

and

$$Z_{n,k,-} \leq Z \leq Z_{n,k,+},$$

Moreover, the sequences $\{Z_{n,k,-}, k \geq n\}$ and $\{Z_{n,k,+}, k \geq n\}$ are decreasing in k. Hence, they converge to some limits $\{Z_{n,-}, k \geq n\}$ and $\{Z_{n,+}, k \geq n\}$, respectively, which are increasing in n. Let us denote by $\{Z_-, k \geq n\}$ and $\{Z_+, k \geq n\}$ the limit of these sequences. We have

$$Z_- \leq X, Z \leq Z_+. \tag{3.29}$$

The coefficients $\bigwedge_{j=n}^k \tilde{F}_j(t)$ and $f_{n,k}$ converge almost everywhere to 0 and f, respectively, as k and n tend to infinity. Consequently, applying Corollary 3.4.2 we get that the processes Z_- and Z_+ solve (3.18). But from Girsanov's theorem, the law of (3.18) is unique. Hence, from (3.29) we deduce that $Z_- = X = Z = Z_+$ which completes the proof of the uniqueness.

□

This approach has been used to derive existence and uniqueness of the solution and approximation of the implicit approximation scheme for the unidimensional heat equation perturbed by a space-time white noise. Let us describe the results obtained for these equations.

Let $\{W(s,t) = W([0,s] \times [0,t]), s,t, \in [0,1]\}$ be a two-parameter Wiener process. Let us consider the stochastic partial differential equation

$$\frac{\partial u}{\partial t} = \frac{\partial^2 u}{\partial x^2} + f(t,x,u(t,x)) + g(t,x,u(t,x))\frac{\partial^2 W}{\partial t\partial x} \qquad (3.30)$$

with Dirichlet boundary conditions

$$u(t,0) = u(t,1) = 0, \quad t \in [0,1]; \qquad (3.31)$$

and with initial condition

$$u(0,x) = u_0(x), \quad x \in [0,1], \qquad (3.32)$$

where u_0 is a continuous function on $[0,1]$ vanishing at 0 and 1. The coefficients $f(t,x,r)$ and $g(t,x,r)$ are locally bounded Borel functions mapping $[0,1]^2 \times \mathbb{R}$ into $\mathbb{R}$.

The existence and uniqueness of the strong solution is well known when f and g satisfy the linear growth condition and are Lipschitz in r. Consider the following hypotheses on the coefficients of Eq. (3.30):

(H1) g has a Lipschitz continuous derivative, it has linear growth, and it satisfies the nondegeneracy condition $g^2 \geq \varepsilon > 0$. f is satisfies the one-sided linear growth condition $rf(t,x,r) \leq C(1+r^2)$.

Under hypothesis (H) the existence and uniqueness of the solution for equation (3.30) has been established in [5]. The main tool used in this paper is an L^p estimate, obtained using Malliavin calculus, similar to Eq. (3.20). This existence and uniqueness result is improved in [36], where the smoothness condition on g is replaced by the local Lipschitzness of g in r.

Consider the implicit approximation scheme for the equation (3.30). For every integer $n \geq 1$ we construct a random field u^n in the following way:

$$u^n(0,x) = u_0(x)$$

$$u^n(t_{i+1}^n, x) = \left(I - \frac{1}{n}\Delta\right)^{-1} u^n(t_i^n, x)$$

$$+ \int_{t_i^n}^{t_{i+1}^n} G_n(x,y) f(s,x,u^n(t_i^n,x)) ds$$

$$+ \int_{t_i^n}^{t_{i+1}^n} \int_0^1 G_n(x,y) g(s,y,u^n(t_i^n,y)) W(ds,dy), \quad i \geq 0,$$

where $t_i^n := \frac{i}{n}$, $0 \leq i \leq n$, G_n is the kernel of the operator $\left(I - \frac{1}{n}\Delta\right)^{-1}$, and $\Delta := \frac{\partial^2}{\partial x^2}$. When $t \in (t_i^n, t_{i+1}^n)$ we define $u^n(t,x)$ by the polygonal approximation

$$u^n(t,x) = u^n(t_i^n,x) + n(t - t_i^n)\left(u^n(t_{i+1}^n,x) - u^n(t_i^n,x)\right).$$

The sequence u_n is called an implicit approximation scheme for the equation (3.30) because we can write:

$$u^n(t_{i+1}^n) = u^n(t_i^n) + \frac{1}{n}\Delta u^n(t_{i+1}^n) + \int_{t_i^n}^{t_{i+1}^n} f(s,u^n(t_i^n)) ds$$

$$+ \int_{t_i^n}^{t_{i+1}^n} \int_0^1 g(s,u^n(t_i^n)) W(ds,dy).$$

Under hypothesis (H) one can show (see [37]) that

$$\lim_n P\left(\sup_{x,t\in[0,1]} |u^n(t,x) - u(t,x)| > \epsilon\right) = 0,$$

for all $\epsilon > 0$. The main ingredient of the proof is the following L^p estimate for the density $p_{t,x}^n$ of the law of $u^n(t,x)$. We obtain it using Lemma 3.1.1:

$$\sup_n \int_0^1 \int_0^1 \int_{\mathbb{R}} p_{t,x}^n(r)^\alpha \, \varphi(t,x) \, dr \, dx \, dt < \infty$$

for any function $\varphi \in C^\infty((0,1)^2)$ with compact support, and for all $\alpha > 1$.

Bibliographical notes: For an m-dimensional nondegenerate random vector F, Watanabe has precised the order (s,p) of the negative Sobolev space $\mathbb{D}^{s,p}$ which contains the distribution $\delta_x(F)$ (cf. [109]). The asymptotic expansion of $X_t(\epsilon\omega)$ using Malliavin calculus has been studied in [108].

Chapter 4

Support theorems

In this chapter we apply the stochastic calculus of variations to study the properties of the support of the law of a random vector. We also discuss the characterization of the support of the law using the so-called skeleton approximation, and the application of the Malliavin calculus to the proof of Varadhan-type estimates.

4.1 Properties of the support

Given a random variable $F : \Omega \to S$ with values on a Polish space S, the topological support of the probability distribution of F is defined as the set of points $x \in S$ such that $P(d(F, x) < \epsilon) > 0$ for all $\epsilon > 0$.

The connected property of the support of the law of a finite-dimensional random variable vector was established by Fang [28]:

Proposition 4.1.1 *Let* $F \in (\mathbb{D}^{1,p})^m$ *for some* $p > 1$. *Then the topological support of the law of* F *is a closed connected subset of* $\mathbb{R}^m$.

Proof: Suppose that $\operatorname{supp} P \circ F^{-1}$ is not connected. There exists two nonempty disjoint closed sets A and B such that $\operatorname{supp} P \circ F^{-1} = A \cup B$.

For each integer $M \geq 2$ let $\psi_M : \mathbb{R}^m \to \mathbb{R}$ be an infinitely differentiable function such that $0 \leq \psi_M \leq 1$, $\psi_M(x) = 0$ if $|x| \geq M$, $\psi_M(x) = 1$ if $|x| \leq M - 1$, and $\sup_{x,M} |\nabla \psi_M(x)| < \infty$.

Set $A_M = A \cap \{|x| \leq M\}$ and $B_M = B \cap \{|x| \leq M\}$. For M large enough we have $A_M \neq \emptyset$ and $B_M \neq \emptyset$, and we can find an infinitely differentiable function f_M such that $0 \leq f_M \leq 1$, $f_M = 1$ in a neighborhood of A_M, and $f_M = 0$ in a neighborhood of B_M.

The sequence $(f_M \psi_M)(F)$ converges a.s. and in L^p to $\mathbf{1}_{\{F \in A\}}$ as M tends to infinity. On the other hand, we have

$$
\begin{aligned}
D[(f_M \psi_M)(F)] &= \sum_{i=1}^{m} (\psi_M \partial_i f_M)(F) DF^i + \sum_{i=1}^{m} (f_M \partial_i \psi_M)(F) DF^i \\
&= \sum_{i=1}^{m} (f_M \partial_i \psi_M)(F) DF^i.
\end{aligned}
$$

Hence,

$$
\sup_M \|D[(f_M \psi_M)(F)]\|_H \leq \sum_{i=1}^{m} \sup_M \|\partial_i \psi_M\|_\infty \|DF^i\|_H \in L^p.
$$

By Lemma 2.4.2 we obtain that $1_{\{F\in A\}}$ belongs to $\mathbb{D}^{1,p}$ and, due to Lemma 1.4.2, this is contradictory because $0 < P(F \in A) < 1$. $\qquad\square$

As a consequence, the support of the law of a random variable $F \in \mathbb{D}^{1,p}$, $p > 1$ is a closed interval. The next results provides sufficient conditions for the density of F to be nonzero in the interior of the support.

Proposition 4.1.2 *Let $F \in \mathbb{D}^{1,p}$, $p > 2$, and suppose that F possesses a locally Lipschitz density $p(x)$. Let a be a point in the interior of the support of the law of F. Then $p(a) > 0$.*

Proof: Suppose $p(a) = 0$. Set $r = \frac{2p}{p+2} > 1$. By Lemma 1.4.2 we know that $1_{\{F>a\}} \notin \mathbb{D}^{1,r}$ because $0 < P(F > a) < 1$. Fix $\epsilon > 0$ and set

$$\varphi_\epsilon(x) = \int_{-\infty}^{x} \frac{1}{2\epsilon} 1_{[a-\epsilon,a+\epsilon]}(y)dy.$$

Then $\varphi_\epsilon(F)$ converges to $1_{\{F>a\}}$ in $L^r(\Omega)$, as $\epsilon \downarrow 0$. Moreover, $\varphi_\epsilon(F) \in \mathbb{D}^{1,r}$ and

$$D(\varphi_\epsilon(F)) = \frac{1}{2\epsilon} 1_{[a-\epsilon,a+\epsilon]}(F)DF.$$

We have

$$E\left(\|D(\varphi_\epsilon(F))\|_H^r\right) \le \left(E(\|DF\|_H^p)\right)^{\frac{2}{p+2}} \left(\frac{1}{(2\epsilon)^2}\int_{a-\epsilon}^{a+\epsilon} p(x)dx\right)^{\frac{p}{p+2}}.$$

The local Lipschitz property of p implies that $p(x) \le K|x - a|$, and we obtain

$$E\left(\|D(\varphi_\epsilon(F))\|_H^r\right) \le \left(E(\|DF\|_H^p)\right)^{\frac{2}{p+2}} 2^{-r} K^{\frac{p}{p+2}}.$$

By Lemma 2.4.2 this implies $1_{\{F>a\}} \in \mathbb{D}^{1,r}$ and we are in contradiction. $\qquad\square$

The following example shows that, unlike the one-dimensional case, in dimension $m > 1$ the density of a nondegenerate random vector can vanish in the interior of the support.

Example: Let h_1 and h_2 be two orthonormal elements of H, and define

$$\begin{aligned} X_1 &= \arctan W(h_1), \\ X_2 &= \arctan W(h_2). \end{aligned}$$

Then $X_1, X_2 \in \mathbb{D}^\infty$ and

$$DX_i = (1 + W(h_i)^2)^{-1} h_i,$$

for $i = 1, 2$, and

$$\det \gamma_X = (1 + W(h_1)^2)^{-2}(1 + W(h_2)^2)^{-2},$$

where γ_X denotes the Malliavin matrix of the nondegenerate random vector $X = (X_1, X_2)$. Notice that the support of the law of X is the rectangle $[-\frac{\pi}{2}, \frac{\pi}{2}]^2$, and the density of X is strictly positive in the interior of the support. Now consider the following vector

$$
\begin{aligned}
Y_1 &= (X_1 + \frac{3\pi}{2}) \cos(2X_2 + \pi) \\
Y_2 &= (X_1 + \frac{3\pi}{2}) \sin(2X_2 + \pi).
\end{aligned}
$$

We have that $Y \in (\mathbb{D}^\infty)^2$ and

$$
\det \gamma_Y = 4(X_1 + \frac{3\pi}{2})^2 (1 + W(h_1)^2)^{-2}(1 + W(h_2)^2)^{-2}.
$$

This implies that Y is a nondegenerate vector. Its support is the set $\{(x,y) : \pi^2 \leq x^2 + y^2 \leq 4\pi^2\}$, and the density of Y vanishes on the points (x,y) in the support such that $\pi < y < 2\pi$, and $x = 0$.

For a smooth and nondegenerate random vector when the density vanishes, then all its partial derivatives also vanish (see Ben Arous and Léandre [9]).

Proposition 4.1.3 *Let $F = (F^1, \ldots, F^m) \in (\mathbb{D}^\infty)^m$ be a nondegenerate random vector in the sense of Definition 3.2.1, and denote its density by $p(x)$. Then $p(x) = 0$ implies $\partial_\alpha p(x) = 0$ for any multiindex α.*

Proof: Suppose that $p(x) = 0$. We know that $p(x) = \langle \delta_x(F), 1 \rangle$. We have that $\langle \delta_x(F), R \rangle \geq 0$ for any smooth random variable R because

$$
\langle \delta_x(F), R \rangle = \lim_{\epsilon \downarrow 0} E(\psi_\epsilon(x - F)R),
$$

where ψ_ϵ is an approximation of the identity. So, from Lemma 2.4.3 there exists a measure η_x on the σ-field $\mathcal{G}$ generated by $\{W(e_i), i \geq 1\}$ such that $\langle \delta_x(F), R \rangle = \int_\Omega R \, d\eta_x$ for any $\mathcal{G}$-measurable and smooth random variable R. Notice that $p(x) = \eta_x(\Omega)$. Therefore, $\eta_x = 0$, which implies that $\delta_x(F) = 0$ as an element of $\mathbb{D}^{-\infty}$. For any multiindex α we have

$$
\partial_\alpha p(x) = \partial_\alpha \langle \delta_x(F), 1 \rangle = \langle (\partial_\alpha \delta_x)(F), 1 \rangle.
$$

Hence, it suffices to show that $(\partial_\alpha \delta_x)(F)$ vanishes. Suppose first that $\alpha = (i)$. We can write

$$
D(\delta_x(F)) = \sum_{i=1}^m (\partial_i \delta_x)(F) DF^i,
$$

as elements of $\mathbb{D}^{-\infty}$, which implies

$$
(\partial_i \delta_x)(F) = \sum_{j=1}^m \langle D[\delta_x(F)], DF^j \rangle_H (\gamma_F^{-1})^{ji} = 0,
$$

because $D[\delta_x(F)] = 0$. The general case follows by recurrence. $\square$

In the case of a diffusion processes, the characterization of the support of the law is obtained by means of the notion of skeleton. A general notion of skeleton is provided by the following definition.

In the sequel we will assume that $\{\Pi^N, N \geq 1\}$ is a sequence of orthogonal projections on H of finite-dimensional range, which converges strongly to the identity. If $\{e_1, \ldots, e_N\}$ is an orthonormal basis of the image on Π^N we set $\Pi^N W = \sum_{i=1}^N W(e_i)e_i$.

Definition 4.1.1 *Let $F : \Omega \to S$ be a random variable taking values on a Polish space space (S, d). We will say that a measurable function $\Phi : H \to S$ is a skeleton of F if the following two conditions are satisfied:*

(i) For all $\epsilon > 0$ we have

$$\lim_N P\left\{d(\Phi(\Pi^N W), F) > \epsilon\right\} = 0.$$

(ii) For all $h \in H$, there exists a sequence of measurable transformations $T_N^h : \Omega \to \Omega$ such that $P \circ (T_N^h)^{-1}$ is absolutely continuous with respect to P, and for every $\epsilon > 0$

$$\limsup_N P\left\{d(F \circ T_N^h, \Phi(h)) < \epsilon\right\} > 0.$$

Proposition 4.1.4 *Suppose that Φ is a skeleton of F in the sense of Definition 4.1.1. Then the support of the law of F is the closure in S of the set $\{\Phi(h), h \in H\}$.*

Proof:

Step 1: Let us first show that condition (i) implies the inclusion

$$\operatorname{supp}\left(P \circ F^{-1}\right) \subset \overline{\Phi(H)}.$$

It suffices to show that $P(F \in \overline{\Phi(H)}) = 1$, and this follows from

$$P\left\{d(F, \overline{\Phi(H)}) \le \epsilon\right\} \ge P\left\{d(\Phi(\Pi^N W), F) \le \epsilon\right\} \to 1,$$

as N tends to infinity. Letting ϵ tend to zero yields the result.

Step 2: We have to check that for each $h \in H$ and each $\epsilon > 0$,

$$P\{d(F, \Phi(h)) < \epsilon\} > 0.$$

Since $P \circ (T_N^h)^{-1}$ is absolutely continuous with respect to P it suffices to show that

$$P\left\{d(F \circ T_N^h, \Phi(h)) < \epsilon\right\} > 0$$

for some $N \ge 1$, and this follows from condition (ii). $\quad\square$

Remarks: Both parts of the proof are independent, in the sense that Step 1 uses only assumption (i) and Step 2 uses (ii), and we could have taken different skeletons Φ_1 and Φ_2 in both assumptions. Also we can replace H by a dense subspace H_0 such that $\Pi^N W$ takes values in H_0, and $\Pi^N W$ by a sequence of random variables ξ_N taking values on H_0.

4.2   Strict positivity of the density and skeleton

Under stronger hypotheses, and assuming that F is a finite-dimensional random variable, one can characterize the set of points where the density is strictly positive in terms of the skeleton. The following propositions are devoted to this problem. More precisely we want to establish the equivalence between the following conditions:

(a) The density p of a random vector F satisfies $p(y) > 0$ at some point $y \in \mathbb{R}^m$.

(b) Fix $y \in \mathbb{R}^m$. There exists an element $h \in H$ such that $\Phi(h) = y$ and $\det \gamma_\Phi(h) > 0$, where $\Phi : H \to \mathbb{R}^m$ is a differentiable mapping and $\gamma_\Phi^{ij} = \langle D\Phi^i, D\Phi^j \rangle_H$.

Proposition 4.2.1 *Let $F = (F^1, \ldots, F^m)$ be a nondegenerate vector in $(\mathbb{D}^\infty)^m$. Suppose that $\Phi : H \to \mathbb{R}^m$ is an infinitely differentiable function such that Φ and all its derivatives have polynomial growth (i.e., $\|D^{(k)}\Phi(h)\| \le C(1 + \|h\|_H^{\nu(k)})$, for all $h \in H$, $k \ge 0$). We will also assume that the following condition holds:*

(H1) $\lim_{N \to \infty} \Phi(\Pi^N W) = F$, *in the norm* $\| \cdot \|_{k,p}$ *for all* $k \ge 0, p > 1$.

Then for each $y \in \mathbb{R}^m$ (a) implies (b).

Proof: Fix $y \in \mathbb{R}^m$. We assume that $p(y) = E[\delta_y(F)] > 0$. For every $M \ge 1$ we consider a function $\alpha_M \in C^\infty(\mathbb{R})$ such that $0 \le \alpha_M \le 1$, $\alpha_M(x) = 0$ if $|x| \le \frac{1}{M}$, and $\alpha_M(x) = 1$ if $|x| \ge \frac{2}{M}$. The fact that F is nondegenerate implies that

$$\lim_{M \to \infty} \alpha_M(\det \gamma_F) = 1,$$

in the norm $\| \cdot \|_{k,p}$ for all k, p. Consequently,

$$0 < E[\delta_y(F)] = \lim_{M \to \infty} E[\delta_y(F) \alpha_M(\det \gamma_F)],$$

and we can find a positive integer M such that

$$E[\delta_y(F) \alpha_M(\det \gamma_F)] > 0. \tag{4.1}$$

Our assumptions imply that $\Phi(\Pi^N W) \in \mathbb{D}^\infty$ for every N, and for every $k \ge 1$ we have

$$D^k(\Phi(\Pi^N W)) = (\Pi^N)^{\otimes k}(D^k \Phi)(\Pi^N W).$$

We have

$$E[\delta_y(F) \alpha_M(\det \gamma_F)] = \lim_{N \to \infty} E[\delta_y(\Phi(\Pi^N W)) \alpha_M \left(\det \gamma_{\Phi(\Pi^N W)} \right)] \tag{4.2}$$

in all the norms $\| \cdot \|_{k,p}$. This convergence follows from hypothesis (H1) and the following integration by parts formulas (see (3.14)):

$$E[\delta_y(F) \alpha_M(\det \gamma_F)] = E[\mathbf{1}_{\{y < F\}} H_{(1,2,\ldots,m)}(F, \alpha_M(\det \gamma_F))]$$
$$E[\delta_y(\Phi(\Pi^N W)) \alpha_M \left(\det \gamma_{\Phi(\Pi^N W)} \right)] = E[\mathbf{1}_{\{y < \Phi(\Pi^N W)\}}$$
$$\times H_{(1,2,\ldots,m)}(\Phi(\Pi^N W), \alpha_M \left(\det \gamma_{\Phi(\Pi^N W)} \right))].$$

Therefore, from (4.1) and (4.2) we can find a positive integer N such that

$$E\left[\delta_y(\Phi(\Pi^N W)) \alpha_M \left(\det \gamma_{\Phi(\Pi^N W)} \right) \right] > 0.$$

Now consider a function $\beta_K \in C^\infty(\mathbb{R})$ such that $0 \le \beta_K \le 1$, $\beta_K(x) = 0$ if $|x| \ge K$, and $\beta_K(x) = 1$ if $|x| \le K - 1$. Then $\beta_K(\|\Pi^N W\|_H^2)$ converges to 1 in $\mathbb{D}^\infty$, and we can find an integer $K \ge 1$ such that

$$E\left[\delta_y(\Phi(\Pi^N W)) \alpha_M \left(\det \gamma_{\Phi(\Pi^N W)} \right) \beta_K(\|\Pi^N W\|_H^2) \right] > 0.$$

This implies that

$$P\left\{|\Phi(\Pi^N W)) - y| < \epsilon, \det \gamma_{\Phi(\Pi^N W)} \geq \frac{1}{M}, \|\Pi^N W\|_H^2 \leq K\right\} > 0, \qquad (4.3)$$

for every $\epsilon > 0$.

Notice that

$$\det \gamma_{\Phi(\Pi^N W)} \leq \det \gamma_\Phi(\Pi^N W),$$

because for every $t \in \mathbb{R}^m$ we have

$$\sum_{i,j=1}^m t^i \langle D(\Phi^i(\Pi^N W)), D(\Phi^j(\Pi^N W)) \rangle_H t^j$$

$$= \sum_{i,j=1}^m t^i \langle \Pi^N[(D\Phi^i)(\Pi^N W)], \Pi^N[D(\Phi^j)(\Pi^N W)] \rangle_H t^j$$

$$= \|\Pi^N(\sum_{i=1}^m t^i (D\Phi^i)(\Pi^N W))\|_H^2 \leq \|\sum_{i=1}^m t^i (D\Phi^i)(\Pi^N W)\|_H^2$$

$$= \sum_{i,j=1}^m t^i \langle (D\Phi^i)(\Pi^N W), (D(\Phi^j)(\Pi^N W) \rangle_H t^j.$$

Hence, (4.3) implies

$$P\left\{|\Phi(\Pi^N W) - y| < \epsilon, \det \gamma_\Phi(\Pi^N W) \geq \frac{1}{M}, \|\Pi^N W\|_H^2 \leq K\right\} > 0. \qquad (4.4)$$

Finally, from (4.4) we can find a sequence of elements $h_k \in \text{Im}(\Pi^N)$ such that $|\Phi(h_k) - y| < \frac{1}{k}$ for every k, $\|h_k\|_H^2 \leq K$, and $\det \gamma_\Phi(h_k) \geq \frac{1}{M}$. Bounded and closed sets in the image of Π^N are compact, so we can select subsequence converging to some element $h \in H$ which verifies the desired properties. $\qquad \square$

In order to formulate and prove the converse implication (that is, that (b) implies (a)) we need some preliminaries. We will denote by $B_a(x)$ the ball of $\mathbb{R}^m$ with center x and radius $a > 0$. The following lemma is a somewhat quantitative version of the classical inverse function theorem. For a function $g : B_1(0) \to \mathbb{R}^m$ which is twice continuous differentiable we will denote by $\|g\|_{C^2}$ the norm

$$\|g\|_{C^2} = \sup_{z \in B_1(0)} \{|g(z)| + |g'(z)| + |g''(z)|\}. \qquad (4.5)$$

Lemma 4.2.1 *For each $\beta > 1$ there exist constants $c_\beta \in (0, \frac{1}{\beta})$, $\delta_\beta > 0$ such that any mapping $g : B_1(0) \to \mathbb{R}^m$ verifying $g(0) = 0$,*

$$\|g\|_{C^2} \leq \beta, \quad \text{and} \quad |\det g'(0)| \geq \frac{1}{\beta}$$

is diffeomorphic from $B_{c_\beta}(0)$ into a neighborhood of $B_{\delta_\beta}(0)$.

Consider a random vector $F \in (\mathbb{D}^\infty)^m$. Given m elements $h_1, \ldots, h_m \in H$ and $z \in \mathbb{R}^m$ we define the shifted Gaussian process

$$(T_z W)(h) = W(h) + \sum_{j=1}^m z_j \langle h, h_j \rangle_H, \quad h \in H. \qquad (4.6)$$

By an elementary change of probability argument we know that for any (integrable or nonnegative) random variable F we have

$$E(F) = E(F(T_zW)J_z), \qquad (4.7)$$

where

$$J_z = \exp\left(-\sum_{j=1}^m z_j W(h_j) - \frac{1}{2}\left\|\sum_{j=1}^m z_j h_j\right\|_H^2\right).$$

We set $\underline{h} = (h_1, \ldots, h_m)$, and note that T_zW and J_z depend on $\underline{h}$. We also set, given $p > m$ and $k \geq 0$

$$\mathcal{R}_{\underline{h},k,p}F = \int_{\{|z|\leq 1\}} \left\|(D^kF)(T_zW)\right\|_{H^{\otimes k}}^p dz.$$

Proposition 4.2.2 *Consider a nondegenerate random vector $F \in (\mathbb{D}^\infty)^m$. Let $\Phi : H \to \mathbb{R}^m$ be a C^1 mapping. Suppose that the following condition holds:*

(H2) For all $h \in H$, there exists a sequence of measurable transformations $T_N^h : \Omega \to \Omega$ such that $P \circ (T_N^h)^{-1}$ is absolutely continuous with respect to P, and for every $\epsilon > 0$,

$$\lim_N P\left\{|F \circ T_N^h - \Phi(h)| > \epsilon\right\} = 0$$

$$\lim_N P\left\{\|(DF) \circ T_N^h - (D\Phi)(h)\|_{H^m} > \epsilon\right\} = 0$$

$$\lim_{M\to\infty} \sup_N P\left\{(\mathcal{R}_{D\Phi(h),k,p}F) \circ T_N^h > M\right\} = 0$$

for some $p > m$, and for $k = 0,1,2,3$.

Then (b) implies (a).

Proof: We fix a point $x \in \mathbb{R}^m$ and an element $h \in H$ such that $\Phi(h) = x$ and $\det \gamma_\Phi(h) > 0$. Consider the elements of H given by $h_j = (D\Phi^j)(h)$, $j = 1, \ldots, m$. Using these elements we can introduce the random mapping $g : \mathbb{R}^m \to \mathbb{R}^m$ defined by $g(z) = F(T_zW) - F$, where T_zW is defined in (4.6). Notice that the random function g has an infinitely differentiable version and for each multi-index $\alpha = (\alpha_1, \ldots \alpha_k) \in \{1, 2, \ldots, m\}^k$ its derivatives are given by

$$(\partial_\alpha g)(z) = \left\langle(D^kF)(T_zW), h_{\alpha_1} \otimes \cdots \otimes h_{\alpha_k}\right\rangle_{H^{\otimes k}}. \qquad (4.8)$$

By Sobolev's inequality, if $p > m$ is the exponent appearing in hypothesis (H2), we have

$$\sup_{|z|\leq 1} |g(z)| \leq c_p \left(\int_{\{|z|\leq 1\}} |g(z)|^p dz + \int_{\{|z|\leq 1\}} \sum_{j=1}^m |\partial_j g(z)|^p dz\right)^{\frac{1}{p}}. \qquad (4.9)$$

Consequently, using (4.8) and (4.9) we can estimate the norm (4.5) in the following way

$$\|g\|_{C^2} \leq c_p \left(\sum_{k=0}^3 \int_{\{|z|\leq 1\}} \sum_{\alpha:|\alpha|=k} |\partial_\alpha g(z)|^p dz\right)^{\frac{1}{p}} \leq G,$$

where G is the random variable defined by

$$G = c_p \left(|F|^p + \sum_{k=0}^{3} (\sum_{j=1}^{m} \|h_j\|_H^p)^k \int_{\{|z| \leq 1\}} \|(D^k F)(T_z W)\|_{H^{\otimes k}}^p dz \right)^{\frac{1}{p}}.$$

For any $\beta > 0$ we will denote by α_β a continuous function such that $0 \leq \alpha_\beta \leq 1$, $\alpha_\beta(x) = 0$ if $|x| \leq \frac{1}{\beta}$, and $\alpha_\beta(x) = 1$ if $|x| \geq \frac{2}{\beta}$. Also k_β will be a continuous function such that $0 \leq k_\beta \leq 1$, $k_\beta(x) = 0$ if $|x| \geq \beta$, and $k_\beta(x) = 1$ if $|x| \leq \beta - 1$. Set

$$H_\beta = k_\beta(G) \alpha_\beta(|\det \langle DF^i, D\Phi^j \rangle_H|).$$

Suppose that $\rho : \mathbb{R}^m \to \mathbb{R}$ is a strictly positive continuous function such that $\int_{\mathbb{R}^m} \rho(z) dz = 1$. Let $f : \mathbb{R}^m \to \mathbb{R}$ be a nonnegative continuous and bounded function. We can write, applying (4.7)

$$\begin{aligned}
E[f(F)] &= \int_{\mathbb{R}^m} E[f(F)] \rho(z) dz \\
&= \int_{\mathbb{R}^m} E\left[f(F(T_z W)) J_z \right] \rho(z) dz \\
&\geq E \left[H_\beta \int_{\mathbb{R}^m} f(F(T_z W)) J_z \rho(z) dz \right]
\end{aligned}$$

Fix $\beta > 1$ and let c_β and δ_β the constants provided by Lemma 4.2.1. We can apply this lemma to the function $g(z) = F(T_z W) - F$ (notice that this function verifies the hypotheses of the lemma if $H_\beta \neq 0$, because $\|g\|_{C^2} \leq G$, and $\det g'(0) = \det \langle DF^i, D\Phi^j \rangle_H$). Hence, making the change of variable $y = g(z)$ we obtain

$$\begin{aligned}
&E \left[H_\beta \int_{\mathbb{R}^m} f(F(T_z W)) J_z \rho(z) dz \right] \\
&\geq E \left[H_\beta \int_{\{|z| \leq c_\beta\}} f(g(z) + F) J_z \rho(z) dz \right] \\
&\geq E \left[H_\beta \int_{\{|y| \leq \delta_\beta\}} f(y + F) J_{g^{-1}(y)} \rho(g^{-1}(y)) |\det \partial_j g^i(g^{-1}(y))| dy \right].
\end{aligned}$$

From the above inequalities we deduce

$$p(x) \geq E \left[H_\beta \mathbf{1}_{\{|F-x| \leq \delta_\beta\}} J_{g^{-1}(x-F)} \rho(g^{-1}(x-F)) |\det \partial_j g^i(g^{-1}(x-F))| \right]$$

Notice that if $H_\beta \neq 0$ and $|x - F| \leq \delta_\beta$ then

$$J_{g^{-1}(x-F)} \rho(g^{-1}(x-F)) |\det \partial_j g^i(g^{-1}(x-F))| > 0.$$

Hence, in order to deduce $p(x) > 0$ it only remains to show that we can choose β in such a way that

$$P \left\{ |F - x| \leq \delta_\beta, G \leq \beta, |\det \langle DF^i, D\Phi^j \rangle_H| \geq \frac{1}{\beta} \right\} > 0.$$

By the absolute continuity of $P \circ (T_N^h)^{-1}$ with respect to P it suffices to show that

$$P \left\{ |F \circ T_N^h - x| \leq \delta_\beta, G \circ T_N^h \leq \beta, |\det \langle (DF^i) \circ T_N^h, D\Phi^j \rangle_H| \geq \frac{1}{\beta} \right\} > 0$$

for some $N \geq 1$. Finally we can write

$$P\left\{|F \circ T_N^h - x| \leq \delta_\beta, G \circ T_N^h \leq \beta, |\det\langle(DF^i) \circ T_N^h, D\Phi^j\rangle_H| \geq \frac{1}{\beta}\right\}$$
$$\geq 1 - P\left\{|F \circ T_N^h - x| > \delta_\beta\right\} - P\left\{G \circ T_N^h > \beta\right\}$$
$$-P\left\{|\det\langle(DF^i) \circ T_N^h, D\Phi^j\rangle_H| \geq \frac{1}{\beta}\right\},$$

and letting N tend to infinity and using hypothesis (H2) we complete the proof. $\square$

4.3 Skeleton and support for diffusion processes

Let us now describe the application of the above results to the case of a diffusion process. In the sequel we will assume that the underlying Gaussian process is a d-dimensional Brownian motion $\{W(t), t \in [0,1]\}$, defined in the canonical probability space $\Omega = C_0([0,1], \mathbb{R}^d)$. Moreover, for each positive integer N, and for any element $h \in H = L^2([0,1]; \mathbb{R}^d)$ we set

$$\Pi^N(h) = \sum_{i=1}^{2^N-1} 2^N \left(\int_{[(i-1)2^{-N}, i2^{-N}]} h(s)ds\right) \mathbf{1}_{(i2^{-N}, (i+1)2^{-N})}.$$

With this definition $(\Pi^N W)(\omega) = \dot{\omega}^N$, where for any continuous function $\omega \in \Omega$ we denote by ω^N the element of Ω such that on each interval $(i2^{-N}, (i+1)2^{-N})$, $1 \leq i \leq 2^N - 1$, has a constant derivative equal to $(\omega(i2^{-N}) - \omega((i-1))2^N$.

We will denote by i the isometry between H and the Cameron-Martin space H^1. That is, for each $h \in H$ we have $i(h)(t) = \int_0^t h(s)ds$.

Let $A_j, B : \mathbb{R}^m \to \mathbb{R}^m$, $j = 1, \ldots, d$ a system of functions such that A_j is of class C_b^2 (i.e., the partial derivatives of first and second order are continuous and bounded), and B is Lipshcitz. Consider the following stochastic differential equation on $\mathbb{R}^m$:

$$X_t = x_0 + \int_0^t B(X_s)ds + \sum_{j=1}^d \int_0^t A_j(X_s)dW_s^j. \tag{4.10}$$

Let $\Phi(h)$ be the solution of

$$\Phi(h)_t = x_0 + \int_0^t A_0(\Phi(h)_s)ds + \sum_{j=1}^d \int_0^t A_j(\Phi(h)_s)h_s^j ds, \tag{4.11}$$

where

$$A_0^i = B_0^i - \frac{1}{2}\sum_{k=1}^m \sum_{j=1}^d \partial_k A_j^i A_j^k.$$

Notice that A_0 drift of the diffusion process X_t if we write the stochastic differential equation (4.10) in the Stratonovich form.

Then one can show that the mapping $\Phi : H \to C([0,1]; \mathbb{R}^m)$ is a skeleton of the process X in the sense of Definition 4.1.1. That is, the following convergences hold ([71]):

$$\lim_N P\left(\sup_{t \in [0,1]} |\Phi(\Pi^N W)_t - X_t| > \epsilon\right) = 0,$$

and

$$\lim_N P\left(\sup_{t\in[0,1]}|X_t\circ T_N^h-\Phi(h)_t|>\epsilon\right)=0,$$

where $T_N^h(\omega)=\omega-\omega^N+i(h)$. Actually, the above convergences hold if we replace $C([0,1];\mathbb{R}^m)$ by the space $C^\alpha([0,1];\mathbb{R}^m)$ of α-Hölder continuous functions equipped with the α-Hölder norm

$$\|x\|_\alpha=\sup_{t\in[0,1]}|x_t|+\sup_{s\neq t}\frac{|x_t-x_s|}{|t-s|^\alpha},$$

for $\alpha\in[0,\frac{1}{2})$.

If the coefficients A_j, $j=0,\dots,m$ are of class C^∞ with bounded partial derivatives of all orders, then, for any fixed $t\in[0,1]$ the convergences of $\Phi(\Pi^N W)_t$ to X_t and $X_t\omega-\omega^N+i(h)$ to $\Phi(h)_t$ holds in the topology of $\mathbb{D}^\infty$. Hence, if the nondegeneracy condition (H) holds then we can apply Propositions 4.2.1 and 4.2.2 to the random vector X_t for all $t\in(0,1]$ and we deduce the following characterization of the points where the density of X_t is strictly positive (see [11]):

Proposition 4.3.1 *The density p_t of the diffusion process X_t satisfies $p_t(y)>0$ at some point $y\in\mathbb{R}^m$ if and only if there exists an element $h\in H$ such that $\Phi(h)_t=y$ and $\det\langle D\Phi^i(h)_t,D\Phi^j(h)_t\rangle_H>0$.*

Example: Consider the following example (cf. [1]). Suppose $\alpha(x)$, $\beta(t,x)$, $\gamma(x,y)$ are smooth functions with bounded derivatives of all orders such that

$$\alpha(x)=x^2\quad\text{if}\quad x\in[-1,1],$$
$$\beta(t,x)\neq0\quad\text{iff}\quad(t,x)\in[0,\frac{1}{2})\times(-1,1),$$
$$\gamma(x,y)\neq0\quad\text{iff}\quad(x,y)\notin[-1,0]\times[-1,1].$$

Consider the process $\{X_t,t\in[0,1]\}$ solution to the equation

$$\begin{aligned}dX_t^1&=\gamma(X_t^1,X_t^2)\circ dW_t^2+\alpha(X_t^2)dt\\dX_t^2&=\beta(t,X_t^2)\circ dW_t^1+\gamma(X_t^1,X_t^2)\circ dW_t^3\\X_0&=0.\end{aligned}$$

Hörmander's condition is satisfied in this example, and, applying the techniques of the Malliavin calculus for the case of time dependent coefficients (cf. [31]), one can show that the vector (X_1^1,X_1^2) is nondegenerate. We have $\{\Phi_1(h),h\in H\}=\mathbb{R}^2$, and the support of the law of X_1 is $\mathbb{R}^2$. However at the point 0 the density vanish because the unique element $h\in H$ such that $\Phi_1(h)=0$ is $h_0=0$, and $\det\langle D\Phi_1^i(h_0),D\Phi_1^j(h_0)\rangle=0$.

4.4 Varadhan estimates

Fix a small parameter $\epsilon\in[0,1]$, and consider the solution X_t^ϵ to the stochastic differential equation

$$X_t^\epsilon=x+\int_0^t B(X_s^\epsilon)ds+\epsilon\sum_{j=1}^d\int_0^t A_j(X_s^\epsilon)dW_s^j.\qquad(4.12)$$

That is, $X_t^\epsilon = X_t(\epsilon W)$. Suppose that the coefficients A_i and B are infinitely differentiable with bounded derivatives of all orders, and Hörmander's condition (H) holds. We can introduce the following functions on $\mathbb{R}^m$, which depend on the skeleton $\Phi_t(h)$:

$$d^2(y) = \inf_{\Phi_1(h)=y} \|h\|_H^2,$$

and

$$d_R^2(y) = \inf_{\Phi_1(h)=y, \det \gamma_{\Phi_1}(h)>0} \|h\|_H^2.$$

Then the following theorem holds (cf. Léandre [58], Ben Arous and Léandre [11] and Léandre and Russo [59]):

Theorem 4.4.1 *Let us denote by $p_\epsilon(y)$ the density of X_1^ϵ. Then*

$$\liminf_{\epsilon \downarrow 0} 2\epsilon^2 \log p_\epsilon(y) \geq -d_R^2(y), \tag{4.13}$$

and

$$\limsup_{\epsilon \downarrow 0} 2\epsilon^2 \log p_\epsilon(y) \leq -d^2(y). \tag{4.14}$$

Moreover, if $\inf_{\Phi_1(h)=y, \det \gamma_{\Phi_1}(h)>0} \det \gamma_{\Phi_1}(h) > 0$ then

$$\lim_{\epsilon \downarrow 0} 2\epsilon^2 \log p_\epsilon(y) = -d_R^2(y). \tag{4.15}$$

We will sketch the proof of this theorem which provides the asymptotic behaviour for the density of a perturbed dynamical system when the noise is small. In order to show the limit results stated in the theorem we will consider a more general case which contains the diffusion case as a particular example.

In the sequel we will work in the framework of an arbitrary Gaussian family $\{W(h), h \in H\}$. Let us first proof the minoration inequality.

Proposition 4.4.1 *Consider a family $\{F_\epsilon, 0 < \epsilon \leq 1\}$ of nondegenerate random vectors, and a function $\Phi \in C_p^\infty(H; \mathbb{R}^m)$ such that:*

$$\lim_{\epsilon \downarrow 0} \frac{1}{\epsilon}\left(F_\epsilon(W + \frac{h}{\epsilon}) - \Phi(h)\right) = Z(h),$$

in the topology of $\mathbb{D}^\infty$, for each $h \in H$, where $Z(h)$ is an m-dimensional random vector in the first Wiener chaos with variance $\gamma_\Phi(h)$. Define

$$d_R^2(y) = \inf_{\Phi(h)=y, \det \gamma_\Phi(h)>0} \|h\|_H^2, \quad y \in \mathbb{R}^m.$$

Then

$$\liminf_{\epsilon \downarrow 0} 2\epsilon^2 \log p_\epsilon(y) \geq -d_R^2(y). \tag{4.16}$$

Notice that the hypothesis of Proposition 4.4.1 implies

$$\lim_{\epsilon \downarrow 0} F_\epsilon(W + \frac{h}{\epsilon}) = \Phi(h),$$

in the topology of $\mathbb{D}^\infty$, for each $h \in H$.

Proof: Fix $y \in \mathbb{R}^m$, and $\eta > 0$. If $d_R^2(y) = \infty$ there is nothing to say. Suppose $d_R^2(y) < \infty$. Let $h \in H$ be such that $\Phi(h) = y$, $\det \gamma_\Phi(h) > 0$, and $\|h\|_H^2 \leq d_R^2(y) + \eta$. Let $f \in C_0^\infty(\mathbb{R}^m)$. By Girsanov's theorem

$$E(f(F_\epsilon)) = e^{-\frac{\|h\|_H^2}{2\epsilon^2}} E\left(f(F_\epsilon(W + \frac{h}{\epsilon}))e^{-\frac{W(h)}{\epsilon}} \right).$$

Consider a function $\chi \in C^\infty(\mathbb{R})$, $0 \leq \chi \leq 1$, such that $\chi(t) = 0$ if $t \notin [-2\eta, 2\eta]$, and $\chi(t) = 1$ if $t \in [-\eta, \eta]$. Then, if $f \geq 0$, we have

$$E(f(F_\epsilon)) \geq e^{-\frac{\|h\|_H^2 + 4\eta}{2\epsilon^2}} E\left(\chi(\epsilon W(h))f(F_\epsilon(W + \frac{h}{\epsilon})) \right).$$

This implies that

$$2\epsilon^2 \log p_\epsilon(y) \geq -(\|h\|_H^2 + 4\eta) + 2\epsilon^2 \log E\left(\chi(\epsilon W(h))\delta_y(F_\epsilon(W + \frac{h}{\epsilon})) \right).$$

Hence, it suffices to show that

$$\lim_{\epsilon \downarrow 0} \epsilon^2 \log E\left(\chi(\epsilon W(h))\delta_y(F_\epsilon(W + \frac{h}{\epsilon})) \right) = 0.$$

We have

$$E\left(\chi(\epsilon W(h))\delta_y(F_\epsilon(W + \frac{h}{\epsilon})) \right) = \epsilon^{-m} E\left(\chi(\epsilon W(h))\delta_0(\frac{F_\epsilon(W + \frac{h}{\epsilon}) - \Phi(h)}{\epsilon}) \right).$$

Then, using hypothesis (ii) the expectation

$$E\left(\chi(\epsilon W(h))\delta_0(\frac{F_\epsilon(W + \frac{h}{\epsilon}) - \Phi(h)}{\epsilon}) \right)$$

converges, as ϵ tends to zero, to $E(\delta_0(Z(h)))$, and this completes the proof of the proposition. $\quad\square$

The majoration estimate requires large deviation assumptions:

Proposition 4.4.2 *Consider a family $\{F_\epsilon, 0 < \epsilon \leq 1\}$ of nondegenerate random vectors, and a function $\Phi \in C_p^\infty(H; \mathbb{R}^m)$ such that:*

(i') $\sup_{\epsilon \in (0,1]} \|F_\epsilon\|_{k,p} < \infty$, *for each $k \geq 1$ and $p > 1$.*

(ii') $\|(\gamma_{F_\epsilon})^{-1}\|_k \leq \epsilon^{-N(k)}$ *for any integer $k \geq 1$.*

(iii) The family $\{F_\epsilon, 0 < \epsilon \leq 1\}$ satisfies a large deviation principle with rate function

$$d^2(y) = \inf_{\Phi(h)=y} \|h\|_H^2, \quad y \in \mathbb{R}^m.$$

Then

$$\limsup_{\epsilon \downarrow 0} 2\epsilon^2 \log p_\epsilon(y) \leq -d^2(y). \tag{4.17}$$

Proof: Fix a point $y \in \mathbb{R}^m$ and consider a function $\chi \in C_0^\infty(\mathbb{R}^m)$, $0 \le \chi \le 1$ such that χ is equal to one in a neighborhood of y. The density of F_ϵ at point y is given by

$$p_\epsilon(y) = E(\chi(F_\epsilon)\delta_y(F_\epsilon)).$$

Using Proposition 3.2.1 we can write

$$\begin{aligned}
E(\chi(F_\epsilon)\delta_y(F_\epsilon)) &= E\left(1_{\{F_\epsilon > y\}} H_{(1,2,\dots,m)}(F_\epsilon, \chi(F_\epsilon))\right)\\
&\le E(|H_{(1,2,\dots,m)}(F_\epsilon, \chi(F_\epsilon))|) = E(|H_{(1,2,\dots,m)}(F_\epsilon, \chi(F_\epsilon))|1_{\{F_\epsilon \in \mathrm{supp}\chi\}})\\
&\le (P(F_\epsilon \in \mathrm{supp}\chi))^{\frac{1}{q}}\|H_{(1,2,\dots,m)}(F_\epsilon, \chi(F_\epsilon))\|_p,
\end{aligned}$$

where $\frac{1}{p} + \frac{1}{q} = 1$. By Proposition 3.2.2 we know that

$$\|H_{(1,2,\dots,m)}(F_\epsilon, \chi(F_\epsilon))\|_p \le C(p)\|\gamma_{F_\epsilon}^{-1}\|_k\|F_\epsilon\|_{a,b}\|\chi(F_\epsilon)\|_{d',b'},$$

for some constants b, d, b', d'. Thus, hypothesis (ii') implies that

$$\lim_{\epsilon \downarrow 0} \epsilon^2 \log \|H_{(1,2,\dots,m)}(F_\epsilon, \chi(F_\epsilon))\|_p = 0.$$

Finally, the large deviation principle for F_ϵ ensures that for ϵ small enough we have

$$(P(F_\epsilon \in \mathrm{supp}\chi))^{\frac{1}{q}} \le e^{-\frac{1}{2q\epsilon^2}(\inf_{y \in \mathrm{supp}\chi} d^2(y))}.$$

$\square$

Proposition 4.4.3 *Consider a family $\{F_\epsilon, 0 < \epsilon \le 1\}$ of nondegenerate random vectors, and a function $\Phi \in C_p^\infty(H; \mathbb{R}^m)$ such that:*

(i') $\sup_{\epsilon \in (0,1]} \|F_\epsilon\|_{k,p} < \infty$, *for each $k \ge 1$ and $p > 1$.*

(ii') $\|(\gamma_{F_\epsilon})^{-1}\|_k \le \epsilon^{-N(k)}$ *for any integer $k \ge 1$.*

(iii') *The family $\{(F_\epsilon, \gamma_{F_\epsilon}), 0 < \epsilon \le 1\}$ satisfies a large deviation principle with rate function*
$$\lambda^2(y,a) = \inf_{\Phi(h)=y, \gamma_\Phi(h)=a} \|h\|_H^2, \quad (y,a) \in \mathbb{R}^m \times \mathbb{R}^{m^2}.$$

(iv) $\lim_{N\to\infty} \Phi(\epsilon\Pi^N W) = F_\epsilon$, *in the norm $\|\cdot\|_{k,p}$ for all $k \ge 0$, $p > 1$, where $\{\Pi^N, N \ge 1\}$ is a sequence of orthogonal projections of H of finite-dimensional rangle strongly convergent to the identity.*

Define $d_R^2(y) = \inf_{\Phi(h)=y, \det \gamma_\Phi(h)>0} \|h\|_H^2$, $y \in \mathbb{R}^m$ and suppose that

$$\gamma := \inf_{\Phi(h)=y, \det \gamma_\Phi(h)>0} \det \gamma_\Phi(h) > 0.$$

Then

$$\limsup_{\epsilon \downarrow 0} 2\epsilon^2 \log p_\epsilon(y) \le -d_R^2(y). \tag{4.18}$$

Proof: The proof is similar to that of Proposition 4.4.2. Consider a function $g \in C^\infty(\mathbb{R})$, $0 \le g \le 1$, such that $g(u) = 1$ if $|u| < \frac{1}{4}\gamma$, and $g(u) = 0$ if $|u| > \frac{1}{2}\gamma$. Set $G_\epsilon = g(\det \gamma_{F_\epsilon})$. With the notations of the proof of Proposition 4.4.2 we have

$$E(\chi(F_\epsilon)\delta_y(F_\epsilon)) = E(G_\epsilon\chi(F_\epsilon)\delta_y(F_\epsilon)) + E((1 - G_\epsilon)\chi(F_\epsilon)\delta_y(F_\epsilon)).$$

We have $E(G_\epsilon\chi(F_\epsilon)\delta_y(F_\epsilon)) = 0$, because, otherwise, using condition (i') and applying the method of the proof of Proposition 4.2.1, one can find an element $\epsilon h \in H$ such that $\Phi(\epsilon h) = y$ and $0 < \det \gamma_\Phi(\epsilon h) < \frac{\gamma}{2}$, and this is in contradiction with the definition of γ.

Proceding as in the Proposition 4.4.2 we obtain

$$E((1 - G_\epsilon)\chi(F_\epsilon)\delta_y(F_\epsilon)) = E\left(1_{F_\epsilon > y}H_{(1,2,\ldots,m)}(F_\epsilon, (1 - G_\epsilon)\chi(F_\epsilon))\right)$$
$$\le E(|H_{(1,2,\ldots,m)}(F_\epsilon, (1 - G_\epsilon)\chi(F_\epsilon))|)$$
$$E(|H_{(1,2,\ldots,m)}(F_\epsilon\chi(F_\epsilon)|1_{F_\epsilon \in \mathrm{supp}\chi, \det \gamma_{F_\epsilon} \ge \frac{1}{4}\gamma})$$
$$\le \left(P\left(F_\epsilon \in \mathrm{supp}\chi, \det \gamma_{F_\epsilon} \ge \frac{1}{4}\gamma\right)\right)^{\frac{1}{q}}\|H_{(1,2,\ldots,m)}(F_\epsilon, \chi(F_\epsilon))\|_p.$$

Finally, hypotheses (ii') and (iii') imply that for any $q > 1$ we have

$$\limsup_{\epsilon \downarrow 0} 2\epsilon^2 \log p_\epsilon(y) \le \exp\left(-\frac{1}{2q\epsilon^2}\left(\inf_{\Phi(h) \in \mathrm{supp}\chi, \det \gamma_\Phi(h) \ge \frac{1}{4}\gamma} \|h\|_H^2\right)\right)$$
$$\le \exp\left(-\frac{1}{2q\epsilon^2}\left(\inf_{y \in \mathrm{supp}\chi} d_R^2(y)\right)\right).$$

$\square$

When the random variable F_ϵ is X_1^ϵ, conditions (i),(ii),(iii), (iv), (i'), (ii') and (iii') are satisfied, and in this way one can show Proposition 4.4.1. More precisely, condition (ii') is proved by Kusuoka and Stroock in [57], the large deviation principle for both the diffusion X_t^ϵ and the Malliavin matrix $\gamma_{X_{t\epsilon}}$ are known, and the convergences (i), (ii) and (iv) are easy to check.

Bibliographical notes The ideas used in this chapter in order to characterize the support of the law of a diffusion go back to the works of Mackevičius [61] and Gyöngy [35]. The support theorem for the Hölder norms has been studied in [8], [71] and [38]. The results on the skeleton are based on the main references [1] and [11]. The characterization of the support of the law of hyperbolic stochastic partial differential equations has been established by Millet and Sanz-Solé in [70], and the characterization of the points of positive density for the solution to such equations has been done by these authors in [72]. The support theorem for parabolic stochastic partial differential equations is proved by Bally, Millet, and Sanz-Solé in [6].

Chapter 5

Anticipating stochastic calculus

We have seen in Chapter 1 that the divergence operator δ on the classical Wiener space is an extension of the Itô stochastic integral, in the sense that the class L_a^2 of square integrable and adapted process is included in Dom δ, and the operator δ restricted to L_a^2 coincides with the Itô integral. Actually, the operator δ coincides with an extension of the Itô integral introduced by Skorohod in [98] using the Wiener chaos expansion. One can develop a stochastic calculus for both the Skorohod integral and the generalized Stratonovich integral. In this chapter we will present the basic facts of the stochastic calculus for anticipating integrals.

5.1 Skorohod integral processes

In this section we will assume that $W = \{W(t), t \in [0,1]\}$ is a Brownian motion defined on the canonical probability space $(\Omega, \mathcal{F}, P)$. In order to simplify the presentation we will assume that W is one-dimensional, but most of the results can be easily generalized to the multidimensional case.

We will say that a square integrable process $u = \{u(t), 0 \le t \le 1\}$ is Skorohod integrable if u belongs to Dom δ, and we will write

$$\delta(u) = \int_0^1 u_t \, dW_t.$$

It may happen that $u \in$ Dom δ, but $u\mathbf{1}_{[0,t]}$ is not Skorohod integrable for some $t \in [0,1]$.

Example: Consider the process

$$u_t = \mathbf{1}_{[0,\frac{1}{2}]}(t)\mathbf{1}_{\{W_1>0\}} - \mathbf{1}_{(\frac{1}{2},1]}(t)\mathbf{1}_{\{W_1>0\}}.$$

The process u is Skorohod integrable and

$$\delta(u) = (2W_{\frac{1}{2}} - W_1)\mathbf{1}_{\{W_1>0\}}. \tag{5.1}$$

Indeed, if we denote by φ_k a smooth nonnegative function such that $0 \le \varphi_k \le 1$, $\varphi_k(x) = 0$ for $x \le 0$, and $\varphi_k(x) = 1$ for $x \ge \frac{1}{k}$, then the sequence of processes

$$u^k(t) = \mathbf{1}_{[0,\frac{1}{2}]}(t)\varphi_k(W_1) - \mathbf{1}_{(\frac{1}{2},1]}(t)\varphi_k(W_1),$$

converges in $L^2([0,1] \times \Omega)$ to u, and

$$\delta(u^k) = (2W_{\frac{1}{2}} - W_1)\varphi_k(W_1).$$

converges in $L^2(\Omega)$ to the right-hand side of (5.1). Hence, u is Skorohod integrable and (5.1) holds. Nevertheless the process $u\mathbf{1}_{[0,\frac{1}{2}]}$ does not belong to Dom δ. In fact, the divergence of this process is the element of $\mathbb{D}^{-\infty}$ given by

$$\delta(u\mathbf{1}_{[0,\frac{1}{2}]}) = W_{\frac{1}{2}}\mathbf{1}_{\{W_1>0\}} - \frac{1}{2}\delta_0(W_1).$$

Let us denote by $\mathbb{L}^s$ the set of processes u such that $u\mathbf{1}_{[0,t]}$ is Skorohod integrable for any $t \in [0, 1]$. Notice that the space $\mathbb{D}^{1,2}(H)$ is included into $\mathbb{L}^s$. Suppose that u belongs to $\mathbb{L}^s$ and define

$$X(t) = \delta(u\mathbf{1}_{[0,t]}) = \int_0^t u_s dW_s. \tag{5.2}$$

The process $X = \{X_t, t \in [0,1]\}$ is not adapted, and its increments satisfy the following orthogonality property:

Lemma 5.1.1 *For any process $u \in \mathbb{L}^s$ we have*

$$E\left(\int_s^t u_r dW_r \big| \mathcal{F}_{[s,t]^c}\right) = 0, \tag{5.3}$$

for all $s < t$, where, as usual, $\mathcal{F}_{[s,t]^c}$ denotes the σ-field generated by the increments of the Brownian motion in the complement of the interval $[s,t]$.

Proof: To show (5.3) it suffices to take an arbitrary $\mathcal{F}_{[s,t]^c}$-measurable random variable F belonging to the space $\mathbb{D}^{1,2}$, and check that

$$E\left(F\int_s^t u_r dW_r\right) = E\left(\int_0^1 u_r D_r F \mathbf{1}_{[s,t]}(r) dr\right) = 0,$$

which holds due to the duality relation (1.7) and Lemma 1.5.1. $\qquad\square$

The Skorohod integral process X is continuous in L^2 for any $u \in \mathbb{L}^s$. However, there exist processes u in $\mathbb{L}^s$ such that the indefinite integral $\int_0^t u_s dW_s$ does not have a continuous version.

Example: Consider the process

$$u(t) = \text{sign}(W_1 - t)\exp\left(W_t - \frac{t}{2}\right).$$

This process belongs to the space $\mathbb{L}^s$, and

$$X_t := \int_0^t u_s dW_s = \text{sign}(W_1 - t)\exp\left(W_t - \frac{t}{2}\right) - \text{sign}W_1. \tag{5.4}$$

As a consequence, the process X has a discontinuity on a point t such that $W_1 = t$. We can show (5.4) by means of a duality argument. In fact, let $G \in \mathcal{S}$ be a smooth random variable. Consider the family of transformations of the Wiener space given by $(T_tW)_s = W_s + s \wedge t, t \in [0,1]$. Notice that if $G = g(W(h_1), \ldots, W(h_n)), g \in C_b(\mathbb{R}^n)$ then

$$
\begin{aligned}
\frac{d}{ds}G(T_sW) &= \frac{d}{ds}g\left(W(h_1) + \int_0^s h_1(r)dr, \ldots, W(h_n) + \int_0^s h_n(r)dr\right) \\
&= \sum_{i=1}^n (\partial_i g)\left(W(h_1) + \int_0^s h_1(r)dr, \ldots, W(h_n) + \int_0^s h_n(r)dr\right) h_i(s) \\
&= (D_sG)(T_sW).
\end{aligned}
$$

Using Girsanov's theorem we can write

$$E(\int_0^t D_s G u_s ds) = E\left(\int_0^t D_s G \text{sign}(W_1 - s)e^{W_s - \frac{s}{2}}ds\right)$$

$$= E\left(\int_0^t (D_s G)(T_s W)\text{sign}(W_1)ds\right) = E\left(G(T_t W)\text{sign}W_1 - G\text{sign}W_1\right)$$

$$= E\left(G\left(\text{sign}(W_1 - t)\exp(W_t - \frac{t}{2}) - \text{sign}W_1\right)\right),$$

and (5.4) follows.

A useful tool in proving the existence of a continuous version for stochastic processes is the real analysis lemma of Garsia, Rodemich and Rumsey [32]. Let us recall this lemma:

Lemma 5.1.2 *Let* $p, \Psi : \mathbb{R}_+ \to \mathbb{R}_+$ *be continuous and strictly increasing functions vanishing at zero and such that* $\lim_{t \uparrow \infty} \Psi(t) = \infty$. *Suppose that* $\phi : \mathbb{R}^m \to \mathbb{B}$ *is a measurable function with values in a separable Banach space* $(\mathbb{B}, \|\cdot\|)$. *Let* $B_K = \{x \in \mathbb{R}^m, |x| \le K\}$. *Suppose that*

$$\Gamma = \int_{B_K} \int_{B_K} \Psi\left(\frac{\|\phi(t) - \phi(s)\|}{p(|t - s|)}\right) ds dt < \infty.$$

Then, there is a set $N \subset B_K$ *of measure zero such that, for all* $s, t \in B_K - N$,

$$\|\phi(t) - \phi(s)\| \le 8 \int_0^{2|t-s|} \Psi^{-1}(\lambda_m \Gamma u^{-2m})p(du),$$

where λ_m *is a universal constant depending only on* m.

Now suppose that $X = \{X(t), t \in B_K\}$ is a stochastic process with values on $\mathbb{B}$, such that the following estimate holds:

$$E(\|X(t) - X(s)\|^\gamma) \le H|t - s|^\alpha$$

for some constants $H > 0$, $\gamma > 0$, $\alpha > m$, and for all $s, t \in B_K$. Then taking $\Psi(x) = x^\gamma$ and $p(x) = x^{\frac{d+2m}{\gamma}}$, with $0 < d < \alpha - m$, one obtains for all $s, t \in B_K - N_\omega$,

$$\|X(t) - X(s)\|^\gamma \le C_1 |t - s|^d \Gamma, \tag{5.5}$$

where N_ω is a subset of B_K of zero Lebesgue measure depending on ω, and $E(\Gamma) \le HC_2$ for some constants C_1 and C_2 depending only on γ, d, and m. In particular, the above inequality implies that the process X has a continuous version (this is the classical Kolmogorov continuity criterion).

For every $p > 1$ and any positive integer k we will denote by $\mathbb{L}^{k,p}$ the space $L^p([0, 1]; \mathbb{D}^{k,p})$. Notice that $\mathbb{L}^{1,2} = \mathbb{D}^{1,2}(H)$, and $L^p([0, 1]; \mathbb{D}^{1,p}) \subset \mathbb{D}^{1,p}(H)$ for $p \ge 2$.

Proposition 5.1.1 *Let* u *be a process in the class* $\mathbb{L}^{1,2}$. *Suppose that* $E \int_0^1 \|Du_t\|_H^p dt < \infty$ *for some* $p > 2$. *Then the integral process* $\{\int_0^t u_s dW_s, 0 \le t \le 1\}$ *has a continuous version.*

Proof: We can assume that $E(u_t) = 0$ for each $t \in [0, 1]$ because the Gaussian process $\int_0^t E(u_s) dW_s$ has a continuous version. Applying the estimate (2.22) we obtain

$$E(|X_t - X_s|^p) \le c_p E\left(\left|\int_s^t \int_0^1 (D_\theta u_r)^2 d\theta dr\right|^{\frac{p}{2}}\right) \le c_p |t - s|^{\frac{p}{2} - 1} \int_s^t E\left(\left|\int_0^1 (D_\theta u_r)^2 d\theta\right|^{\frac{p}{2}}\right) dr.$$

Set

$$A_r = |\int_0^1 (D_\theta u_r)^2 d\theta|^{\frac{p}{2}}.$$

Fix an exponent $2 < \alpha < 1 + \frac{p}{2}$, and assume that p is close to 2. Applying Fubini's theorem we can write

$$E\left(\int_0^1 \int_0^1 \frac{|X_t - X_s|^p}{|t - s|^\alpha} ds dt\right) \le c_p E\left(\int_0^1 \int_0^1 |t - s|^{\frac{p}{2} - 1 - \alpha} \int_s^t A_r dr ds dt\right)$$

$$= \frac{2c_p}{\frac{p}{2} - \alpha} E\left(\int_{\{s < r\}} (|1 - s|^{\frac{p}{2} - \alpha} - |r - s|^{\frac{p}{2} - \alpha}) A_r dr ds\right)$$

$$= \frac{2c_p}{(\alpha - \frac{p}{2})(\frac{p}{2} + 1 - \alpha)} \int_0^1 (r^{\frac{p}{2} + 1 - \alpha} + |1 - r|^{\frac{p}{2} + 1 - \alpha} - 1) E(A_r) dr < \infty.$$

Hence, the random variable defined by

$$\Gamma = \int_0^1 \int_0^1 \frac{|X_t - X_s|^p}{|t - s|^\alpha} ds dt$$

is finite almost surely, and by Lemma 5.1.2 we obtain

$$|X_t - X_s| \le c_{p,\alpha} \Gamma^{\frac{1}{p}} |t - s|^{\frac{\alpha - 2}{p}},$$

for some constant $c_{p,\alpha}$. □

Note that the above proposition implies that for a process u in the space $\mathbb{L}_{loc}^{1,p}$, with $p > 2$, the Skorohod integral $\int_0^t u_s dW_s$ has a continuous version. Furthermore, if $u \in \mathbb{L}^{1,p}$, $p > 2$, we have

$$E(\sup_{t \in [0,1]} |\int_0^t u_s dW_s|^p) < \infty.$$

The next result will show the existence of a nonzero quadratic variation for the indefinite Skorohod integral.

Theorem 5.1.1 *Suppose that u is a process of the space $\mathbb{L}_{loc}^{1,2}$. Then*

$$\sum_{i=0}^{n-1} \left(\int_{t_i}^{t_{i+1}} u_s dW_s\right)^2 \to \int_0^1 u_s^2 ds, \tag{5.6}$$

in probability, as $|\pi| \to 0$, where π runs over all finite partitions $\{0 = t_0 < t_1 < \cdots < t_n = 1\}$ of $[0, 1]$, and $|\pi| = \sup_{0 \le i \le n-1} |t_{i+1} - t_i|$. Moreover, the convergence is in $L^1(\Omega)$ if u belongs to $\mathbb{L}^{1,2}$.

Proof: We will describe the details of the proof only for the case $u \in \mathbb{L}^{1,2}$. The general case would be deduced by an easy argument of localization.

For any process u in $\mathbb{L}^{1,2}$ and for any partition $\pi = \{0 = t_0 < t_1 < \cdots < t_n = 1\}$ we define

$$V^\pi(u) = \sum_{i=0}^{n-1} \left(\int_{t_i}^{t_{i+1}} u_s dW_s \right)^2.$$

Suppose that u and v are two processes in $\mathbb{L}^{1,2}$. Then we have

$$E\left(|V^\pi(u) - V^\pi(v)|\right) \leq \left(E \sum_{i=0}^{n-1} \left(\int_{t_i}^{t_{i+1}} (u_s - v_s) dW_s \right)^2 \right)^{\frac{1}{2}}$$

$$\times \left(E \sum_{i=0}^{n-1} \left(\int_{t_i}^{t_{i+1}} (u_s + v_s) dW_s \right)^2 \right)^{\frac{1}{2}}$$

$$\leq \|u - v\|_{\mathbb{L}^{1,2}} \|u + v\|_{\mathbb{L}^{1,2}}. \tag{5.7}$$

Therefore, it suffices to show the result for a class of processes u which is dense in $\mathbb{L}^{1,2}$. So we can assume that

$$u_t = \sum_{j=0}^{m-1} F_j \mathbf{1}_{(s_j, s_{j+1})},$$

where for each j, F_j is a smooth and bounded random variable, and $0 = s_0 < \cdots < s_m = 1$. We can assume that the partition π contains the points $\{s_0, \ldots, s_m\}$, because

$$\lim_{|\pi| \downarrow 0} E(|V^{\pi \vee \{s_0, \ldots, s_m\}}(u) - V^\pi(u)|) = 0.$$

If π contains the points s_j we can write

$$V^\pi(u) = \sum_{j=0}^{m-1} \sum_{\{i: s_j \leq t_i < s_{j+1}\}} \left(F_j(W(t_{i+1}) - W(t_i)) - \int_{t_i}^{t_{i+1}} D_s F_j ds \right)^2$$

$$= \sum_{j=0}^{m-1} \left[\sum_{\{i: s_j \leq t_i < s_{j+1}\}} F_j^2 (W(t_{i+1}) - W(t_i))^2 \right.$$

$$\left. - 2(W(t_{i+1}) - W(t_i)) \int_{t_i}^{t_{i+1}} D_s F_j ds + \left(\int_{t_i}^{t_{i+1}} D_s F_j ds \right)^2 \right].$$

Using the properties of the quadratic variation of the Brownian motion this converges in $L^1(\Omega)$ to

$$\sum_{j=0}^{m-1} F_j^2 (s_{j+1} - s_j) = \int_0^1 u_s^2 ds,$$

as $|\pi|$ tends to zero. □

As a consequence of these results, if $u \in \mathbb{L}_{loc}^{1,p}$, $p > 2$, is a process such that the Skorohod integral $\int_0^t u_s dW_s$ has bounded variation paths, then $u = 0$.

Let $X \in \mathbb{L}^{k,p}$, $p > 1$, and let $q \in [1, p]$. We will denote by $D^+ X$ (resp. $D^- X$) the element of $L^q([0,1] \times \Omega)$ defined by

$$\lim_n \int_0^1 \sup_{s < t \leq (s + \frac{1}{n}) \wedge 1} E(|D_s X_t - (D^+ X)_s|^q) ds = 0. \tag{5.8}$$

(resp.

$$\lim_n \int_0^1 \sup_{(s-\frac{1}{n})\vee 0 \leq t < s} E(|D_s X_t - (D^- X)_s|^q) ds = 0) \tag{5.9}$$

provided that these limit exist. We will denote by $\mathbb{L}_{q+}^{k,p}$ (resp. $\mathbb{L}_{q-}^{k,p}$) the class of processes in $\mathbb{L}^{k,p}$ such that the limit (5.8) (resp. (5.9)) exist. For each $p > 1$, and $q \in [1,p]$ we set $\mathbb{L}_q^{k,p} = \mathbb{L}_{q+}^{k,p} \cap \mathbb{L}_{q-}^{k,p}$. For $u \in \mathbb{L}_q^{k,p}$ we will write $(\nabla X)_t = (D^+ X)_t + (D^- X)_t$.

Notice that a process X of the form

$$X_t = X_0 + \int_0^t u_s dW_s + \int_0^t v_s ds,$$

where $X_0 \in \mathbb{D}^{1,2}$, $u \in \mathbb{L}^{2,2}$, and $v \in \mathbb{L}^{1,2}$ belongs to the class $\mathbb{L}_2^{1,2}$. Indeed, we have

$$D_s X_t = u_s \mathbf{1}_{\{s \leq t\}} + D_s X_0 + \int_0^t D_s v_r dr + \int_0^t D_s u_r dW_r,$$

and this implies that $(D^+ X)_t$ and $(D^- X)_t$ exist and

$$(D^+ X)_t = u_t + D_t X_0 + \int_0^t D_t v_r dr + \int_0^t D_t u_r dW_r,$$

$$(D^- X)_t = D_t X_0 + \int_0^t D_t v_r dr + \int_0^t D_t u_r dW_r.$$

In fact, we have

$$\lim_n \int_0^1 \sup_{s < t \leq (s+\frac{1}{n})\wedge 1} E(|D_s X_t - (D^+ X)_s|^2) ds$$

$$\leq \frac{1}{n} \int_0^1 \int_s^{(s+\frac{1}{n})\wedge 1} E(|D_s v_r|^2) dr ds + \int_0^1 \int_s^{(s+\frac{1}{n})\wedge 1} E(|D_s u_r|^2) dr ds$$

$$+ \int_0^1 \int_0^1 \int_s^{(s+\frac{1}{n})\wedge 1} E(|D_\theta D_s u_r|^2) dr ds d\theta,$$

and this converges to zero as n tends to infinity.

One can show a change-of-variables formula similar to Itô formula for Skorohod integral processes.

Theorem 5.1.2 *Consider a process of the form* $X_t = X_0 + \int_0^t u_s dW_s + \int_0^t v_s ds$, *where* $X_0 \in \mathbb{D}_{loc}^{1,2}$, $u \in (\mathbb{L}^{2,2} \cap \mathbb{L}^{1,4})_{loc}$, *and* $v \in \mathbb{L}_{loc}^{1,2}$. *Let* $F : \mathbb{R} \to \mathbb{R}$ *be a twice continuously differentiable function. Then we have*

$$F(X_t) = F(X_0) + \int_0^t F'(X_s) dX_s + \frac{1}{2} \int_0^t F''(X_s) u_s^2 ds + \int_0^t F''(X_s)(D^- X)_s u_s ds. \tag{5.10}$$

Notice that if the process X is adapted then $(D^- X)_s$ vanishes, and we obtain the classical Itô formula.

Proof: Suppose that $(\Omega^{n,1}, X_0^n)$, $(\Omega^{n,2}, u^n)$ and $(\Omega^{n,3}, v^n)$ are localizing sequences for X_0, u, and v, respectively. For each positive integer k let ψ_k be a smooth function such that $0 \leq \psi_k \leq 1$, $\psi_k(x) = 0$ if $|x| \geq k+1$, and $\psi_k(x) = 1$ if $|x| \leq k$. Define

$$u_t^{n,k} = u_t^n \psi_k \left(\int_0^1 (u_s^n)^2 ds \right).$$

Set $X_t^{n,k} = X_0^n + \int_0^t u_s^{n,k} dW_s + \int_0^t v_s^n ds$, and consider the family of sets

$$G^{n,k} = \Omega^{n,1} \cap \Omega^{n,2} \cap \Omega^{n,3} \cap \{\sup_{t \in [0,1]} |X_t| \leq k\} \cap \{\int_0^1 (u_s^n)^2 ds \leq k\}.$$

Then it suffices to show the result for the processes X_0^n, $u^{n,k}$, and v^n, and for the function $F^n := F\psi_n$. In this way we can assume that $X_0 \in \mathbb{D}^{1,2}$, $u \in \mathbb{L}^{2,2} \cap \mathbb{L}^{1,4}$, $v \in \mathbb{L}^{1,2}$, $\int_0^1 u_s^2 ds \leq k$, and that the functions F, F' and F'' are bounded.

Set $t_i^n = \frac{i t}{2^n}$, $0 \leq i \leq 2^n$. Applying Taylor development up to the second order we obtain

$$F(X_t) = F(X_0) + \sum_{i=0}^{2^n-1} F'(X(t_i^n))(X(t_{i+1}^n) - X(t_i^n)) + \sum_{i=0}^{2^n-1} \frac{1}{2} F''(\bar{X}_i)(X(t_{i+1}^n) - X(t_i^n))^2,$$

(5.11)

where $\bar{X}_i$ denotes a random intermediate point between $X(t_i^n)$ and $X(t_{i+1}^n)$. Now the proof will be decomposed in several steps.

Step 1. Let us show that

$$\sum_{i=0}^{2^n-1} F''(\bar{X}_i^n)(X(t_{i+1}^n) - X(t_i^n))^2 \to \int_0^t F''(X_s) u_s^2 ds,$$

(5.12)

in probability (actually in $L^1(\Omega)$), as n tends to infinity.

The increment $(X(t_{i+1}^n) - X(t_i^n))^2$ can be decomposed into

$$\left(\int_{t_i^n}^{t_{i+1}^n} u_s dW_s\right)^2 + \left(\int_{t_i^n}^{t_{i+1}^n} v_s ds\right)^2 + 2\left(\int_{t_i^n}^{t_{i+1}^n} u_s dW_s\right)\left(\int_{t_i^n}^{t_{i+1}^n} v_s ds\right).$$

Using the continuity of $F''(X_t)$ we can show that the contribution of the last two terms to the limit (5.12) is zero. Therefore, it suffices to show that

$$\sum_{i=0}^{2^n-1} F''(\bar{X}_i)\left(\int_{t_i^n}^{t_{i+1}^n} u_s dW_s\right)^2 \to \int_0^t F''(X_s) u_s^2 ds.$$

(5.13)

Suppose that $n \geq m$, and for any $i = 1, \ldots, 2^n$ let us denote by $t_i^{(m)}$ the point of the mth partition which is closer to t_i^n from the left. Then we have

$$\left| \sum_{i=0}^{2^n-1} F''(\bar{X}_i)\left(\int_{t_i^n}^{t_{i+1}^n} u_s dW_s\right)^2 - \int_0^t F''(X_s) u_s^2 ds \right|$$

$$\leq \left| \sum_{i=0}^{2^n-1} [F''(\bar{X}_i) - F''(X(t_i^{(m)}))]\left(\int_{t_i^n}^{t_{i+1}^n} u_s dW_s\right)^2 \right|$$

$$+ \left| \sum_{j=0}^{2^m-1} F''(X(t_j^m)) \sum_{i:t_i^n \in [t_j^m, t_{j+1}^m)} \left[\left(\int_{t_i^n}^{t_{i+1}^n} u_s dW_s\right)^2 - \int_{t_i^n}^{t_{i+1}^n} u_s^2 ds \right] \right|$$

$$+ \left| \sum_{j=0}^{2^m-1} F''(X(t_j^m)) \int_{t_j^m}^{t_{j+1}^m} u_s^2 ds - \int_0^t F''(X_s) u_s^2 ds \right|$$

$$= b_1 + b_2 + b_3.$$

The term b_3 can be bounded by

$$\sup_{|s-r|\leq t2^{-m}} |F''(X_s) - F''(X_r)| \int_0^t u_s^2 ds,$$

which converges to zero as m tends to infinity by the continuity of the process X_t. In the same way the term b_1 is bounded by

$$\sup_{|s-r|\leq t2^{-m}} |F''(X_s) - F''(X_r)| \sum_{i=0}^{2^n-1} (\int_{t_i^n}^{t_{i+1}^n} u_s dW_s)^2,$$

which again tends to zero in probability as m tends to infinity uniformly in n. Finally, the term b_2 converges to zero as n tends to infinity, for any fixed m, due to Theorem 5.1.1.

Step 2. Clearly the term

$$\sum_{i=0}^{2^n-1} F'(X(t_i^n))(\int_{t_i^n}^{t_{i+1}^n} v_s ds)$$

converges a.s. and in $L^1(\Omega)$ to $\int_0^t F'(X_s)v_s ds$, as n tends to infinity.

Step 3. From property (1.12) of the Skorohod integral we deduce

$$
\begin{aligned}
F'(X(t_i^n)) \int_{t_i^n}^{t_{i+1}^n} u_s dW_s &= \int_{t_i^n}^{t_{i+1}^n} F'(X(t_i^n))u_s dW_s + \int_{t_i^n}^{t_{i+1}^n} D_s[F'(X(t_i^n))]u_s ds \\
&= \int_{t_i^n}^{t_{i+1}^n} F'(X(t_i^n))u_s dW_s + F''(X(t_i^n)) \int_{t_i^n}^{t_{i+1}^n} D_s X(t_i^n)u_s ds \\
&= c_1 + c_2.
\end{aligned}
$$

Notice that the process $F'(X_t)u_t$ belongs to $\mathbb{L}^{1,2}$ because $u \in \mathbb{L}^{2,2} \cap \mathbb{L}^{1,4}$, $v \in \mathbb{L}^{1,2}$, and the processes $F'(X_t)$, $F''(X_t)$, and $\int_0^1 u_s^2 ds$ are uniformly bounded. In fact, we have

$$
\begin{aligned}
D_s[F'(X_t)u_t] &= F''(X_t)\left(u_s 1_{\{s\leq t\}} + D_s X_0 + \int_0^t D_s u_r dW_r + \int_0^t D_s v_r dr\right)u_t \\
&+ F'(X_t)D_s u_t,
\end{aligned}
$$

and all the terms in the right-hand side of the above expression are square integrable. For the second term we use the duality relationship of the Skorohod integral:

$$
\begin{aligned}
&\int_0^1 \int_0^1 E\left(\left|u_t \int_0^t D_s u_r dW_r\right|^2\right) dsdt \\
&= E\left\{ \int_0^1 \int_0^1 \int_0^t D_s u_r \left[2u_t D_r u_t \left(\int_0^t D_s u_\theta dW_\theta\right) + u_t^2 D_s u_r\right.\right. \\
&\left.\left.+ u_t^2 \left(\int_0^t D_r D_s u_\theta dW_\theta\right)\right] drdsdt \right\} \\
&\leq cE\left((\|Du\|_{L^2([0,1]^2)}^2 + \|D^2 u\|_{L^2([0,1]^3)})^2 \right).
\end{aligned}
$$

Then, for any smooth random variable $G \in \mathcal{S}$ we have

$$\lim_n E\left(G \sum_{i=0}^{2^n-1} \int_{t_i^n}^{t_{i+1}^n} F'(X(t_i^n))u_s dW_s \right) = E\left(G \int_0^t F'(X_s)u_s dW_s \right). \tag{5.14}$$

This convergence is easily checked by duality and in this way obtain the convergence of $E(Gc_1)$ to $E(G \int_0^t F'(X_s)u_s dW_s)$. One can also show the convergence in L^2 of c_1 to $\int_0^t F'(X_s)u_s dW_s$.

On the other hand, the term c_2 converges in L^1 to $\int_0^t F''(X_s)(D^- X)_s u_s ds$. In fact, we have

$$\left| \sum_{i=0}^{2^n-1} F''(X(t_i^n)) \int_{t_i^n}^{t_{i+1}^n} D_s X(t_i^n) u_s ds - \int_0^t F''(X_s)(D^- X)_s u_s ds \right|$$

$$\leq \left| \sum_{i=0}^{2^n-1} F''(X(t_i^n)) \int_{t_i^n}^{t_{i+1}^n} [D_s X(t_i^n) - (D^- X)_s] u_s ds \right|$$

$$+ \left| \sum_{i=0}^{2^n-1} \int_{t_i^n}^{t_{i+1}^n} [F''(X(t_i^n)) - F''(X_s)](D^- X)_s u_s ds \right|.$$

Consequently, we obtain

$$E\left(\left| \sum_{i=0}^{2^n-1} F''(X(t_i^n)) \int_{t_i^n}^{t_{i+1}^n} D_s X(t_i^n) u_s ds - \int_0^t F''(X_s)(D^- X)_s u_s ds \right| \right)$$

$$\leq \|F''\|_\infty \left\{ \int_0^1 E(u_s^2) ds \int_0^1 \sup_{(s-2^{-n})^+ \leq t \leq s} E(|D_s X_t - (D^- X)_s|^2) ds \right\}^{\frac{1}{2}}$$

$$+ E\left(\sup_{|s-t| \leq 2^{-n}} |F''(X_t) - F''(X_s)| \int_0^1 |(D^- X)_s u_s| ds \right),$$

which converges to zero as n tends to infinity. $\qquad \square$

Remarks:

1. Notice that for a process $u \in \mathbb{L}^{2,2}$, the condition $E(\int_0^1 \|Du_t\|^2 dt) < \infty$ is only required to insure that X has a continuous version, and to show that $F'(X_t)u_t$ belongs to $\mathbb{L}^{1,2}$. One can show the Itô formula under different type of hypotheses. More precisely, one can impose some conditions on X and modify the assumptions on X_0, u, and v. For instance, one can assume either

(a) $u \in \mathbb{L}^{1,4}_{\text{loc}}$, $v \in L^1([0,1])$ a.s., and $X \in (\mathbb{L}^{1,2}_2 \cap \mathbb{L}^{1,4})_{\text{loc}}$, or

(b) $u \in (\mathbb{L}^{1,2} \cap L^\infty([0,1] \times \Omega))_{\text{loc}}$, $v \in L^1([0,1])$ a.s., $X \in \mathbb{L}^{1,2}_{2^-,\text{loc}}$, and X possesses a continuous version.

2. Suppose that X, u, v are stochastic processes verifying one of the hypotheses (a) or (b). Then if there is a set $G \in \mathcal{F}$ such that for almost all $\omega \in G$ we have

$$X_t = X_0 + \int_0^t u_s dW_s + \int_0^t v_s ds, \quad t \in [0,1], \tag{5.15}$$

then (5.10) holds a.s. on G. In fact, it suffices to check that

$$\sum_{i=0}^{2^n-1} F'(X(t_i^n)) \left(\int_{t_i^n}^{t_{i+1}^n} u_s dW_s + \int_{t_i^n}^{t_{i+1}^n} v_s ds \right)$$

$$+ \sum_{i=0}^{2^n-1} \frac{1}{2} F''(\bar{X}_i) \left(\int_{t_i^n}^{t_{i+1}^n} u_s dW_s + \int_{t_i^n}^{t_{i+1}^n} v_s ds \right)^2,$$

converges in probabiblity as $|\pi|$ tends to zero to the right-hand side of (5.10).

3. Using the operator ∇ we can write Eq. (5.10) in the following way:

$$F(X_t) = F(X_0) + \int_0^t F'(X_s)dX_s + \frac{1}{2}\int_0^t F''(X_s)(\nabla X)_s u_s ds. \qquad (5.16)$$

The following result contains a multidimensional and local version of the change-of-variables formula for the Skorohod integral. We will make the usual convention of summation over repeated indices.

Theorem 5.1.3 *Let $W = \{W_t, t \in [0,1]\}$ be an N-dimensional Wiener process. Suppose that $X_0^i \in \mathbb{D}^{1,2}$, $u^{ij} \in \mathbb{L}^{2,2} \cap \mathbb{L}^{1,4}$, and $v^i \in \mathbb{L}^{1,2}$, $1 \le i \le M$, $1 \le j \le N$, are processes such that on a set $G \in \mathcal{F}$ we have a.s.*

$$X_t^i = X_0^i + \int_0^t u_s^{ij} dW_s^i + \int_0^t v_s^i ds, \quad 0 \le t \le 1.$$

Let $F : \mathbb{R}^M \to \mathbb{R}$ be a twice continuously differentiable function. Then we have on G a.s.

$$\begin{aligned}
F(X_t) &= F(X_0) + \int_0^t (\partial_i F)(X_s)dX_s^i + \frac{1}{2}\int_0^t (\partial_i \partial_j F)(X_s) u_s^{ik} u_s^{jk} ds \\
&\quad + \int_0^t (\partial_i \partial_j F)(X_s)((D^k)^- X^j)_s u_s^{ik} ds,
\end{aligned}$$

where D^k denotes the derivative with respect to the kth component of the Wiener process.

5.2 Extended Stratonovich integral

In the sequel, π will denote a partition of the interval $[0,1]$ of the form $\pi = \{0 = t_0 < t_1 < \cdots < t_n = 1\}$. The mesh of π is defined by $|\pi| = \sup_i(t_{i+1} - t_i)$. Given a measurable process $u = \{u_t, t \in [0,1]\}$ such that $\int_0^1 |u_t|dt < \infty$ a.s., we can introduce the following Riemann-type sums

$$S^\pi = \sum_{i=0}^{n-1} \frac{1}{t_{i+1} - t_i}\left(\int_{t_i}^{t_{i+1}} u_s ds\right)(W(t_{i+1}) - W(t_i)).$$

Notice that

$$S^\pi = \int_0^1 u_s W_s^\pi ds,$$

where

$$W^\pi = \sum_{i=0}^{n-1}\left(\frac{W(t_{i+1}) - W(t_i)}{t_{i+1} - t_i}\right)\mathbf{1}_{[t_i,t_{i+1}]}.$$

Definition 5.2.1 *We say that a measurable process $u = \{u_t, t \in [0,1]\}$ such that $\int_0^1 |u_t|dt < \infty$ a.s. is Stratonovich integrable if the family S^π converges in probability as $|\pi| \downarrow 0$, and in this case the limit will be denoted by $\int_0^1 u_t \circ dW_t$.*

We know that if the process u is a continuous semimartingale, then the Stratonovich integral of u exists and it coincides with the Itô integral plus the correction term $\frac{1}{2}\langle u, W\rangle_t$. Let us now describe the relationship between the extended Stratonovich integral and the Skorohod integral. A basic result on the approximation of the Skorohod integral by Riemann-type sums is the following. For any process $u \in L^2([0,1] \times \Omega)$ we define

$$\hat{S}^\pi = \sum_{i=0}^{n-1} \frac{1}{t_{i+1} - t_i} \left(\int_{t_i}^{t_{i+1}} E(u_s | \mathcal{F}_{[t_i, t_{i+1}]^c} ds) \right) (W(t_{i+1}) - W(t_i)).$$

Notice each term in the summation is the product of two independent factors. Furthermore, from Lemma 1.5.2 the processes

$$\hat{u}^\pi(t) = \sum_{i=0}^{n-1} \frac{1}{t_{i+1} - t_i} \left(\int_{t_i}^{t_{i+1}} E(u_s | \mathcal{F}_{[t_i, t_{i+1}]^c} ds) \right) \mathbf{1}_{(t_i, t_{i+1}]}(t)$$

are Skorohod integrable, and

$$\hat{S}^\pi = \delta(\hat{u}^\pi).$$

Proposition 5.2.1 *Let $u \in L^2([0,1] \times \Omega)$. If there exist a sequence of partitions $\{\pi(n)\}$ of $[0,1]$ whose mesh tends to zero such that $\hat{S}^{\pi(n)}$ converges in $L^2(\Omega)$ as n tends to infinity, then u is Skorohod integrable, and*

$$\delta(u) = \lim_n \hat{S}^{\pi(n)}.$$

Moreover, this convergence always holds if $u \in \mathbb{L}^{1,2}$.

Proof: The first part of the proof follows from the fact that the operator δ is closed and $\hat{u}_n^\pi$ converges to u in $L^2([0,1] \times \Omega)$. The second part follows from the convergence of $\hat{u}^\pi$ to u as $|\pi|$ tends to zero, in the norm of $\mathbb{L}^{1,2}$ (see [76] for the details of the proof). $\qquad\square$

Concerning the extended Stratonovich integral we can show the following result:

Theorem 5.2.1 *Let $u \in \mathbb{L}_{1,\mathrm{loc}}^{1,2}$. Then u is Stratonovich integrable and*

$$\int_0^1 u_t \circ dW_t = \int_0^1 u_t dW_t + \frac{1}{2} \int_0^1 (\nabla u)_t dt. \tag{5.17}$$

Proof: By a localization argument we can assume that $u \in \mathbb{L}_1^{1,2}$. We have, for any partition π on $[0,1]$

$$\delta(u^\pi) = S^\pi - \sum_{i=0}^{n-1} \frac{1}{t_{i+1} - t_i} \int_{t_i}^{t_{i+1}} \int_{t_i}^{t_{i+1}} D_s u_t \, ds \, dt, \tag{5.18}$$

where

$$u^\pi(t) = \sum_{i=0}^{n-1} \frac{1}{t_{i+1} - t_i} \left(\int_{t_i}^{t_{i+1}} u_s \, ds \right) \mathbf{1}_{(t_i, t_{i+1}]}(t).$$

The processes u^π converge to u as $|\pi|$ tends to zero in the norm of $\mathbb{L}^{1,2}$. Hence, $\delta(u^\pi)$ converges to $\delta(u)$ in $L^2(\Omega)$ as $|\pi| \downarrow 0$. Therefore, it suffices to show the following convergences:

$$\lim_{|\pi| \downarrow 0} E \left(\left| \sum_{i=0}^{n-1} \frac{1}{t_{i+1} - t_i} \int_{t_i}^{t_{i+1}} dt \int_t^{t_{i+1}} (D_t u_s) ds - \frac{1}{2} \int_0^1 (D^+ u)_t dt \right| \right) = 0, \quad (5.19)$$

$$\lim_{|\pi| \downarrow 0} E \left(\left| \sum_{i=0}^{n-1} \frac{1}{t_{i+1} - t_i} \int_{t_i}^{t_{i+1}} dt \int_{t_i}^t (D_t u_s) ds - \frac{1}{2} \int_0^1 (D^- u)_t dt \right| \right) = 0. \quad (5.20)$$

We will only show (5.19). We can write

$$E \left(\left| \sum_{i=0}^{n-1} \frac{1}{t_{i+1} - t_i} \int_{t_i}^{t_{i+1}} dt \int_t^{t_{i+1}} (D_t u_s) ds - \frac{1}{2} \int_0^1 (D^+ u)_t dt \right| \right)$$

$$\leq E \left(\left| \sum_{i=0}^{n-1} \frac{1}{t_{i+1} - t_i} \int_{t_i}^{t_{i+1}} \left(\int_t^{t_{i+1}} [D_t u_s - (D^+ u)_t] ds \right) dt \right| \right)$$

$$+ E \left(\left| \sum_{i=0}^{n-1} \int_{t_i}^{t_{i+1}} \frac{t_{i+1} - t}{t_{i+1} - t_i} (D^+ u)_t dt - \frac{1}{2} \int_0^1 (D^+ u)_t dt \right| \right)$$

$$\leq \int_0^1 \sup_{t \leq s \leq (t + |\pi|) \wedge 1} E(|D_t u_s - (D^+ u)_t|) dt$$

$$+ E \left(\left| \int_0^1 (D^+ u)_t \left(\sum_{i=0}^{n-1} \frac{t_{i+1} - t}{t_{i+1} - t_i} 1_{(t_i, t_{i+1}]}(t) - \frac{1}{2} \right) dt \right| \right).$$

The first summand in the above converges to zero due to the definition of the class $\mathbb{L}_{1^+}^{1,2}$. For the second term we will use the convergence of the functions $\sum_{i=0}^{n-1} \frac{t_{i+1} - t}{t_{i+1} - t_i} 1_{(t_i, t_{i+1}]}(t)$ to the constant $\frac{1}{2}$ in the weak topology of $L^2([0,1])$. This weak convergence implies that

$$\int_0^1 (D^+ u)_t \left(\sum_{i=0}^{n-1} \frac{t_{i+1} - t}{t_{i+1} - t_i} 1_{(t_i, t_{i+1}]}(t) - \frac{1}{2} \right) dt$$

converges a.s. to zero as $|\pi| \to 0$. Finally, the convergence in $L^1(\Omega)$ follows by dominated convergence, using the definition of the space $\mathbb{L}_1^{1,2}$. $\qquad \square$

Itô's formula for the Skorohod integral allows us to deduce a change-of-variables formula for the Stratonovich integral. Let us first introduce the following classes of processes:

Set $\mathbb{L}_S^{2,4} = \{ u \in \mathbb{L}_1^{2,4} : \nabla u \in \mathbb{L}^{1,2}, u \text{ is continuous in } L^2(\Omega) \}$.

Theorem 5.2.2 Let F be a real-valued, twice continuously differentiable function. Consider a process of the form $X_t = X_0 + \int_0^t u_s \circ dW_s + \int_0^t v_s ds$, where $X_0 \in \mathbb{D}_{loc}^{1,2}$, $u \in \mathbb{L}_{S,loc}^{2,4}$ and $v \in \mathbb{L}_{loc}^{1,2}$. Then we have

$$F(X_t) = F(X_0) + \int_0^t F'(X_s) v_s ds + \int_0^t [F'(X_s) u_s] \circ dW_s. \quad (5.21)$$

Proof: As in the proof of the change-of-variables formula for the Skorohod integral we can assume that the functions F, F', and F'' are bounded, $\int_0^1 u_s^2 ds$ is bounded,

$X_0 \in \mathbb{D}^{1,2}$, $u \in \mathbb{L}_S^{2,4}$, and $v \in \mathbb{L}^{1,2}$. We know that the process X_t has the following decomposition:

$$X_t = X_0 + \int_0^t u_s dW_s + \int_0^t v_s ds + \frac{1}{2} \int_0^t (\nabla u)_s ds.$$

This process verifies the assumptions of Theorem 5.1.2. Consequently, we can apply Itô's formula to X and obtain

$$\begin{aligned} F(X_t) &= F(X_0) + \int_0^t F'(X_s) v_s ds + \frac{1}{2} \int_0^t F'(X_s)(\nabla u)_s ds \\ &+ \int_0^t F'(X_s) u_s dW_s + \int_0^t F''(X_s)(D^- X)_s u_s ds + \frac{1}{2} \int_0^t F'(X_s) u_s^2 ds. \end{aligned}$$

The process $\{F'(X_t) u_t, t \in [0,1]\}$ belongs to $\mathbb{L}_1^{1,2}$. In fact, notice first that as in the proof of Theorem 5.1.2, the boundedness of F', F'', $\int_0^1 u_s^2 ds$, and the fact that $u \in \mathbb{L}^{2,4}$, $v, \nabla u \in \mathbb{L}^{1,2}$ and $X_0 \in \mathbb{D}^{1,2}$ imply that this process belongs to $\mathbb{L}^{1,2}$ and

$$D_s[F'(X_t) u_t] = F'(X_t) D_s u_t + F''(X_t) u_t D_s X_t.$$

On the other hand, using that $u \in \mathbb{L}_1^{1,2}$, u is continuous in L^2, and $X \in \mathbb{L}_2^{1,2}$ we deduce that $\{F'(X_t) u_t, t \in [0,1]\}$ belongs to $\mathbb{L}_1^{1,2}$ and that

$$(\nabla(F'(X) u))_t = F'(X_t)(\nabla u)_t + F''(X_t) u_t(\nabla X)_t.$$

Hence, applying Theorem 5.2.1 we can write

$$\begin{aligned} \int_0^t [F'(X_s) u_s] \circ dW_s &= \int_0^t F'(X_s) u_s dW_s + \frac{1}{2} \int_0^t (\nabla(F'(X) u))_s ds \\ &= \int_0^t F'(X_s) u_s dW_s + \frac{1}{2} \int_0^t F'(X_s)(\nabla u)_s ds + \frac{1}{2} \int_0^t F''(X_s) u_s(\nabla X)_s ds. \end{aligned}$$

Finally, notice that
$$(\nabla X)_t = 2(D^- X)_t + u_t.$$

This completes the proof of the theorem. $\qquad\qquad \square$

The above theorem is still valid if we assume $X \in \mathbb{L}_2^{1,4}$, $u \in (\mathbb{L}_1^{1,4} \cap C^2([0,1], L^2(\Omega))$, and $v \in L^1([0,1])$ a.s. In the next theorem we state a multidimensional version of the change-of-variables formula for the Stratonovich integral, using these type of hypotheses.

Theorem 5.2.3 *Let $W = \{W_t, t \in [0,1]\}$ be an N-dimensional Wiener process. Suppose that we have processes $X^i \in \mathbb{L}_2^{1,4}$, $u^{ij} \in \mathbb{L}_1^{1,4} \cap C^2([0,1], L^2(\Omega))$, and $v^i \in L^1([0,1])$ a.s., $1 \le i \le M$, $1 \le j \le N$, such that on some set $G \in \mathcal{F}$ a.s. we have*

$$X_t^i = X_0^i + \int_0^t u_s^{ij} \circ dW_s^j + \int_0^t v_s^i ds, \quad 0 \le t \le 1.$$

Let $F : \mathbb{R}^M \to \mathbb{R}$ be a twice continuously differentiable function. Then we have on G

$$F(X_t) = F(X_0) + \int_0^t (\partial_i F)(X_s) u_s^{ij} \circ dW_s^j + \int_0^t (\partial_i F)(X_s) v_s^i ds.$$

5.3 Substitution formulas

In this section we will consider the following problem. Suppose that $u = \{u_t(x), 0 \leq t \leq 1\}$ is a stochastic process parametrized by $x \in \mathbb{R}^m$, which is square integrable and adapted for each $x \in \mathbb{R}^m$. For each x we can define the Itô integral

$$\int_0^1 u_t(x) dW_t.$$

Assume now that the resulting random field is a.s. continuous in x, and let F be an m-dimensional random variable. Then we can evaluate the stochastic integral at $x = F$, that is, we can define the random variable

$$\int_0^1 u_t(x) dW_t|_{x=F}. \tag{5.22}$$

A natural question is under which conditions is the nonadapted process $\{u_t(F), 0 \leq t \leq 1\}$ Skorohod integrable, and what is the relationship between the Skorohod integral of this process and the random variable defined by (5.22). A similar question can be asked for the extended Stratonovich integral.

Let us first consider the following preliminary result.

Lemma 5.3.1 *Suppose that $\{Y_n(\theta), \theta \in \mathbb{R}^m\}$, $n \geq 1$ is a sequence of processes which converges in probability to a random field $\{Y(\theta), \theta \in \mathbb{R}^m\}$ for each $\theta \in \mathbb{R}^m$. Suppose that*

$$E(|Y_n(\theta) - Y_n(\theta')|^p) \leq c_{p,K}|\theta - \theta'|^\alpha, \tag{5.23}$$

for all $|\theta|, |\theta'| \leq K$, $n \geq 1$, $K > 0$ and for some constants $p > 0$ and $\alpha > m$. Then, for any m-dimensional random variable F one has

$$\lim_n Y_n(F) = Y(F),$$

in probability. Moreover, the convergence is in L^p if F is bounded.

Proof: Fix $K > 0$. Replacing F by $F_K := F 1_{\{|F| \leq K\}}$ we can assume that F is bounded by K. Fix $\epsilon > 0$ and consider a random variable F_ϵ wich takes finitely many values and such that $|F_\epsilon| \leq K$ and $\|F - F_\epsilon\|_\infty \leq \epsilon$. We can write

$$|Y_n(F) - Y(F)| \leq |Y_n(F) - Y_n(F_\epsilon)| + |Y_n(F_\epsilon) - Y(F_\epsilon)| + |Y(F_\epsilon) - Y(F)|.$$

Take $0 < m' < \alpha - m$. By (5.5) there exist constants C_1, C_2, and random variables Γ, Γ_n such that

$$|Y_n(\theta) - Y_n(\theta')|^p \leq C_1|\theta - \theta'|^{m'}\Gamma_n,$$
$$E(\Gamma_n) \leq C_2 c_{p,K},$$

and

$$|Y(\theta) - Y(\theta')|^p \leq C_1|\theta - \theta'|^{m'}\Gamma,$$
$$E(\Gamma) \leq C_2 c_{p,K}.$$

Notice that Eq. (5.23) is also satisfied by the random field $Y(\theta)$. Hence,

$$E(|Y_n(F) - Y(F)|^p) \le c_p \left(2C_1C_2c_{p,K}\epsilon^{m'} + E(|Y_n(F_\epsilon) - Y(F_\epsilon)|^p) \right),$$

and the desired convergence follows by taking first the limit as $n \to \infty$ and then the limit as $\epsilon \downarrow 0$. $\qquad\square$

Consider a random field $u = \{u_t(x), 0 \le t \le 1, x \in \mathbb{R}^m\}$ satisfying the following conditions:

(h1) For each $x \in \mathbb{R}^m$, and $t \in [0,1]$, $u_t(x)$ is $\mathcal{F}_t$-measurable.

(h2) There exist constants $p \ge 2$ and $\alpha > m$ such that

$$E(|u_t(x) - u_t(x')|^p) \le C_{t,K}|x - x'|^\alpha,$$

for all $|x|, |x'| \le K$, $K > 0$, where $\int_0^1 C_{t,K}dt < \infty$. Moreover $\int_0^1 E(|u_t(0)|^2)dt < \infty$.

Notice that under the above conditions for each $t \in [0,1]$ the random field $\{u_t(x), x \in \mathbb{R}^m\}$ possesses a continuous version, and the Itô integral $\int_0^1 u_t(x)dW_t$ possess a continuous version in (t, x). In fact, for all $|x|, |x'| \le K$, $K > 0$, we have

$$E\left(\sup_{t \in [0,1]} \left| \int_0^t (u_s(x) - u_s(x'))dW_s \right|^p \right) \le c_p E\left(\left| \int_0^1 |u_s(x) - u_s(x')|^2 ds \right|^{\frac{p}{2}} \right)$$
$$\le c_p \left(\int_0^1 C_{t,K}dt \right)|x - x'|^\alpha.$$

The following theorem provides the relationship between the evaluated integral $\int_0^1 u_t(x)dW_t|_{x=F}$ and the Skorohod integral $\delta(u(F))$. We will need the following hypothesis which is stronger that **(h2)**:

(h3) For each (t, ω) the mapping $x \to u_t(x)$ is continuously differentiable, and for each $K > 0$.

$$\int_0^1 E\left(\sup_{|x| \le K} |u_t'(x)|^q \right)dt < \infty,$$

where u' denotes the gradient of u and $q \ge 4$, $q > m$. Moreover $\int_0^1 E(|u_t(0)|^2)dt < \infty$.

Theorem 5.3.1 *Suppose that $u = \{u_t(x), 0 \le t \le 1, x \in \mathbb{R}^m\}$ is a random field satisfying conditions (h1) and (h3). Let $F : \Omega \to \mathbb{R}^m$ be a bounded random variable such that $F^i \in \mathbb{D}^{1,4}$ for $1 \le i \le m$. Then the composition $u(F) = \{u_t(F), 0 \le t \le 1\}$ belongs to the domain of δ and*

$$\delta(u(F)) = \int_0^1 u_t(x)dW_t\Big|_{x=F} - \sum_{j=1}^m \int_0^1 (\partial_j u_t)(F)D_tF^j dt. \tag{5.24}$$

Proof: Consider the approximation of the process u given by

$$u_t^n(x) = \sum_{i=1}^{n-1} n \left(\int_{\frac{i-1}{n}}^{\frac{i}{n}} u_s(x) ds \right) 1_{(\frac{i}{n}, \frac{i+1}{n}]}(t). \tag{5.25}$$

Notice that $u_t(F)$ belongs to $L^2([0,1] \times \Omega)$ and the sequence of processes $u_t^n(F)$ converges to $u_t(F)$ in $L^2([0,1] \times \Omega)$. Indeed, if F is bounded by K we have

$$E\left(\int_0^1 u_t(F)^2 dt \right) \leq \int_0^1 E\left(\sup_{|x| \leq K} |u_t(x)|^2 dt \right)$$

$$\leq 2 \int_0^1 E(|u_t(0)|^2) dt + 2K^2 \int_0^1 E(\sup_{|x| \leq K} |u_t'(x)|^2) dt < \infty.$$

The convergence of $u^n(F)$ to $u(F)$ is obtained by first approximating $u(F)$ by an element of $C([0,1]; L^2(\Omega))$.

From Lemma 1.5.3 we deduce that $u_t^n(F)$ belongs to Dom δ and

$$\delta(u^n(F)) = \sum_{i=1}^{n-1} n \left(\int_{\frac{i-1}{n}}^{\frac{i}{n}} u_s(F) ds \right) \left(W(\frac{i+1}{n}) - W(\frac{i}{n}) \right)$$

$$- \sum_{i=1}^{n-1} \sum_{j=1}^{m} n \left(\int_{\frac{i-1}{n}}^{\frac{i}{n}} (\partial_j u_s)(F) ds \right) \left(\int_{\frac{i}{n}}^{\frac{i+1}{n}} D_r F^j dr \right). \tag{5.26}$$

The Itô stochastic integrals $\int_0^1 u_t^n(x) dW_t$ satisfy

$$E\left(\left| \int_0^1 u_t^n(x) dW_t - \int_0^1 u_t^n(x') dW_t \right|^q \right) \leq c_q |x - x'|^q.$$

Hence, by Lemma 5.3.1 the sequence $\int_0^1 u_t^n(x) dW_t \big|_{x=F}$ converges in L^q to the random variable $\int_0^1 u_t(x) dW_t \big|_{x=F}$. On the other hand, the second summand in the right-hand side of (5.26) converges to $\sum_{j=1}^m \int_0^1 (\partial_j u_t)(F) D_t F^j dt$ in L^2, as it follows from the estimate

$$\left| \int_0^1 (\partial_j u_t)(F) D_t F^j dt - \int_0^1 (\partial_j u_t^n)(F) D_t F^j dt \right|$$

$$\leq \|DF^j\|_H \left(\int_0^1 |(\partial_j u_t)(F) - (\partial_j u_t^n)(F)|^2 dt \right)^{\frac{1}{2}}.$$

The operator δ being closed the result follows. $\qquad\square$

The preceding theorem can be localized as follows:

Theorem 5.3.2 *Suppose that $u = \{u_t(x), 0 \leq t \leq 1, x \in \mathbb{R}^m\}$ is a random field that satisfies (h1) and (h3). Let $F : \Omega \to \mathbb{R}^m$ be a random variable such that $F^i \in \mathbb{D}_{loc}^{1,4}$ for $1 \leq i \leq m$. Then the composition $u(F)$ is locally in the domain of δ and (5.24) holds.*

Notice that in the above theorem the Skorohod integral $\delta(u(F))$ may depend on the particular localizing sequence, because we do not know if the operator δ is local in Dom δ.

The Stratonovich integral also satisfies a commutativity relationship but in this case we do not need a complementary term. This fact is coherent with the the general behavior of the Stratonovich integral as an ordinary integral.

Let us consider the following condition which is stronger that (h2):

(h4) There exist constants $p \geq 2$ and $\alpha > m$ such that

$$E(|u_t(x) - u_t(x')|^p) \leq c_K|x - x'|^\alpha,$$
$$E(|u_t(x) - u_t(x') - u_s(x) + u_s(x')|^p) \leq c_K|x - x'|^\alpha|s - t|^{\frac{p}{2}},$$

for all $|x|, |x'| \leq K$, $K > 0$, and $s, t \in [0, 1]$. Moreover $\int_0^1 E(|u_t(0)|^2)dt < \infty$, and $t \to u_t(x)$ is continuous in L^2 for each x.

Theorem 5.3.3 *Let* $u = \{u_t(x), 0 \leq t \leq 1, x \in \mathbb{R}^m\}$ *be a random field satisfying hypothesis (h1) and (h4). Suppose that for each* $x \in \mathbb{R}^m$ *the Stratonovich integral* $\int_0^1 u_t(x) \circ dW_t$ *exists. Consider an arbitrary m-dimensional random variable F. Then $u(F)$ is Stratonovich integrable, and we have*

$$\int_0^1 u_t(F) \circ dW_t = \int_0^1 u_t(x) \circ dW_t \bigg|_{x=F}. \tag{5.27}$$

Proof: Fix a partition $\pi = \{0 = t_0 < t_1 < \ldots < t_n = 1\}$. We can write

$$\begin{aligned}
S^\pi &= \sum_{i=0}^{n-1} \frac{1}{t_{i+1} - t_i} \left(\int_{t_i}^{t_{i+1}} u_s(F)ds \right) (W(t_{i+1}) - W(t_i)) \\
&= \sum_{i=0}^{n-1} u_{t_i}(F)(W(t_{i+1}) - W(t_i)) \\
&\quad + \sum_{i=0}^{n-1} \frac{1}{t_{i+1} - t_i} \left(\int_{t_i}^{t_{i+1}} [u_s(F) - u_{t_i}(F)]ds \right) (W(t_{i+1}) - W(t_i)) \\
&= a_1 + a_2.
\end{aligned}$$

The continuity of $t \to u_t(x)$ in L^2 implies that for each $x \in \mathbb{R}^m$ the Riemann sums

$$\sum_{i=0}^{n-1} u_{t_i}(x)(W(t_{i+1}) - W(t_i))$$

converge in L^2, as $|\pi|$ tends to zero, to the Itô integral $\int_0^1 u_t(x)dW_t$. Moreover, we have for $|x|, |x'| \leq K$,

$$E\left(\left| \sum_{i=0}^{n-1} [u_{t_i}(x) - u_{t_i}(x')](W(t_{i+1}) - W(t_i)) \right|^p \right) \leq c_p c_K|x - x'|^\alpha.$$

Hence, by Lemma 5.3.1 we deduce that a_1 converges in probability to

$$\int_0^1 u_t(x)dW_t \bigg|_{x=F}.$$

Now we will study the convergence of the term a_2. For each fixed $x \in \mathbb{R}^m$, the sums

$$R^\pi(x) := \sum_{i=0}^{n-1} \frac{1}{t_{i+1} - t_i} \left(\int_{t_i}^{t_{i+1}} [u_s(x) - u_{t_i}(x)]ds \right) (W(t_{i+1}) - W(t_i))$$

converge in probability, as $|\pi|$ tends to zero, to the difference

$$V(x) := \int_0^1 u_t(x) \circ dW_t - \int_0^1 u_t(x)dW_t.$$

So it suffices to show that $R^\pi(F)$ converges in probability, as $|\pi|$ tends to zero, to $V(F)$. This convergence follows from Lemma 5.3.1 and the following estimations where $0 < \epsilon < 1$ verifies $\epsilon\alpha > m$, and $|x|, |x'| \le K$,

$$\|R^\pi(x) - R^\pi(x')\|_{p\epsilon}$$

$$\le \sum_{i=0}^{n-1} \frac{1}{t_{i+1} - t_i} \int_{t_i}^{t_{i+1}} \|[u_s(x) - u_{t_i}(x) - u_s(x') + u_{t_i}(x')](W(t_{i+1}) - W(t_i))\|_{p\epsilon} ds$$

$$\le c_{p,\epsilon} \sum_{i=0}^{n-1} |t_{i+1} - t_i|^{\frac{1}{2}} \sup_{s \in [t_i, t_{i+1}]} [E(|u_s(x) - u_{t_i}(x) - u_s(x') + u_{t_i}(x')|^p)]^{\frac{1}{p}}$$

$$\le c_{p,\epsilon} c_K |x - x'|^{\frac{\alpha}{p}}.$$

$\square$

It is also possible to establish substitution formulas similar to those appearing in theorems 5.3.1 and 5.3.3 for random fields $u_t(x)$ which are not adapted. In this case additional regularity assumptions are required. More precisely we need regularity in x of $D_s u_t(x)$ (for the Skorohod integral) and of $(\nabla u(x))_t$ for the Stratonovich integral.

One can also compute the differentials of processes of the form $F_t(X_t)$ where $\{X_t, t \in [0,1]\}$ and $\{F_t(x), t \in [0,1], x \in \mathbb{R}^m\}$ are generalized continuous semimartingales, i.e., they have a bounded variation component and an indefinite Skorohod (or Stratonovich) integral component. This type of change-of-variables formula are known as Itô-Ventzell formulas. Below we show a change-of-variables of this type which will be used in the next chapter (see [85]).

Let us first state the following differentiation rule (see Ocone and Pardoux [85, Lemma 2.3]) .

Lemma 5.3.2 Let $F = (F^1, \ldots, F^m)$ be a random vector whose components belong to $\mathbb{D}^{1,p}$, $p > 1$ and suppose that $|F| \le K$. Consider a measurable random field $u = \{u(x), x \in \mathbb{R}^m\}$ with continuously differentiable paths, such that for any $x \in \mathbb{R}^m$, $u(x) \in \mathbb{D}^{1,r}$, $r > 1$, and the derivative $Du(x)$ has a continuous version as an H-valued process. Suppose that we have

$$E\left(\sup_{|x| \le K} [|u(x)|^r + \|Du(x)\|_H^r]\right) < \infty,$$

$$E\left(\sup_{|x| \le K} |u'(x)|^q\right) < \infty,$$

where $\frac{1}{q} + \frac{1}{p} = \frac{1}{r}$. Then the composition $G = u(F)$ belongs to $\mathbb{D}^{1,r}$, and we have

$$DG = \sum_{i=1}^k (\partial_i u)(F)DF^i + (Du)(F).$$

Sketch of the proof: Approximate the composition $u(F)$ by the integral

$$\int_{\mathbb{R}^m} u(x)\psi_\epsilon(F-x)dx,$$

where ψ_ϵ is an approximation of the identity. □

Let W be an N-dimensional Wiener process. Consider two stochastic processes of the form:

$$X_t^i = X_0^i + \int_0^t u_s^{ij} \circ dW_s^j + \int_0^t v_s^i ds, \quad 1 \le i \le m, \tag{5.28}$$

and

$$F_t(x) = F_0(x) + \int_0^t H_s^i(x) \circ dW_s^i + \int_0^t G_s(x)ds, \tag{5.29}$$

We introduce the following set of hypotheses:

(A1) For all $1 \le i \le N$, $1 \le j \le m$, $u^{ij}, X^j \in \mathbb{L}_2^{1,4}$, $v^j \in L^1([0,1])$ a.s., and the processes X and u are continuous and bounded by constants K and K_1 respectively.

(A2) $x \to F_t(x)$ is of class C^2 for all (t,ω); the processes $F_t(x)$, $F_t'(x)$, and $F_t''(x)$ are continuous in (t,x); $F(x) \in \mathbb{L}^{1,2}$ and there exists a version of $D_sF_t(x)$ which is differentiable in x and for each fixed $s \in [0,1]$ the functions $D_sF_t(x)$, $(D_sF_t)'(x)$ are continuous in the region $\{s \le t \le 1, |x| \le K\}$ (resp. $\{0 \le t \le s, |x| \le K\}$).

(A3) For all $|x| \le K$, $1 \le i \le N$, $H^i(x) \in \mathbb{L}^{1,2}$; the processes $H_s(x)$, and $H_s'(x)$ are continuous in (s,x); and there exists a version of $D_sH_t(x)$ which is differentiable in x and for each fixed $s \in [0,1]$ the functions $D_sH_t(x)$, $(D_sH_t)'(x)$ are continuous in the region $\{s \le t \le 1, |x| \le K\}$ (resp. $\{0 \le t \le s, |x| \le K\}$).

(A4) $x \to G_t(x)$ is continuous.

(A5) The following estimates hold:

$$E\left(\sup_{|x| \le K, t \in [0,1]} \left(|F_t(x)|^4 + |F_t'(x)|^4 + |F_t''(x)|^4 \right. \right.$$
$$\left. \left. + |H_t(x)|^4 + |H_t'(x)|^4 + |G_t(x)^4| \right) \right) < \infty,$$

$$\int_0^1 E\left(\sup_{|x| \le K, t \in [0,1]} \left(|D_sF_t(x)|^4 + |(D_sF_t)'(x)|^4 \right. \right.$$
$$\left. \left. + |D_sH_t(x)|^4 + |(D_sH_t)'(x)|^4 \right) \right) ds < \infty.$$

Theorem 5.3.4 *Assume that the processes X_t and $F_t(x)$ satisfy the above hypotheses. Then $\langle F_t'(X_t), u_t^i \rangle$ and $H_t^i(X_t)$ are elements of $\mathbb{L}_1^{1,2}$, $i = 1, \ldots, m$, and*

$$F_t(X_t) = F_0(X_0) + \int_0^t \langle F_s'(X_s), u_s^i \rangle \circ dW_s^i + \int_0^t \langle F_s'(X_s), v_s \rangle ds$$
$$+ \int_0^t H_s^i(X_s) \circ dW_s^i + \int_0^t G_s(X_s)ds. \tag{5.30}$$

Moreover if (5.28) and (5.29) hold a.s. for all $t \in [0,1]$ on some set $G \in \mathcal{F}$ then (5.30) holds for all $t \in [0,1]$ a.s. on G.

Sketch of the proof: To simplify we will assume $N = 1$. The proof will be done in several steps.

Step 1: Consider an approximation of the identity ψ_ϵ on $\mathbb{R}^m$. For each $x \in \mathbb{R}^m$ we can apply the multidimensional change-of-variables for the Stratonovich integral (Theorem 5.2.3) to the process $F_t(x)\psi_\epsilon(X_t - x)$ and we obtain

$$F_t(x)\psi_\epsilon(X_t - x) = F_0(x)\psi_\epsilon(X_0 - x)$$
$$+ \int_0^t H_s(x)\psi_\epsilon(X_s - x) \circ dW_s + \int_0^t G_s(x)\psi_\epsilon(X_s - x)ds$$
$$+ \int_0^t F_s(x)(\partial_i\psi_\epsilon)(X_s - x)u_s^i \circ dW_s + \int_0^t F_s(x)(\partial_i\psi_\epsilon)(X_s - x)v_s^i ds.$$

If we express the above Stratonovich integrals in terms of the corresponding Skorohod integrals we obtain

$$F_t(x)\psi_\epsilon(X_t - x) = F_0(x)\psi_\epsilon(X_0 - x)$$
$$+ \int_0^t H_s(x)\psi_\epsilon(X_s - x)dW_s + \int_0^t [G_s(x) + \frac{1}{2}(\nabla H)_s(x)]\psi_\epsilon(X_s - x)ds$$
$$+ \frac{1}{2}\int_0^t H_s(x)(\partial_i\psi_\epsilon)(X_s - x)(\nabla X^i)_s ds$$
$$+ \int_0^t F_s(x)(\partial_i\psi_\epsilon)(X_s - x)u_s^i dW_s$$
$$+ \frac{1}{2}\int_0^t (\nabla F)_s(x)(\partial_i\psi_\epsilon)(X_s - x)u_s^i ds + \int_0^t F_s(x)(\partial_i\psi_\epsilon)(X_s - x)(\nabla u^i)_s ds$$
$$+ \int_0^t F_s(x)(\partial_j\partial_i\psi_\epsilon)(X_s - x)(\nabla X^j)_s u_s^i ds$$
$$+ \int_0^t F_s(x)(\partial_i\psi_\epsilon)(X_s - x)v_s^i ds.$$

Notice that all the terms appearing in the above expression have continuous versions in (t, x). Grouping the Skorohod integral terms and the Lebesgue integral terms we can write

$$F_t(x)\psi_\epsilon(X_t - x) = F_0(x)\psi_\epsilon(X_0 - x) + \int_0^t \alpha_\epsilon(s, x)dW_s + \int_0^t \beta_\epsilon(s, x)ds.$$

Step 2: Consider the processes $H_s(X_s)$ and $\langle F_s'(X_s), u_s \rangle$. We claim that these processes belong to the space $\mathbb{L}^{1,2}$. In fact, first notice that for each x, $F'(x)$ belongs to $\mathbb{L}^{1,2}$ and $D_s(F_t'(x)) = (D_sF_t)'(x)$. This follows from hypotheses (A2) and (A5). On the other hand, $X, u \in \mathbb{L}^{1,4}$. Then we can apply Lemma 5.3.2 (suitably extended to stochastic processes) to $H_t(x)$ and X_t, and also to $\langle F_t'(x), u_t \rangle$ and X_t taking into account the following estimates that

$$\int_0^1 E(|H_s(X_s)|^2)ds \leq E\int_0^1 \sup_{|x|\leq K} |H_s(x)|^2 ds < \infty,$$

$$E\int_0^1 \sup_{|x|\leq K} \|DH_s(x)\|_H^2 ds < \infty,$$

$$E\int_0^1 \sup_{|x|\leq K} |H_s'(x)|^4 ds < \infty,$$

$$\int_0^1 E(|\langle F_s'(X_s), u_s\rangle|^2)ds \le \|u\|_\infty^2 E \int_0^1 \sup_{|x|\le K} |F_s'(x)|^2 ds < \infty,$$

$$\int_0^1 E(|F_s'(X_s)|^2 \|Du_s\|_H^2)ds \le \left\{ E \int_0^1 \sup_{|x|\le K} |F_s'(x)|^4 ds E \int_0^1 \|Du_s\|_H^4 ds \right\}^{\frac{1}{2}} < \infty,$$

$$\int_0^1 E(|F_s''(X_s)|^2 \|DX_s\|_H^2 u_s^2)ds \le \|u\|_\infty^2$$

$$\times \left\{ E \int_0^1 \sup_{|x|\le K} |F_s''(x)|^4 ds E \int_0^1 \|DX_s\|_H^4 ds \right\}^{\frac{1}{2}} < \infty,$$

$$\int_0^1 E(\|(DF_s)'(X_s)\|_H^2 u_s^2)ds \le \|u\|_\infty^2 E \int_0^1 \sup_{|x|\le K} \|(DF_s)'(x)\|_H^2 ds < \infty,$$

The derivatives of these processes are then given by

$$D_s[H_t(X_t)] = \langle H_t'(X_t), D_s X_t \rangle + (D_s H_t)(X_t),$$

and

$$D_s[\langle F_t'(X_t), u_t\rangle] = \langle F_t''(X_t), D_s X_t \otimes u_t \rangle + \langle (D_s F_t)'(X_t), u_t \rangle + \langle F_t'(X_t), D_s u_t \rangle.$$

Then from our hypotheses it follows that the processes $H_s(X_s)$ and $\langle F_s'(X_s), u_s\rangle$ belong to $\mathbb{L}_1^{1,2}$, and

$$(\nabla[H_t(X_t)])_t = \langle H_t'(X_t), (\nabla X)_t \rangle + (\nabla H)_t(X_t),$$
$$(\nabla[\langle F' \cdot (X.), u.\rangle])_t = \langle F_t''(X_t), (\nabla X)_t \otimes u_t \rangle + \langle (\nabla F)_t'(X_t), u_t \rangle + \langle F_t'(X_t), (\nabla u)_t \rangle.$$

Step 3: We claim that

$$\int_{\mathbb{R}^m} \left(F_t(x)\psi_\epsilon(X_t - x) - F_0(x)\psi_\epsilon(X_0 - x) - \int_0^t \beta_\epsilon(s, x)ds \right) dx \tag{5.31}$$

converges in $L^1(\Omega)$ to the process

$$\begin{aligned} \Phi_t &= F_t(X_t) - F_0(X_0) - \int_0^t \Big[G_s(X_s) + \frac{1}{2}(\nabla H)_s(X_s) \\ &\quad + \frac{1}{2}(\partial_i H)_s(X_s)(\nabla X^i)_s + \frac{1}{2}\partial_i(\nabla F)_s(X_s) u_s^i \\ &\quad + (\partial_i F)_s(X_s)(\nabla u^i)_s + (\partial_j \partial_i F)_s(X_s)(\nabla X^j)_s u_s^i \\ &\quad + (\partial_i F)_s(X_s)v_s^i \Big] ds. \end{aligned}$$

In fact, the convergence for each (s, ω) follows from the continuity assumption on x of the processes $G_s(x)$, $(\nabla H)_s(x)$, $\partial_i(\nabla F)_s(x)$, and the fact that $H_s(x)$ is of class C^1, and $F_s(x)$ is of class C^2. The $L^1(\Omega)$ convergence follows by dominated convergence because we have

$$E \int_0^1 \sup_\epsilon \left(\int_{\mathbb{R}^m} |\beta_\epsilon(s, x)|dx \right) ds < \infty.$$

From Step 2 it follows that

$$\Phi_t = F_t(X_t) - F_0(X_0) - \int_0^t [G_s(X_s) + \langle F_s'(X_s), v_s \rangle] ds$$
$$- \int_0^t [H_s(X_s) + \langle F_s'(X_s), u_s \rangle] \circ dW_s$$
$$+ \int_0^t [H_s(X_s) + \langle F_s'(X_s), u_s \rangle] dW_s.$$

Step 4: Finally, we have

$$\int_{\mathbb{R}^m} \alpha_\epsilon(s, x) dx = \int_{\mathbb{R}^m} H_s(x) \psi_\epsilon(X_s - x) dx$$
$$+ \int_{\mathbb{R}^m} F_s(x)(\partial_i \psi_\epsilon)(X_s - x) u_s^i dx$$
$$= \int_{\mathbb{R}^m} \left[H_s(x) + (\partial_i F_s)(x) u_s^i \right] \psi_\epsilon(X_s - x) dx.$$

The integrals $\int_{\mathbb{R}^m} \alpha_\epsilon(s, x) dx$ converge as ϵ tends to zero in the norm of $\mathbb{L}^{1,2}$ to $H_s(X_s) + \langle F_s'(X_s), u_s \rangle$. Hence, the Skorohod integral

$$\int_0^t \left(\int_{\mathbb{R}^m} \alpha_\epsilon(s, x) dx \right) dW_s \tag{5.32}$$

converges in $L^2(\Omega)$ to

$$\int_0^t [H_s(X_s) + \langle F_s'(X_s), u_s \rangle] dW_s.$$

By Fubini's theorem (5.32) coincides with expression (5.31). Consequently, we get

$$\Phi_t = \int_0^t [H_s(X_s) + \langle F_s'(X_s), u_s \rangle] dW_s,$$

which completes the proof. □

Bibliographical notes: There are different methods to define stochastic integrals of non adapted integrands with respect to the Wiener process. One approach, developed by Ogawa [87, 88, 89] and Rosinski [94] is based on the orthogonal development of the integrand in $L^2([0, 1])$. Another possibility is to use Riemann sums that converge to the so-called forward integral, which is equal to $\delta(u) + \int_0^1 D_t^- u_t dt$. This stochastic integral is an extension of the Itó integral, and it follows the same change-of-variable rule. Berger and Mizel [13] introduced this integral in order to solve stochastic Volterra equations. Russo and Vallois [95] introduced the forward integral considering the approximation of the Brownian path by its convolution with a rectangular function. In [56] Kuo and Russek study anticipating stochastic integrals in the context of the white noise analysis. Different versions of the change-of-variables formula for the Skorohod integral can be found in Sevljakov [97], Hitsuda [43], Üstünel [105], and Sekiguchi and Shiota [96]. The indefinite Skorohod integral has properties similar to that of a semimartingale. In particular, one has established the existence of a continuous local time and a Tanaka's type formula in [47] and [48].

Chapter 6

Anticipating stochastic differential equations

The anticipating stochastic calculus developed in Chapter 5 can be used to formulate stochastic differential equations where the solutions are nonadapted processes. Such type of equations appear, for instance, when the initial condition is not independent of the Wiener process, or when we impose a condition relating the values of the process at the initial and final times.

6.1 Stochastic differential equations in the Stratonovich sense

In this section we will present examples of stochastic differential equations formulated using the extended Stratonovich integral. These equations can be solved taking into account that this integral satisfies the usual differentiation rules.

Stochastic differential equations with random initial condition

Suppose that $W = \{W(t), t \in [0,1]\}$ is a d-dimensional Brownian motion on the canonical probability space $(\Omega, \mathcal{F}, P)$. Let $A_i : \mathbb{R}^m \to \mathbb{R}^m$, $0 \le i \le d$, be continuous functions. Consider the stochastic differential equation

$$X_t = X_0 + \sum_{i=1}^{d} \int_0^t A_i(X_s) \circ dW_s^i + \int_0^t A_0(X_s) ds, \qquad t \in [0,1]. \tag{6.1}$$

The initial condition is an arbitrary random variable $X_0 \in L^0(\Omega; \mathbb{R}^m)$.

Suppose that the coefficients A_i, $1 \le i \le d$, are continuously differentiable with bounded derivatives. Define $B = A_0 + \frac{1}{2} \sum_{k=1}^{m} \sum_{i=1}^{d} A_i^k \partial_k A_i$, and suppose moreover that B is Lipschitz.

Let $\{\varphi_t(x), t \in [0,1]\}$ be the solution to the following stochastic differential equation starting at point $x \in \mathbb{R}^m$:

$$\varphi_t(x) = x + \sum_{i=1}^{d} \int_0^t A_i(\varphi_s(x)) dW_s^i + \int_0^t B(\varphi_s(x)) ds. \tag{6.2}$$

In terms of the Stratonovich stochastic integral we can write

$$\varphi_t(x) = x + \sum_{i=1}^{d} \int_0^t A_i(\varphi_s(x)) \circ dW_s^i + \int_0^t A_0(\varphi_s(x)) ds.$$

We know (see, for instance, Kunita [54]) that there exists a version of $\varphi_t(x)$ such that $(t, x) \hookrightarrow \varphi_t(x)$ is continuous, and we have

$$E(|\varphi_s(x) - \varphi_t(x')|^p) \leq C_{p,K}(|t - s|^{\frac{p}{2}} + |x - x'|^p).$$

for all $s, t \in [0, 1]$, $|x|, |x'| \leq K$, $p \geq 2$, $K > 0$. Then we can establish the following result

Theorem 6.1.1 *Assume that the coefficients A_i, $1 \leq i \leq d$ are of class C^3, and A_i, $1 \leq i \leq d$, B have bounded partial derivatives of first order. Then for any random vector X_0, the process $X = \{\varphi_t(X_0), t \in [0, 1]\}$ satisfies the anticipating stochastic differential equation (6.1).*

Proof: Under the assumptions of the theorem we know that for any $x \in \mathbb{R}^m$ Eq. (6.2) has a unique solution $\varphi_t(x)$. Set $X = \{\varphi_t(X_0), 0 \leq t \leq 1\}$. By a localization argument we can assume that the coefficients A_i, $1 \leq i \leq d$, and B have compact support. It suffices to show that for any $i = 1, \ldots, d$, the process $u_i(t, x) = A_i(\varphi_t(x))$ satisfies the hypotheses of Theorem 5.3.3, suitably extended to a d-dimensional Brownian motion.

Condition (h1) is obvious. On the other hand, using Itô's formula one can easily check condition (h4). This completes the proof. $\square$

The uniqueness is a more difficult problem. We are going to show that there is a unique solution in the class of processes $\mathbb{L}_{2,\text{loc}}^{1,4}(\mathbb{R}^m)$, assuming some additional regularity conditions on the initial condition X_0. We will denote by $L(\mathbb{R}^m)$ the space of m-dimensional processes (or random variables) whose components belong to the class L.

Lemma 6.1.1 *Suppose that $X_0 \in \mathbb{D}_{\text{loc}}^{1,p}(\mathbb{R}^m)$ for some $p > q \geq 2$, and assume that the coefficients A_i, $1 \leq i \leq d$, and B are of class C^2 with bounded partial derivatives of first order, then we claim that $\{\varphi_t(X_0), 0 \leq t \leq 1\}$ belongs to $\mathbb{L}_{2,\text{loc}}^{1,q}(\mathbb{R}^m)$.*

Proof: Let (Ω^n, X_0^n) be a localizing sequence for X_0 in $\mathbb{D}_{\text{loc}}^{1,p}(\mathbb{R}^m)$. Set

$$G^{n,k,M} = \Omega^n \cap \{|X_0^n| \leq k\} \cap \{ \sup_{|x| \leq k, t \in [0,1]} |\varphi_t(x)| \leq M\}.$$

On the set $G^{n,k,M}$ the process $\varphi_t(X_0)$ coincides with $\varphi_t^M(X_0^n \beta_k(X_0^n))$, where β_k is a smooth function such that $0 \leq \beta_k \leq 1$, $\beta_k(x) = 1$ if $|x| \leq k$, and $\beta_k(x) = 0$ if $|x| \geq k + 1$, and $\varphi_t^M(x)$ is the stochastic flow associated with the coefficients $\beta_M A_i$, $1 \leq i \leq d$, and $\beta_M B$. Hence, we can assume that the coefficients A_i, $1 \leq i \leq d$, and B are bounded and have bounded derivatives up to the second order, $X_0 \in \mathbb{D}^{1,p}(\mathbb{R}^m)$, and X_0 is bounded by k. Now we can apply Lemma 5.3.2. The following estimates hold:

$$E \left(\sup_{|x| \leq k, t \in [0,1]} [|\varphi_t(x)|^r + \|D[\varphi_t(x)]\|_H^r] \right) < \infty, \tag{6.3}$$

$$E \left(\sup_{|x| \leq k, t \in [0,1]} |\varphi_t'(x)|^r \right) < \infty, \tag{6.4}$$

for any $r \geq 2$. These estimate (6.3) and (6.4) follow from our assumptions on the coefficients A_i, $1 \leq i \leq d$, and B, taking into account that

$$
\begin{aligned}
D_s^j[\varphi_t(x)] \;=\; & A_j(\varphi_s(x)) + \int_s^t (\partial_k A_l)(\varphi_r(x)) D_s^j[\varphi_r^k(x)] dW_r^l \\
& + \int_s^t (\partial_k B)(\varphi_r(x)) D_s^j[\varphi_r^k(x)] dr,
\end{aligned}
$$

for $1 \leq j \leq d$ and $0 \leq s \leq t \leq 1$. Hence, from Lemma 5.3.2 we obtain that $\{\varphi_t(X_0), 0 \leq t \leq 1\}$ belongs to $\mathbb{L}^{1,q}(\mathbb{R}^m)$ and that

$$
D_s[\varphi_t(X_0)] = \varphi_t'(X_0) D_s X_0 + [D_s \varphi_t](X_0).
$$

Finally, from the above expression for the derivative of $\varphi_t(X_0)$ it is not difficult to show that $\{\varphi_t(X_0), 0 \leq t \leq 1\}$ belongs to $\mathbb{L}_2^{1,q}(\mathbb{R}^m)$, and

$$
(D[\varphi(X_0)])_s^{+,-} = \varphi_s'(X_0) D_s X_0 + (D\varphi)_s^{+,-}(X_0).
$$

$\square$

Theorem 6.1.2 *Assume that the coefficients A_i, $1 \leq i \leq d$ and B are of class C^4, with bounded partial derivatives of first order. The initial condition X_0 belongs to $\mathbb{D}_{loc}^{1,p}(\mathbb{R}^m)$ for some $p > 4$. Then the process $X = \{\varphi_t(X_0), t \in [0,1]\}$ is the unique solution to Eq. (6.1) in $\mathbb{L}_{2,loc}^{1,4}(\mathbb{R}^m)$ which is continuous.*

Proof: By Lemma 6.1.1 we already know that X belongs to $\mathbb{L}_{2,loc}^{1,4}(\mathbb{R}^m)$. Let Y be another continuous solution in this space. Suppose that (Ω^n, X_0^n) is a localizing sequence for X_0 in $\mathbb{D}_{loc}^{1,p}(\mathbb{R}^m)$, and $(\Omega^{n,1}, Y^n)$ is a localizing sequence for Y in $\mathbb{L}_2^{1,4}(\mathbb{R}^m)$. Set

$$
H^{n,k,M} = \Omega^n \cap \Omega^{n,1} \cap \{|X_0^n| \leq k\} \cap \{ \sup_{|x| \leq k, t \in [0,1]} |\varphi_t(x)| \leq M \} \cap \{ \sup_{t \in [0,1]} |Y_t| \leq M \}.
$$

The processes $Y_t^n \beta_M(Y_t^n)$, $(\beta_M A_i)(Y_t^n)$, and $(\beta_M A_0)(Y_t^n)$ satisfy hypothesis (A1). On the set $H^{n,k,M}$ we have the following equality

$$
Y_t = X_0 + \sum_{i=1}^d \int_0^t A_i(Y_s) \circ dW_s^i + \int_0^t A_0(Y_s) ds, \quad t \in [0,1].
$$

Let us denote by $\varphi_t^{-1}(x)$ the inverse of the mapping $x \to \varphi_t(x)$. By the classical Itô's formula we can write

$$
\varphi_t^{-1}(x) = x - \sum_{i=1}^d \int_0^t (\varphi_s'(\varphi_s^{-1}(x)))^{-1} A_i(x) \circ dW_s^i - \int_0^t (\varphi_s'(\varphi_s^{-1}(x)))^{-1} A_0(x) ds.
$$

We claim that the processes $F_t(x) = \varphi_t^{-1}(x)$, $H_t^i(x) = (\varphi_s'(\varphi_t^{-1}(x)))^{-1} A_i(x)$, and $G_t(x) = (\varphi_s'(\varphi_t^{-1}(x)))^{-1} A_0(x)$ with values in the space of $m \times m$ matrices satisfy hypotheses (A2) to (A5). This follows from the properties of the stochastic flow associated with the coefficients A_j.

Consequently, we can apply Theorem 5.3.4 and we obtain, on the set $H^{n,k,M}$,

$$\varphi_t^{-1}(Y_t) = X_0 + \int_0^t (\varphi_s^{-1})'(Y_s)A_i(Y_s) \circ dW_s^i + \int_0^t (\varphi_s^{-1})'(Y_s)A_0(Y_s)ds$$
$$- \int_0^t (\varphi_s'(\varphi_s^{-1}(Y_s)))^{-1}A_i(Y_s) \circ dW_s^i - \int_0^t (\varphi_s'(\varphi_s^{-1}(Y_s)))^{-1}A_0(Y_s)ds.$$

Notice that

$$(\varphi_s^{-1})'(x) = -(\varphi_s'(\varphi_s^{-1}(x)))^{-1}.$$

Hence, we get $\varphi_t^{-1}(Y_t) = X_0$, that is, $Y_t = \varphi_t(X_0)$ which completes the proof of the uniqueness. □

One-dimensional stochastic differential equations with random coefficients

Consider the one-dimensional equation

$$X_t = X_0 + \int_0^t \sigma(X_s) \circ dW_s + \int_0^t b(X_s)ds, \tag{6.5}$$

where the coefficients σ and b and the initial condition X_0 are random. If the coefficients and the initial condition are deterministic and we assume that the function σ is of class C^2 with bounded first and second derivaties, and that b is Lipschitz, then there is a unique solution X_t. This solution (see Doss [26]) can be written as $X_t = u(W_t, Y_t)$, where $u : \mathbb{R}^2 \to \mathbb{R}$ is the solution of the ordinary differential equation

$$\frac{\partial u}{\partial x} = \sigma(u(x)), \quad u(0, y) = y,$$

and Y_t is the solution to the ordinary differential equation with random parameters

$$Y_t = X_0 + \int_0^t f(W_s, Y_s)ds,$$

where $f(x, y) = b(u(x, y)) \exp\left(-\int_0^x \sigma'(u(z, y))dz\right)$. Notice that

$$\frac{\partial u}{\partial y} = \exp\left(\int_0^x \sigma'(u(z, y))dz\right).$$

Suppose that now the coefficients and the initial condition are random. We claim that in that case the process $X_t = u(W_t, Y_t)$ solves Eq. (6.5).

Consider a partition $\pi = \{0 = t_0 < t_1 < \cdots < t_n = 1\}$ of the interval $[0, 1]$. We can introduce the polygonal approximation of the Brownian motion defined by

$$W_t^\pi = \int_0^t \sum_{i=0}^{n-1} \left(\frac{W(t_{i+1}) - W(t_i)}{t_{i+1} - t_i}\right) \mathbf{1}_{[t_i, t_{i+1}]}(s)ds.$$

Set $X_t^\pi = u(W_t^\pi, Y_t)$. This process satisfies

$$X_t^\pi = X_0 + \int_0^t \sigma(X_s^\pi)\dot{W}_s^\pi ds + \int_0^t \exp\left(\int_{W_s}^{W_s^\pi} \sigma'(u(y, Y_s))dy\right) b(X_s)ds = A_t^1 + A_t^2.$$

The term A_t^1 can be written as

$$A_t^1 = \int_0^t [\sigma(X_s^\pi) - \sigma(X_s)]\dot{W}_s^\pi ds + \int_0^t \sigma(X_s)\dot{W}_s^\pi ds = B_t^1 + B_t^2.$$

Note that B_t^2 is the Riemann sum that approximates the Stratonovich integral of $\sigma(X_s)$. We know that A_t^1 converges in probability to $X_t - X_0 - \int_0^t b(X_s)ds$ as $|\pi|$ tends to zero. As a consequence, if we show that B_t^1 converges in probability to zero as $|\pi|$ tends to zero this will imply $\sigma(X_t)$ is Stratonovich integrable on any interval $[0, t]$ and (6.5) holds.

We have

$$\sigma(X_s^\pi) - \sigma(X_s) = \sigma\sigma'(X_s)(W_s^\pi - W_s) + \frac{1}{2}[\sigma(\sigma')^2 + \sigma^2\sigma''](\xi)(W_s^\pi - W_s)^2,$$

where ξ is a random intermediate point between W_s^π and W_s. The last summand of the above expression does not contribute to the limit of B_t^1. Finally the convergence to zero of B_t^1 follows from Lemma 6.1.2.

Lemma 6.1.2 *Let Φ_t be a continuous process. Then, the integral*

$$\int_0^1 \Phi_s \dot{W}_s^\pi (W_s^\pi - W_s) ds$$

converge in probability to zero as $|\pi|$ tends to zero, where $\pi = \{0 = t_0 < t_1 < \cdots < t_n = 1\}$ runs over all partitions of the interval $[0, 1]$.

Proof: Fix a partition $\pi = \{0 = t_0 < t_1 < \cdots < t_n = 1\}$. Consider another partition of the interval $[0, 1]$ of the form $\{s_j := \frac{j}{m}, 0 \le j \le m\}$. We can write

$$\left| \int_0^1 \Phi_s \dot{W}_s^\pi (W_s^\pi - W_s) ds \right| \le \left| \sum_{j=0}^{m-1} \Phi_{s_j} \int_{s_j}^{s_{j+1}} \dot{W}_s^\pi (W_s^\pi - W_s) ds \right|$$

$$+ \left| \sum_{j=0}^{m-1} \int_{s_j}^{s_{j+1}} (\Phi_s - \Phi_{s_j}) \dot{W}_s^\pi (W_s^\pi - W_s) ds \right| = C_1 + C_2.$$

The term C_1 converges to zero in probability as $|\pi|$ tends to zero, for any fixed value of m. In fact, if the points s_j and s_{j+1} belong to the partition π, we have

$$\int_{s_j}^{s_{j+1}} \dot{W}_s^\pi (W_s^\pi - W_s) ds = \sum_{i:t_i \in [s_j, s_{j+1})} \frac{W(t_{i+1}) - W(t_i)}{t_{i+1} - t_i} \left(\int_{t_i}^{t_{i+1}} (W_s^\pi - W_s) ds \right)$$

$$= -\frac{1}{2} \sum_{i:t_i \in [s_j, s_{j+1})} \frac{1}{t_{i+1} - t_i} \int_{t_i}^{t_{i+1}} [(W_{t_{i+1}} - W_s)^2 - (W_s - W_{t_i})^2] ds,$$

and this converges to $\frac{1}{2}(s_{j+1} - s_j) - \frac{1}{2}(s_{j+1} - s_j) = 0$.

The term C_2 can be bounded as follows:

$$C_2 \le \sup_{|s-t| \le \frac{1}{m}} |\Phi_s - \Phi_t| \sum_{j=0}^{m-1} \int_{s_j}^{s_{j+1}} |\dot{W}_s^\pi (W_s^\pi - W_s)| ds$$

$$\le \sup_{|s-t| \le \frac{1}{m}} |\Phi_s - \Phi_t|$$

$$\times \sum_{j=0}^{m-1} \sum_{i:t_i \in [s_j, s_{j+1})} \frac{1}{2(t_{i+1} - t_i)} \int_{t_{i+1}}^{t_i} [(W_{t_{i+1}} - W_s)^2 + (W_s - W_{t_i})^2] ds.$$

As n tends to infinity the right-hand side of the above inequality converges in probability to $\frac{1}{2}$. Hence, we obtain

$$\limsup_{n} C_2 \leq \frac{1}{2} \sup_{|s-t|\leq\frac{1}{m}} |\Phi_s - \Phi_t|,$$

and this expression converges to zero in probability as m tends to zero, due to the continuity of the process Φ_t. □

The uniqueness requires the use of the Itô-Ventzell formula, and we have to impose regularity assumptions on the coefficients (cf. [51]).

6.2 Stochastic differential equations with boundary conditions

Consider the following Stratonovich differential equation on the time interval $[0, 1]$, where instead of giving the value X_0, we impose a linear relationship between the values of X_0 and X_1:

$$\begin{cases} dX_t = \sum_{i=1}^{d} A_i(X_t) \circ dW_t^i + A_0(X_t)dt \\ h(X_0, X_1) = 0 \end{cases} \tag{6.6}$$

We are interested in proving the existence and uniqueness of a solution for this type of equations. We will discuss two particular cases:

(a) The case where the coefficients A_i, $0 \leq i \leq d$ and the function h are affine (see Ocone and Pardoux [86]).

(b) The one-dimensional case (see Donati-Martin [25]).

Linear stochastic differential equations with boundary conditions

Consider the following stochastic boundary value problem:

$$\begin{cases} X_t = X_0 + \sum_{i=1}^{d} \int_0^t A_i X_s \circ dW_s^i + A_0 \int_0^t X_s ds, \\ H_0 X_0 + H_1 X_1 = h. \end{cases} \tag{6.7}$$

We assume that A_i, $i = 0, \ldots, d$, H_0, and H_1 are $m \times m$ deterministic matrices and $h \in \mathbb{R}^m$. We will also assume that the $m \times 2m$ matrix $H_0 : H_1$ has rank m.

Concerning the boundary condition, two particular cases are interesting:

Two-point boundary-value problem: Let $l \in \mathbb{N}$ be such that $0 < l < m$. Suppose that $H_0 = \begin{pmatrix} H_0' \\ 0 \end{pmatrix}$, $H_1 = \begin{pmatrix} 0 \\ H_1'' \end{pmatrix}$, where H_0' is an $l \times m$ matrix and H_1'' is an $(m - l) \times m$ matrix. Condition $\operatorname{rank}(H_0 : H_1) = m$ implies that H_0' has rank l and that H_1'' has rank $m - l$. If we write $h = \begin{pmatrix} h_0 \\ h_1 \end{pmatrix}$, where $h_0 \in \mathbb{R}^l$ and $h_1 \in \mathbb{R}^{m-l}$, then the boundary condition becomes

$$H_0' X_0 = h_0, \quad H_1' X_1 = h_1.$$

Periodic solutions to SDE's: Suppose $H_0 = -H_1 = I$ and $h = 0$. Then the boundary condition becomes

$$X_0 = X_1.$$

With (6.7) we can associate an $m \times m$ adapted and continuous matrix-valued process Φ solution of

$$\begin{cases} d\Phi_t = \sum_{i=1}^d A_i \Phi_t \circ dW_t^i + B\Phi_t dt, \\ \Phi_0 = I. \end{cases} \tag{6.8}$$

Using the properties of the extended Stratonovich integral, one can obtain an explicit solution to Eq. (6.7). By a solution we mean a continuous stochastic process X such that $A_i(X_s)$ is Stratonovich integrable with respect to W^i on any interval $[0, t]$ and such that (6.7) holds.

Theorem 6.2.1 *Suppose that the random matrix $H_0 + H_1\Phi_1$ is a.s. invertible. Then there exist a solution to the stochastic differential equation (6.7), which is unique among those continuous processes whose components belong to the space $\mathbb{L}_{2,\text{loc}}^{1,4}$.*

Proof: Define

$$X_t = \Phi_t X_0, \tag{6.9}$$

where X_0 is given by

$$X_0 = \left[H_0 + H_1\Phi_1 \right]^{-1} h. \tag{6.10}$$

Then it follows from this expression that X^i belongs to $\mathbb{L}_{2,\text{loc}}^{1,4}$, for all $i = 1, \ldots, m$, and this process satisfies Eq. (6.7).

In order to show the uniqueness, we proceed as follows. Let $Y \in \mathbb{L}_{2,\text{loc}}^{1,4}(\mathbb{R}^m)$ be a solution to (6.7). Then we have

$$\Phi_t^{-1} = I - \sum_{i=1}^d \int_0^t \Phi_s^{-1} A_i \circ dW_s^i - \int_0^t \Phi_s^{-1} A_0 ds.$$

By the change-of-variables formula for the Stratonovich integral (Theorem 5.2.2), we see that $\Phi_t^{-1} Y_t = Y_0$, namely, $Y_t = \Phi_t Y_0$. Therefore, Y satisfies (6.9). But (6.10) follows from (6.9) and the boundary condition $H_0 Y_0 + H_1 Y_1 = h$. Consequently, Y satisfies (6.9) and (6.10), and it must be equal to X. $\square$

Notice that, in general, the solution will not belong to $\mathbb{L}_2^{1,4}(\mathbb{R}^m)$. One can also treat more general linear equations. For instance, suppose that we have the equation:

$$\begin{cases} X_t = X_0 + \sum_{i=1}^d \int_0^t A_i X_s \circ dW_s^i + \int_0^t A_0 X_s ds + V_t, \\ H_0 X_0 + H_1 X_1 = h, \end{cases} \tag{6.11}$$

where V_t is a continuous semimartingale. In that case,

$$X_t = \Phi_t \left[H_0 + H_1\Phi_1 \right]^{-1} \left[h - H_1\Phi_1 \int_0^1 \Phi_s^{-1} \circ dV_s \right] + \Phi_t \int_0^t \Phi_s^{-1} \circ dV_s,$$

is a solution to equation (6.11). The uniqueness in the class of processes $\mathbb{L}_{2,\text{loc}}^{1,4}(\mathbb{R}^m)$ can be established provided the process V_t belongs also to this space.

One-dimensional stochastic differential equations with boundary conditions

Consider the one-dimensional stochastic boundary value problem

$$\begin{cases} X_t = X_0 + \int_0^t \sigma(X_s) \circ dW_s + \int_0^t b(X_s)ds \\ a_0 X_0 + a_1 X_1 = a_2. \end{cases} \tag{6.12}$$

Applying the techniques of the anticipating stochastic calculus we can show the following result.

Theorem 6.2.2 *Suppose that the functions σ and $b_1 := b + \frac{1}{2}\sigma\sigma'$ are of class C^2 with bounded derivatives and $a_0 a_1 > 0$. Then there exists a solution to (6.12). Furthermore if the functions σ and b_1 are of class C^4 with bounded derivatives then the solution is unique in the class $\mathbb{L}^{1,4}_{2,\text{loc}}$.*

Proof: Let $\varphi_t(x)$ be the stochastic flow associated with the coefficients σ and b_1. By Theorem 6.1.1 for any random variable X_0 the process $\varphi_t(X_0)$ satisfies

$$X_t = X_0 + \int_0^t \sigma(X_s) \circ dW_s + \int_0^t b(X_s)ds.$$

Hence, in order to show the existence of a solution it suffices to prove that there is a unique random variable X_0 such that

$$\varphi_1(X_0) = \frac{a_2 - a_0 X_0}{a_1}. \tag{6.13}$$

The mapping $g(x) = \varphi_1(x)$ is strictly increasing and this implies the existence of a unique solution to equation (6.13).

Let us now turn to the proof of the uniqueness. It suffices to show that the unique solution X_0 to equation (6.13) belongs to $\mathbb{D}^{1,p}_{\text{loc}}$ for some $p > 4$. Taking into account that the equation is one dimensional, one can represent the solution $\varphi_t(x)$ by means of the method of Doss. Hence, $\varphi_t(x)$ is Fréchet differentiable in ω. Using this fact, and the implicit function theorem it is not difficult to show that X_0 belongs to the space $\mathbb{D}^{1,p}_{\text{loc}}$. $\qquad\square$

6.3 Stochastic differential equations in the Skorohod sense

Let $W = \{W_t, t \in [0,1]\}$ be a one-dimensional Brownian motion defined on the canonical probability space $(\Omega, \mathcal{F}, P)$. Consider the stochastic differential equation

$$X_t = X_0 + \int_0^t \sigma(s, X_s)dW_s + \int_0^t b(s, X_s)ds, \tag{6.14}$$

$0 \le t \le 1$, where X_0 is $\mathcal{F}_1$-measurable and σ and b are deterministic functions. The stochastic integral $\int_0^t \sigma(s, X_s)dW_s$ is interpreted in the Skorohod sense. First notice that the usual Picard iteration procedure cannot be applied to show that this equation has a unique solution. Indeed, the Sokorohod integral is not a bounded operator in

L^p. In some sense Eq. (6.14) is an infinite dimensional hyperbolic partial differential equation. In fact, this equation can be formally written as

$$X_t = X_0 + \int_0^t \sigma(s, X_s) \circ dW_s + \frac{1}{2}\int_0^t \sigma'(s, X_s)[(D^+X)_s + (D^-X_s)]ds + \int_0^t b(s, X_s)ds.$$

If the diffusion coefficient is linear one can show that there exist a unique global solution. In fact, let us consider the following particular case:

$$X_t = X_0 + \sigma\int_0^t X_s dW_s. \tag{6.15}$$

When X_0 is deterministic, the solution is given by the martingale

$$X_t = X_0 e^{\sigma W_t - \frac{1}{2}\sigma^2 t}.$$

We have seen in (5.4) that if $X_0 = \text{sign}W_1$, then, a solution to Eq. (6.15) is

$$X_t = \text{sign}(W_1 - \sigma t)e^{\sigma W_t - \frac{1}{2}\sigma^2 t}.$$

More generally, one can show (see Theorem 6.3.1 below) that if $X_0 \in L^p(\Omega)$ for some $p > 2$, then

$$X_t = X_0(A_t)e^{\sigma W_t - \frac{1}{2}\sigma^2 t}$$

is a solution to Eq. (6.15) where $A_t(\omega)_s = \omega_s - \sigma(t \wedge s)$. In terms of the Wick product (see [22]) one can write

$$X_t = X_0 \Diamond e^{\sigma W_t - \frac{1}{2}\sigma^2 t} = X_0(A_t)e^{\sigma W_t - \frac{1}{2}\sigma^2 t}.$$

Consider the following equation:

$$X_t = X_0 + \int_0^t \sigma_s X_s dW_s + \int_0^t b(s, X_s)ds, \qquad 0 \le t \le 1, \tag{6.16}$$

where $\sigma \in L^2([0,1])$, X_0 is a random variable, and b is a random function satisfying the following condition:

(H.1) $b : [0,1] \times \mathbb{R} \times \Omega \to \mathbb{R}$ is a measurable function such that there exist an nonnegative function γ_t on $[0,1]$, $\gamma_t \ge 0$, a constant $L > 0$, and a set $N_1 \in \mathcal{F}$ of probability one, verifying

$$|b(t, x, \omega) - b(t, y, \omega)| \le \gamma_t |x - y|, \quad \int_0^1 \gamma_t dt \le L,$$
$$|b(t, 0, \omega)| \le L,$$

for all $x, y \in \mathbb{R}$, $t \in [0,1]$ and $\omega \in N_1$.

Let us introduce some notation. Consider the family of transformations T_t, A_t : $\Omega \to \Omega$, $t \in [0,1]$, given by

$$T_t(\omega)_s = \omega_s + \int_0^{t \wedge s} \sigma_u du,$$
$$A_t(\omega)_s = \omega_s - \int_0^{t \wedge s} \sigma_u du.$$

Note that $T_t A_t = A_t T_t = \text{Identity}$. Define

$$\varepsilon_t = \exp\left(\int_0^t \sigma_s dW_s - \frac{1}{2}\int_0^t \sigma_s^2 ds\right).$$

Then, by Girsanov's theorem $E[F(A_t)\varepsilon_t] = E[F]$ for any random variable $F \in L^1(\Omega)$. For each $x \in \mathbb{R}$ and $\omega \in \Omega$ we denote by $Z_t(\omega, x)$ the solution of the integral equation

$$Z_t(\omega, x) = x + \int_0^t \varepsilon_s^{-1}(T_t(\omega))\, b\big(s,\, \varepsilon_s(T_t(\omega))Z_s(\omega, x),\, T_s(\omega)\big)ds. \qquad (6.17)$$

Notice that for $s \leq t$ we have $\varepsilon_s(T_t) = \exp\left(\int_0^s \sigma_u dW_u + \frac{1}{2}\int_0^s \sigma_u^2 du\right) = \varepsilon_s(T_s)$. Henceforth we will omit the dependence on ω in order to simplify the notation.

Theorem 6.3.1 *Fix an initial condition $X_0 \in L^p(\Omega)$ for some $p > 2$, and define*

$$X_t = \varepsilon_t Z_t\left(A_t, X_0(A_t)\right). \qquad (6.18)$$

Then the process $X = \{X_t, 0 \leq t \leq 1\}$ satisfies $1_{[0,t]}\sigma X \in \text{Dom}\,\delta$ for all $t \in [0,1]$, $X \in L^2([0,1] \times \Omega)$, and X is the unique solution of Eq. (6.16) verifying these conditions.

Proof: We will only prove the existence. The uniqueness can be shown using similar arguments.

Let us prove first that the process X given by (6.18) satisfies the desired conditions. By Gronwall's lemma and using hypothesis (H.1), we have

$$|X_t| \leq \varepsilon_t e^{tL}(|X_0(A_t)| + L\int_0^t \varepsilon_s^{-1}(T_s)ds), \qquad (6.19)$$

which implies $\sup_{t\in[0,1]} E(|X_t|^q) < \infty$, for all $2 \leq q < p$, as it follows easily from Girsanov's theorem and Hölder's inequality. Indeed, we have

$$
\begin{aligned}
E(|X_t|^q) &\leq c_q E\left(\varepsilon_t^q e^{qtL}\left(|X_0(A_t)|^q + L^q \int_0^t \varepsilon_s^{-q}(T_s)ds\right)\right)\\
&\leq c_q e^{qL}\left\{E\left(\varepsilon_t^{q-1}(T_t)|X_0|^q + L^q\varepsilon_t^{q-1}(T_t)\int_0^t \varepsilon_s^{-q}(T_s^2)ds\right)\right\}\\
&\leq C\left\{E\left(|X_0|^p\right)^{\frac{q}{p}} + 1\right\}.
\end{aligned}
$$

Now fix $t \in [0,1]$ and let us prove that $1_{[0,t]}\sigma X \in \text{Dom}\,\delta$ and that (6.16) holds. Let $G \in S$ be a smooth random variable. Using (6.18) and Girsanov's theorem, we obtain

$$
\begin{aligned}
E\left[\int_0^t \sigma_s X_s D_s G ds\right] &= E\left[\int_0^t \sigma_s \varepsilon_s Z_s(A_s, X_0(A_s))D_s G ds\right]\\
&= E\left[\int_0^t \sigma_s Z_s(X_0)[D_s G](T_s)ds\right]. \qquad (6.20)
\end{aligned}
$$

Notice that $\frac{d}{ds}G(T_s) = \sigma_s[D_s G](T_s)$. Therefore, integrating by parts in (6.20) and again applying Girsanov's theorem yield

$$
\begin{aligned}
E\left[\int_0^t Z_s(X_0)\frac{d}{ds}G(T_s)ds\right] &= E\Big[Z_t(X_0)G(T_t) - Z_0(X_0)G\\
&\quad - \int_0^t \varepsilon_s^{-1}(T_t)b(s, \varepsilon_s(T_s)Z_s(X_0), T_s)G(T_s)ds\Big]\\
&= E\left[\varepsilon_t Z_t(A_t, X_0(A_t))G\right] - E\left[Z_0(X_0)G\right]\\
&\quad - \int_0^t E\Big[b\big(s, \varepsilon_s Z_s(A_s, X_0(A_s))\big)G\Big]ds\\
&= E[X_t G] - E[X_0 G] - \int_0^t E[b(s, X_s)G]ds.
\end{aligned}
$$

Because the random variable $X_t - X_0 - \int_0^t b(s, X_s)ds$ is square integrable, we deduce that $\mathbf{1}_{[0,t]}\sigma X$ belongs to the domain of δ and that (6.16) holds. □

When the diffusion coefficient is not linear one can show that there exist a solution up to a random time. Existence and uniqueness results for one-dimensional Skorohod equations with random initial condition are established by Buckdahn in [20]. Let us state the main result of this paper. Its proof is based on Doss representation of the solution to a one-dimensional stochastic differential equation and on the change-of-variable formula for the Skorohod integral. We recall that i denotes the mapping from $H = L^2([0,1])$ into $\Omega = C_0([0,1])$, defined by $i(h)_t = \int_0^t h_s ds$.

Theorem 6.3.2 *Consider the stochastic differential equation in the Skorohod sense*

$$X_t = X_0 + \int_0^t \sigma(X_s)dW_s + \int_0^t b(X_s)ds. \tag{6.21}$$

Suppose that σ and b are functions of class C^2 with bounded derivatives of first and second order. Suppose that the initial condition X_0 is bounded and there exist a process $D_s X_0$ bounded by M such that

$$X_0(\omega + i(h)) = X_0(\omega) + \int_0^1 D_s X_0 h_s ds + o(\|h\|_H),$$
$$|D_s X_0(\omega + i(h)) - D_s X_0(\omega)| \le M\|h\|_H.$$

Then for any bounded ball $B_r(h)$ of radius $r > 0$ around $i(h)$ there exist a $t > 0$ such that Eq. (6.21) has a unique solution X on $[0, t] \times B_{r-3t\|\sigma'\|_\infty}(h)$ which belongs to $L^1([0, t] \times B_r(h))$ and it possesses and extension $\tilde{X}$ in the space $\mathbb{D}_\infty^{1,2}$ (set of elements in $\mathbb{D}^{1,2}$ with a bounded derivative).

Bibliographical notes: There are several papers devoted to the study of the anticipating process $\varphi_t(X_0)$, composition of a stochastic flow with a random initial condition. The large deviation principle has been investigated in [68] and [69], the characterization of the support of the law of the solution in [67], the absolute continuity of the law was studied in [65], and the regularity of the density was treated in [23]. Stochastic differential equations in the Skorohod sense were analyzed by Buckdahn, in the linear case, in connection with the anticipating Girsanov theorem (see [19] and [21]). Other works on the Skorohod stochastic differential equations are [78] (multidimensional Brownian motion) and [22] (linear multidimensional case).

Bibliography

[1] S. Aida, S. Kusuoka, and D. Stroock: On the support of Wiener functionals. In: *Asymptotic problems in probability theory: Wiener functionals and asymptotics*, eds: K. D. Elworthy and N. Ikeda. Pitman Research Notes in Math. Series **284**, 3–34, Longman Scient. Tech. 1993.

[2] H. Airault and P. Malliavin: Intégration géométrique sur l'espace de Wiener. *Bull. Sciences Math.* **112** (1988) 3–52.

[3] D. Bakry: L'hypercontractivité et son utilisation en théorie des semi-groupes. In: *Ecole d'Eté de Probabilités de Saint Flour XXII-1992*, Lecture Notes in Math. **1581** (1994) 1–114.

[4] V. Bally: On the connection between the Malliavin covariance matrix and Hörmander's condition. *J. Functional Anal.* **96** (1991) 219–255.

[5] V. Bally, I. Gyöngy, and E. Pardoux: White noise driven parabolic SPDEs with measurable drift. *J. Functional Anal.* **120** (1994) 484–510.

[6] V. Bally, A. Millet, and M. Sanz-Solé: Approximation and support theorem in Hölder norm for parabolic stochastic partial differential equation. *Annals of Probability* **23** (1995) 178–222.

[7] D. R. Bell: *The Malliavin Calculus*, Pitman Monographs and Surveys in Pure and Applied Math. 34, Longman and Wiley, 1987.

[8] G. Ben Arous, M. Gradinaru and M. Ledoux: Hölder norms and the support theorem for diffusions. *Ann. Institut Henri Poincaré* **30** (1994) 415–436.

[9] G. Ben Arous and R. Léandre: Annulation plate du noyau de la chaleur. *C. R. Acad. Sci. Paris* **312** (1991) 463–464.

[10] G. Ben Arous and R. Léandre: Décroissance exponentielle du noyau de la chaleur sur la diagonale I. *Probab. Theory Rel. Fields* **90** (1991) 175–202.

[11] G. Ben Arous and R. Léandre: Décroissance exponentielle du noyau de la chaleur sur la diagonale II. *Probab. Theory Rel. Fields* **90** (1991) 377–402.

[12] M. A. Berger: A Malliavin–type anticipative stochastic calculus. *Ann. Probab.* **16** (1988) 231–245.

[13] M. A. Berger and V. J. Mizel: An extension of the stochastic integral. *Ann. Probab.* **10** (1982) 435–450.

[14] J. M. Bismut: Martingales, the Malliavin Calculus and hypoellipticity under general Hörmander's condition. *Z. für Wahrscheinlichkeitstheorie verw. Gebiete* **56** (1981) 469–505.

[15] J. M. Bismut: *Large deviations and the Malliavin calculus*, Progress in Math. 45, Birkhäuser, 1984.

[16] N. Bouleau and F. Hirsch: Propriétés d'absolue continuité dans les espaces de Dirichlet et applications aux équations différentielles stochastiques. In: *Seminaire de Probabilités XX*, Lecture Notes in Math. **1204** (1986) 131–161.

[17] N. Bouleau and F. Hirsch: *Dirichlet Forms and Analysis on Wiener Space*. de Gruyter Studies in Math. 14, Walter de Gruyter, 1991.

[18] R. Buckdahn: Quasilinear partial stochastic differential equations without nonanticipation requirement. Preprint 176, Humboldt Universität, Berlin, 1988.

[19] R. Buckdahn: Linear Skorohod stochastic differential equations. *Probab. Theory Rel. Fields* **90** (1991) 223–240.

[20] R. Buckdahn: Skorohod stochastic differential equations of diffusion type. *Probab. Theory Rel. Fields* **93** (1993) 297–324.

[21] R. Buckdahn: *Anticipative Girsanov Transformations and Skorohod Stochastic Differential Equations*, Memoirs of the A.M.S., **533**, 1994.

[22] R. Buckdahn and D. Nualart: Linear stochastic differential equations and Wick products. *Probab. Theory Rel. Fields.* **99** (1994) 501–526.

[23] M.E. Caballero, B. Fernández, D. Nualart: Smoothness of distributions for solutions of anticipating stochastic differential equations. *Stochastics and Stochastics Reports* **52** (1995) 303–322.

[24] J. M. C. Clark: The representation of functionals of Brownian motion by stochastic integrals. *Ann. Math. Statist.* **41** (1970) 1282–1295; **42** (1971) 1778.

[25] C. Donati-Martin: Equations différentielles stochastiques dans ℝ avec conditions au bord. *Stochastics and Stochastics Reports* **35** (1991) 143–173.

[26] H. Doss: Liens entre équations différentielles stochastiques et ordinaires. *Ann. Inst. Henri Poincaré* **13** (1977) 99–125.

[27] N. Dunford, J. T. Schwartz: *Linear Operators, Part II*. Interscience Publishers, 1963.

[28] S. Fang: Une inégalité isopérimétrique sur l'espace de Wiener. *Bull. Sciences Math.* **112** (1988) 345–355.

[29] D. Feyel and A. de La Pradelle: Espaces de Sobolev Gaussiens. *Ann. Institut Fourier* **39** (1989) 875–908.

[30] D. Feyel and A. de La Pradelle: Capacités Gaussiennes. *Ann. Institut Fourier* **41** (1991) 49–76.

[31] P. Florchinger: Malliavin calculus with time-dependent coefficients and application to nonlinear filtering. *Probab. Theory Rel. Fields* **86** (1990) 203–223.

[32] A. Garsia, E. Rodemich, and H. Rumsey: A real variable lemma and the continuity of paths of some Gaussian processes. *Indiana Univ. Math. Journal* **20** (1970/71) 565–578.

[33] B. Gaveau and P. Trauber: L'intégrale stochastique comme opérateur de divergence dans l'espace fonctionnel. *J. Functional Anal.* **46** (1982) 230–238.

[34] L. Gross: Logarithmic Sobolev inequalities. *Amer. J. Math.* **97** (1975) 1061-1083.

[35] I. Gyöngy: On the support of the solutions of stochastic differential equations. (1989) 649–653. *Theory Probability Appl.* **3** (1994) 649–653.

[36] I. Gyöngy: On nondegenerate quasilinear stochastic partial differential equations. *Potential Analysis* **4** (1995) 157–171.

[37] I. Gyöngy, D. Nualart: Implicit scheme for stochastic parabolic partial differential equations drive by space-time white noise. *Potential Analysis*.

[38] I. Gyöngy, D. Nualart, M. Sanz-Solé: Approximation and support theorem in modulus spaces. *Probability Theory and Rel. Fields* **101** (1995) 495–509.

[39] I. Gyöngy and E. Pardoux: On quasi-linear stochastic partial differential equations. *Probab. Theory Rel. Fields* **94** (1993) 413–425.

[40] B. Hajek and E. Wong: Multiple stochastic integrals: Projection and Iteration. *Z. für Wahrscheinlichkeitstheorie verw. Gebiete* **63** (1983) 349-368.

[41] F. Hirsch: Propriété d'absolue continuité pour les équations différentielles stochastiques dépendant du passé. *J. Functional Anal.* **76** (1988) 193–216.

[42] F. Hirsch: Theory of capacity on the Wiener space. In: *Stochastic analysis and related topics V*, eds.: H. Korezlioglu, B. Øksendal and A. S. Üstünel, *Progress in Probability* vol. 38 Birkhäuser (1996) 69–98.

[43] M. Hitsuda: Formula for Brownian partial derivatives. *Publ. Fac. of Integrated Arts and Sciences Hiroshima Univ.* **3** (1979) 1–15.

[44] N. Ikeda and S. Watanabe: An introduction to Malliavin's calculus. In: *Stochastic Analysis, Proc. Taniguchi Inter. Symp. on Stoch. Analysis*, Katata and Kyoto 1982, ed. : K. Itô, Kinokuniya/North-Holland, Tokyo, 1984, 1–52.

[45] N. Ikeda and S. Watanabe: *Stochastic Differential Equations and Diffusion Processes*, second edition, North-Holland, 1989.

[46] P. Imkeller: Regularity of Skorohod integral processes based on integrands in a finite Wiener chaos. *Probab. Theory Rel. Fields* **98** (1994) 137–142.

[47] P. Imkeller, D. Nualart: Continuity of the occupation density for anticipating stochastic integral processes. *Potential Analysis* **2** (1993) 137–155.

[48] P. Imkeller, D. Nualart: Integration by parts on the Wiener space and the existence of occupation densities. *Annals of Probability* **22** (1994) 469–493.

[49] K. Itô: Multiple Wiener integral. *J. Math. Soc. Japan* **3** (1951) 157–169.

[50] I. Karatzas, D. Ocone, and Jinju Li: An extension of Clark's formula. *Stochastics and Stochastics Reports* **37** (1991) 127–131.

[51] A. Kohatsu-Higa and J. León: Anticipating stochastic differential equations of Stratonovich type. *Applied Mathematics and Optimization* **36** (1997) 263–289.

[52] N. V. Krylov: An inequality in the theory of stochastic integrals. *Theory Probab. Appl.* **XVI** (1971) 438–448.

[53] N. V. Krylov: *Controlled Diffusion Processes*. Sringer-Verlag, 1980.

[54] H. Kunita: Stochastic differential equations and stochastic flow of diffeomorphisms. In: *Ecole d'Eté de Probabilités de Saint Flour XII*, 1982, Lecture Notes in Math. **1097** (1984) 144–305.

[55] H. H. Kuo: *Gaussian Measures in Banach Spaces*, Lecture Notes in Math. **463**, Springer-Verlag, 1975.

[56] H. H. Kuo and A. Russek: White noise approach to stochastic integration. *Journal Multivariate Analysis* **24** (1988) 218–236.

[57] S. Kusuoka and D. W. Stroock: Application of the Malliavin calculus II. *J. Fac. Sci. Univ. Tokyo Sect IA Math.* **32** (1985) 1–76.

[58] R. Léandre: Intégration dans la fibre associée à une diffusion dégénérée. *Probab. Theory Rel. Fields* **76** (1987) 341–358.

[59] R. Léandre and F. Russo: Estimation de Varadhan pour les diffusions à deux paramètres. *Probab. Theory Rel. Fields* **84** (1990) 429–451.

[60] Jin Ma, Ph. Protter, and J. San Martín: Anticipating integrals for a class of martingales. Preprint.

[61] V. Mackevičius: On the support of the solution of stochastic differential equations. *Lietuvos Matematikow Rinkings* **XXXVI** (1) (1986) 91–98.

[62] P. Malliavin: Stochastic calculus of variations and hypoelliptic operators. In: *Proc. Inter. Symp. on Stoch. Diff. Equations, Kyoto* 1976, Wiley 1978, 195–263.

[63] P. Malliavin: Implicit functions of finite corank on the Wiener space. *Stochastic Analysis, Proc. Taniguchi Inter. Symp. on Stoch. Analysis*, Katata and Kyoto 1982, ed.: K. Itô, Kinokuniya/North-Holland, Tokyo, 1984, 369–386.

[64] P. Malliavin and D. Nualart: Quasi-sure analysis of stochastic flows and Banach space valued smooth functionals on the Wiener space. *J. Functional Anal.* **112** (1993) 287–317.

[65] T. Masuda: Absolute continuity of distributions of solutions of anticipating stochastic differential equations. *J. Functional Anal.* **95** (1991) 414–432.

[66] P. A. Meyer: Transformations de Riesz pour les lois gaussiennes. In: *Seminaire de Probabilités XVIII*, Lecture Notes in Math. **1059** (1984) 179–193.

[67] A. Millet, D. Nualart: Support theorems for a class of anticipating stochastic differential equations. *Stochastics and Stochastics Reports* **39** (1992) 1–24.

[68] A. Millet, D. Nualart, M. Sanz: Composition of large deviation principles and applications. Stochastic Analysis: Liber Amicorum for Moshe Zakai 383–396, Academic Press 1991.

[69] A. Millet, D. Nualart, M. Sanz: Large deviations for a class of anticipating stochastic differential equations. *Annals of Probability* **20** (1992) 1902–1931.

[70] A. Millet and M. Sanz-Solé: The support of an hyperbolic stochastic partial differential equation. *Probability Theory and Related Fields* **98** (1994) 361–387.

[71] A. Millet and M. Sanz-Solé: A simple proof of the support theorem for diffusion processes. In: *Seminaire de Probabilités XXVIII, Lecture Notes in Math.* **1583** (1994) 36–48.

[72] A. Millet and M. Sanz-Solé: Points of positive density for the solution to a hyperbolic SPDE. Preprint.

[73] E. Nelson: The free Markov field. *J. Functional Anal.* **12** (1973) 217–227.

[74] J. Neveu: Sur l'espérance conditionnelle par rapport à un mouvement Brownien. *Ann. Inst. Henri Poincaré* **12** (1976) 105–109.

[75] D. Nualart: *Malliavin Calculus and Related Topics.* Springer-Verlag, 1995.

[76] D. Nualart and E. Pardoux: Stochastic calculus with anticipating integrands. *Probab. Theory Rel. Fields* **78** (1988) 535–581.

[77] D. Nualart and E. Pardoux: Boundary value problems for stochastic differential equations. *Ann. Probab.* **19** (1991) 1118–1144 .

[78] D. Nualart, M. Thieullen: Skorohod stochastic differential equations on random intervals. *Stochastics and Stochastics Reports* **49** (1994) 149–167.

[79] D. Nualart, A.S. Üstünel: Mesures cylindriques et distributions sur l'espace de Wiener. *Lecture Notes in Math.* **1390** (1989) 186–191.

[80] D. Nualart and M. Zakai: Generalized multiple stochastic integrals and the representation of Wiener functionals. *Stochastics* **23** (1988) 311–330.

[81] D. Nualart and M. Zakai: The partial Malliavin calculus. In: *Seminaire de Probabilités XXIII*, Lecture Notes in Math. **1372** (1989) 362–381.

[82] D. Nualart, M. Zakai: Positive and strongly positive Wiener functionals. In: Barcelona Seminar on Stochastic Analysis, *Progress in Probability*, vol. 32, Birkhäuser (1993) 132–146.

[83] D. Ocone: Malliavin calculus and stochastic integral representation of diffusion processes. *Stochastics* **12** (1984) 161–185.

[84] D. Ocone: A guide to the stochastic calculus of variations. In: *Stochastic Analysis and Related Topics*, eds.: H. Korezlioglu and A. S. Üstünel, Lecture Notes in Math. **1316** (1987) 1–79.

[85] D. Ocone and E. Pardoux: A generalized Itô–Ventzell formula. Application to a class of anticipating stochastic differential equations. *Ann. Inst. Henri Poincaré* **25** (1989) 39–71.

[86] D. Ocone and E. Pardoux: Linear stochastic differential equations with boundary conditions. *Probab. Theory Rel. Fields* **82** (1989) 489–526.

[87] S. Ogawa: Quelques propriétés de l'intégrale stochastique de type noncausal. *Japan J. Appl. Math.* **1** (1984) 405–416.

[88] S. Ogawa: The stochastic integral of noncausal type as an extension of the symmetric integrals. *Japan J. Appl. Math.* **2** (1984) 229–240.

[89] S. Ogawa: Une remarque sur l'approximation de l'intégrale stochastique du type noncausal par une suite des intégrales de Stieltjes. *Tohoku Math. J.* **36** (1984) 41–48.

[90] E. Pardoux: Applications of anticipating stochastic calculus to stochastic differential equations. In: *Stochastic Analysis and Related Topics II*, eds.: H. Korezlioglu and A. S. Üstünel, Lecture Notes in Math. **1444** (1990) 63–105.

[91] G. Pisier: Riesz transforms: A simple analytic proof of P. A. Meyer's inequality. In: *Seminaire de Probabilités XXIII*, Lecture Notes in Math. **1321** (1988) 485–501.

[92] J. Potthoff: On positive generalized functionals. *J. Functional Anal.* **74** (1987) 81–95.

[93] C. Rovira and M. Sanz-Solé: Anticipating stochastic differential equations: Regularity of the law. *J. Functional Anal.* **142** (1996).

[94] J. Rosinski: On stochastic integration by series of Wiener integrals. *Appl. Math. Optimization* **19** (1989) 137–155.

[95] F. Russo and P. Vallois: Forward, backward and symmetric stochastic integration. *Probab. Theory Rel. Fields* **97** (1993) 403–421.

[96] T. Sekiguchi and Y. Shiota: L^2-theory of noncausal stochastic integrals. *Math. Rep. Toyama Univ.* **8** (1985) 119–195.

[97] A. Ju. Sevljakov: The Itô formula for the extended stochastic integral. *Theory Prob. Math. Statist.* **22** (1981) 163–174.

[98] A. V. Skorohod: On a generalization of a stochastic integral. *Theory Probab. Appl.* **20** (1975) 219–233.

[99] D. W. Stroock: Some applications of stochastic calculus to partial differential equations. In: *Ecole d'Eté de Probabilités de Saint Flour*, Lecture Notes in Math. **976** (1983) 267–382.

[100] D. W. Stroock: Homogeneous chaos revisited. In: *Seminaire de Probabilités XXI*, Lecture Notes in Math. **1247** (1987) 1–8.

[101] H. Sugita: Sobolev spaces of Wiener functionals and Malliavin's calculus. *J. Math. Kyoto Univ.* **25**-1 (1985) 31–48.

[102] H. Sugita: On a characterization of the Sobolev spaces over an abstract Wiener space. *J. Math. Kyoto Univ.* **25**-4 (1985) 717–725.

[103] H. Sugita: Positive generalized Wiener functionals and potential theory over abstract Wiener spaces. *Osaka J. Math.* **25** (1988) 665–696.

[104] A. S. Üstünel: Representation of the distributions on Wiener space and stochastic calculus of variations. *J. Functional Anal.* **70** (1987) 126–139.

[105] A. S. Üstünel: The Itô formula for anticipative processes with nonmonotonous time via the Malliavin calculus. *Probab. Theory Rel. Fields* **79** (1988) 249–269.

[106] S. R. S. Varadhan: Diffusion processes in a small time interval. *Comm. Pure Appl. Math.* **XX** (1967) 659–685.

[107] S. Watanabe: *Lectures on Stochastic Differential Equations and Malliavin Calculus*, Tata Institute of Fundamental Research, Springer-Verlag, 1984.

[108] S. Watanabe: Analysis of Wiener functionals (Malliavin Calculus) and its applications to heat kernels. *Ann. Probab.* **15** (1987) 1–39.

[109] S. Watanabe: Donsker's δ-functions in the Malliavin calculus. In: *Stochastic Analysis, Liber Amicorum for Moshe Zakai*, eds.: E. Mayer-Wolf, E. Merzbach, and A. Shwartz, Academic Press, 1991, 495–502.

[110] Y. Yokoi: Positive generalized white noise functionals. *Hiroshima Math. J.* **20** (1990) 137–157.

LISTE DES EXPOSES

- Mr ABRAHAM Romain
    Serpent brownien et mesure de sortie du super-mouvement brownien
- Mlle AL-KHACH Rim
    Régularité du temps local d'une diffusion
- Mr ASPANDIJAROV Sanjar
    Sur un problème de grandes déviations pour les processus de Poisson composés
- Mr BALDI Paolo
    Estimations précises pour le théorème de Cramer dans $\mathbb{R}^k$
- Mr BOUREKH Youcef
    Deux résultats sur le problème de SKOROKHOD
- Mr BRIAND Philippe
    Fonctions de Lyapunov et interprétation probabiliste des EDP semi-linéaires
- Mr BUCKDAHN Rainer
    EDS rétrograde avec des sauts. Application à la finance
- Mlle CASTELL Fabienne
    Approximation semi-classique pour l'équation de Schrödinger
- Mr CATONI Olivier
    Une nouvelle borne supérieure pour l'énergie libre du modèle de
Shermington-Kirkpatrick
- Mr CERF Razphaël
    Métastabilité du modèle d'Ising 3D à très petite température
- Mr DECREUSEFOND Roland
    Analyse stochastique du mouvement brownien fractionnaire
- Mr DERMOUNE Azzouz
    The inviscid Burgers equation with initial value of Poissonian type
- Mr DHERSIN Jean-Stéphane
    Serpent brownien et équations aux dérivées partielles
- Mr DELMAS Jean-François
    Supermouvement brownien avec catalyse
- Mr EDDAHBI M'hamed
    Grandes déviations des diffusions à deux paramètres en norme höldérienne
- Melle FRADON Myriam
    Formes de Dirichlet et diffusions singulières éventuellement dégénérées bootstrap
- Mr FRANZ Uwe
    Processus stochastiques multiplicatifs sur les groupes quantiques
- Mr GRISHIN Stas
    Positivity of the Lyapunov exponent
- Mlle JACQUOT Sophie
    Ergodicité d'une classe d'EDPS
- Mr JAKUBOWSKI Jacek
    Itô formula in conuclear spaces
- Mr LACHAL Aimé
    Lois du logarithme itéré pour les primitives successives du mouvement brownien
- Mr LEURIDAN Christophe
    Les théorèmes de Ray-Knight et la mesure d'Itô pour le mouvement brownien
dans $\mathbb{R}/Z$
- Mlle MARSALLE Laurence
    Quelques propriétés des temps de croissance des processus de Lévy stables

– Mme PIETRUSKA-PALUBA Katarzyna

Long-time asymptotics for the surviving brownian motion on the Sierpinski gasket with Poisson obstacles

– Mme THIEULLEN Michèle

Symétries en calcul des variations stochastiques

– Mr TINDEL Samy

EDPS hyperboliques à coefficient croissant

– Mr TOUBOL Alain

Le modèle de Shenington.Kirkpatrick avec couplage

– Mr VIENS Frederi

Borne supérieure précise sur le comportement exponentiel d'une équation aux dérivées partielles stochastique

– Mr WOLF Jochen

Sur des transformées de semimartingales

LISTE DES AUDITEURS

Mr	ABRAHAM Romain	UFR Mathématiques et Informatique, Université René Descartes, PARIS V
Mlle	AL-KHACH Rim	Laboratoire de Probabilités, Université PARIS VI
Mr	AMGHIBECH Said	URA CNRS 1378 Mathématiques, Université de ROUEN
Mr	ASPANDIJAROV Sanjar	Laboratoire de Probabilités, Université PARIS VI
Mr	ASSING Sigurd	
Mr	AZEMA Jacques	Laboratoire de Probabilités, Université PARIS VI
Mr	BALDI Paolo	Département de Mathématiques, Université de ROME (Italie)
Mr	BALLY Vlad	Laboratoire de Probabilités, Université PARIS VI
Mr	BARDINA Xavier	Département de Mathématiques, Université de BARCELONE (Espagne)
Mr	BENASSI Albert	Laboratoire de Mathématiques Appliquées, Université Blaise Pascal, CLERMONT-FD
Mr	BERCU Bernard	Département de Mathématiques, Université Paris-Sud (ORSAY)
Mr	BERNARD Pierre	Laboratoire de Mathématiques Appliquées, Université Blaise Pascal (CLERMONT-FD)
Mr	BODINEAU Thierry	URA CNRS 1321, Université PARIS VII
Mr	BOUGEROL Philippe	Laboratoire de Probabilités, Université PARIS VI
Mr	BOUREKH Youcef	Laboratoire de Probabilités, Université PARIS VI
Mr	BRIAND Philippe	Laboratoire de Mathématiques Appliquées, Université Blaise Pascal, CLERMONT-FD
Mr	BRUNAUD Marc	UFR Mathématiques, Université PARIS VII
Mr	BUCKDAHN Rainer	Département de Mathématiques, Université de BREST
Mr	CAMPILLO Fabien	Unité de Recherche INRIA, SOPHIA ANTIPOLIS
Mlle	CARAMELLINO Lucia	Département de Mathématiques, Université de ROME (Italie)
Mlle	CASTELL Fabienne	Département of Mathématiques, Université Paris-Sud (ORSAY)
Mr	CATONI Olivier	Laboratoire de Mathématiques, Ecole Normale Supérieure, PARIS
Mme	CECI Claudia	Institut de Mathématiques, Université de FIRENZE (Italie)
Mr	CERF Raphaël	Département de Mathématiques, Université Paris-Sud (ORSAY)
Mme	CHALEYAT-MAUREL Mireille	Laboratoire de Probabilités, Université PARIS VI
Mlle	CHEVET Simone	Laboratoire de Mathématiques Appliquées, Université Blaise Pascal, CLERMONT-FD
Mr	COHEN Serge	Département de Mathématiques, Université de VERSAILLES
Mr	DAW Ibrahima	URA CNRS 1378, Université de ROUEN
Mlle	DEACONU Madalina	Institut Elie Cartan, VANDOEUVRE-LES-NANCY
Mr	DECREUSEFOND Laurent	Ecole Supérieure des Télécommunications Département Réseaux, PARIS
Mr	DELMAS Jean-François	CERMICS - ENPC, La Courtine
Mr	DERMOUNE Azzouz	Départant de Mathématiques, Université du MANS
Mr	DHERSIN Jean-Stéphane	Laboratoire de Probabilités, Université PARIS VI
Mme	DONATI-MARTIN	Laboratoire de Probabilités, Université PARIS VI
Mr	DUNLOP François	CPHT, Ecole Polytechnique, PALAISEAU
Mr	EDDAHBI M'hamed	Département de Mathématiques, Université Cadi Ayyad MARRAKECH (Maroc)
Mr	ELKADIRI Mohamed	Département de Mathématiques, Faculté des Sciences de RABAT (Maroc)
Mme	FARRE Mercé	Département de Mathématiques Université de BARCELONE (Espagne)
Mr	FERRANTE Marco	Département de Mathématiques Université de PADOVA (Italie)
Mme	FLORIT Carme	Département de Mathématiques Université de BARCELONE (Espagne)

Mlle	FRADON Myriam	Département de Mathématiques, Université Paris-Sud (ORSAY)
Mr	FRANZ Uwe	Département de Mathématiques, Université de NANCY I
Mr	GALLARDO Léonard	Département de Mathématiques, Université de BREST
Mlle	GANTERT Nina	Fachbereich Mathematik, BERLIN (Allemagne)
Mr	GARNIER Josselin	CMAP, Ecole Polytechnique, PALAISEAU
Mr	GRADINARU Mihai	Département de Mathématiques, Université Paris-Sud (ORSAY)
Mr	GRISIIIN Stas	Department of Mathematics, University of California IRVINE (USA)
Mr	GRORUD Axel	Centre de Mathématiques et Informatique, Université de Provence, MARSEILLE
Mr	GUIMIER Alain	Département de Mathématiques, Université de YAOUNDE (Cameroun)
Mr	GUIOTTO Paolo	Ecole Normale Supérieure, PISE (Italie)
Mr	HU Ying	Laboratoire de Mathématiques Appliquées Université Blaise Pascal (CLERMONT-FD)
Mlle	JACQUOT Sophie	Département de Mathématiques, Université d'ORLEANS
Mr	JAKUBOWSKI Jacek	Institute of Mathematics, University of WARSAW (Pologne)
Mme	JOLIS Maria	Département de Mathématiques, Université de BARCELONE (Espagne)
Mr	LACHAL Aimé	INSA de LYON
Mlle	LAGAIZE Sandrine	Département de Mathématiques, Université d'ORLEANS
Mr	LEDOUX Michel	Laboratoire de Statistique et Probabilités, Université de TOULOUSE
Mr	LE GALL Jean-François	Laboratoire de Probabilités, Université PARIS VI
Mr	LEURIDAN Christophe	Institut Fourier, SAINT MARTIN D'HERES
Mr	LORANG Gérard	Département de Mathématiques, Université de NANCY I
Mlle	MALITA SCARIAT Elena	Laboratoire de Mathématiques, Université de NANCY I
Mr	MARQUEZ David	Département de Mathématiques, Université de BARCELONE (Espagne)
Mlle	MARSALLE Laurence	Laboratoire de Probabilités, Université PARIS VI
Mme	MASTROENI Loretta	Faculté d'Economie, Université de ROME (Italie)
Mr	MATHIEU Pierre	CMI-LATP, Université de Provence, MARSEILLE
Mlle	MAVIRA MANCINO	Institut de Mathématiques, Université de FIRENZE (Italie)
Mr	MAZLIAK Laurent	Laboratoire de Probabilités, Université PARIS VI
Mr	MESNAGER Laurent	Département de Mathématiques, Université Paris-Sud (ORSAY)
Mr	MICLO Laurent	Laboratoire de Statistiques et Probabilités, Université de TOULOUSE
Mme	MILLET Annie	Laboratoire de Probabilités, Université PARIS VI
Mme	MULINACCI Sabrina	Département de Mathématiques, Université de PISE (Italie)
Mlle	NAJA Dania	Institut Elie Cartan, VANDOEUVRE-LES-NANCY
Mme	NEGRI Ilia	Département de Statistique et Mathématiques, Université de MILAN (Italie)
Mr	OUZINA Mostafa	URA CNRS 1378, Université de ROUEN
Mr	PARDOUX Etienne	Département de Mathématiques, Université de Provence, MARSEILLE
Mlle	PAROUX Katy	Ecole Normale Supérieure, LYON I
Mr	PESZAT Szymon	Institut de Mathématiques, KRAKOW (Pologne)
Mr	PIAU Didier	Laboratoire de Probabilités, Université de LYON I
Mr	PICARD Jean	Laboratoire de Mathématiques Appliquées, Université Blaise Pascal, CLERMONT-FD
Mme	PIETRUSKA-PALUBA Katarzyna	Institut de Mathématiques, Université de WARSAW (Pologne)
Mme	PONTIER Monique	Département de Mathématiques, Université d'ORLEANS
Mr	RACHAD Abdelhak	Laboratoire de Mathématiques Appliquées, Université Blaise Pascal, CLERMONT-FD
Mr	ROUX Daniel	Laboratoire de Mathématiques Appliquées, Université Blaise Pascal, CLERMONT-FD
Mr	ROVIRA Carles	Département de Mathématiques, Université de BARCELONE (Espagne)

Mlle	SAADA Diane	UFR Mathématiques de la Décision, Université PARIS IX
Mr	SABOT Christophe	Laboratoire de Probabilités, Université PARIS VI
Mr	SAINT LOUBERT BIE Erwan	Laboratoire de Mathématiques Appliquées, Université Blaise Pascal, CLERMONT-FD
Mlle	SARRA Monica	Département de Mathématiques, Université de BARCELONE (Espagne)
Mlle	SAVONA Catherine	Laboratoire de Mathématiques Appliquées, Université Blaise Pascal, CLERMONT-FD
Mr	SERLET Laurent	Département de Mathématiques, YAMOUSSOUKRO (Côte d'Ivoire)
Mme	THIEULLEN Michèle	Laboratoire de Probabilités, Université PARIS VI
Mr	TINDEL Samy	Département de Statistique, Université de BARCELONE (Espagne)
Mr	TOUBOL Alain	ENPC - CERMICS, La Courtine
Mr	TROUVE Alain	Département de Mathématiques et Informatique, Ecole Normale Supérieure, PARIS
Mme	TROUVE Isabelle	UFR Sciences Economiques, Université Paris-Nord, VILLETANEUSE
Mr	VIENS Frederi	Department of Mathematics, University of California, IRVINE (USA)
Mlle	WANTZ Sophie	Institut Elie Cartan, VANDOEUVRE-LES-NANCY
Mr	WOLF Jochen	Institut für Stochastik Universität JENA (Allemagne)
Mr	YAKOVLEV Andrei	Département de Mathématiques, Université d'ORLEANS
Mr	ZEITOUNI Ofer	Department of Electrical Engineering, Technion Institute, HAIFA (Israel)

LIST OF PREVIOUS VOLUMES OF THE
"Ecole d'Eté de Probabilités"

1977 -   D. DACUNHA-CASTELLE   (LNM 678)

"Vitesse de convergence pour certains problèmes statistiques"

H. HEYER

"Semi-groupes de convolution sur un groupe localement compact
et applications à la théorie des probabilités"

B. ROYNETTE

"Marches aléatoires sur les groupes de Lie"

1978 -   R. AZENCOTT   (LNM 774)

"Grandes déviations et applications"

Y. GUIVARC'H

"Quelques propriétés asymptotiques des produits de matrices
aléatoires"

R.F. GUNDY

"Inégalités pour martingales à un et deux indices : l'espace H^p"

1979 -   J.P. BICKEL   (LNM 876)

"Quelques aspects de la statistique robuste"

N. EL KAROUI

"Les aspects probabilistes du contrôle stochastique"

M. YOR

"Sur la théorie du filtrage"

1980 -   J.M. BISMUT   (LNM 929)

"Mécanique aléatoire"

L. GROSS

"Thermodynamics, statistical mechanics and random fields"

K. KRICKEBERG

"Processus ponctuels en statistique"

1981 -   X. FERNIQUE   (LNM 976)

"Régularité de fonctions aléatoires non gaussiennes"

P.W. MILLAR

"The minimax principle in asymptotic statistical theory"

D.W. STROOCK

"Some application of stochastic calculus to partial
differential equations"

M. WEBER

"Analyse infinitésimale de fonctions aléatoires"

1982 - R.M. DUDLEY   (LNM 1097)
        "A course on empirical processes"
        H. KUNITA
        "Stochastic differential equations and stochastic flow of
        diffeomorphisms"
        F. LEDRAPPIER
        "Quelques propriétés des exposants caractéristiques"
1983 - D.J. ALDOUS   (LNM 1117)
        "Exchangeability and related topics"
        I.A. IBRAGIMOV
        "Théorèmes limites pour les marches aléatoires"
        J. JACOD
        "Théorèmes limites pour les processus"
1984 - R. CARMONA   (LNM 1180)
        "Random Schrödinger operators"
        H. KESTEN
        "Aspects of first passage percolation"
        J.B. WALSH
        "An introduction to stochastic partial differential equations"
1985-87 - S.R.S. VARADHAN   (LNM 1362)
        "Large deviations"
        P. DIACONIS
        "Applications of non-commutative Fourier analysis to
        probability theorems"
        H. FOLLMER
        "Random fields and diffusion processes"
        G.C. PAPANICOLAOU
        "Waves in one-dimensional random media"
        D. ELWORTHY
        "Geometric aspects of diffusions on manifolds"
        E. NELSON
        "Stochastic mechanics and random fields"
1986 - O.E. BARNDORFF-NIELSEN   (LNS 50)
        "Parametric statistical models and likelihood"

1988 -   A. ANCONA   (LNM 1427)
         "Théorie du potentiel sur les graphes et les variétés"
         D. GEMAN
         "Random fields and inverse problems in imaging"
         N. IKEDA
         "Probabilistic methods in the study of asymptotics"
1989 -   D.L. BURKHOLDER   (LNM 1464)
         "Explorations in martingale theory and its applications"
         E. PARDOUX
         "Filtrage non linéaire et équations aux dérivées partielles
         stochastiques associées"
         A.S. SZNITMAN
         "Topics in propagation of chaos"
1990 -   M.I. FREIDLIN   (LNM 1527)
         "Semi-linear PDE's and limit theorems for large deviations"
         J.F. LE GALL
         "Some properties of planar Brownian motion"
1991 -   D.A. DAWSON   (LNM 1541)
         "Measure-valued Markov processes"
         B. MAISONNEUVE
         "Processus de Markov : Naissance, Retournement, Régénération"
         J. SPENCER
         "Nine Lectures on Random Graphs"
1992 -   D. BAKRY   (LNM 1581)
         "L'hypercontractivité et son utilisation en théorie des semigroupes"
         R.D. GILL
         "Lectures on Survival Analysis"
         S.A. MOLCHANOV
         "Lectures on the Random Media"
1993 -   P. BIANE   (LNM 1608)
         "Calcul stochastique non-commutatif"
         R. DURRETT
         "Ten Lectures on Particle Systems"

1994 -   R. DOBRUSHIN   (LNM 1648)
         "Perturbation methods of the theory of Gibbsian fields"
         P. GROENEBOOM
         "Lectures on inverse problems"
         M. LEDOUX
         "Isoperimetry and gaussian analysis"
1995 -   M.T. BARLOW   (LNM 1690)
         "Diffusions on fractals"
         D. NUALART
         "Analysis on W ner space and anticipating stochastic calculus"
1996 -   E. GINE   (LNM 1665)
         "Decoupling anf limit theorems for U-statistics and U-processes"
         "Lectures on some aspects theory of the bootstrap"
         G. GRIMMETT
         "Percolation and disordered systems"
         L. SALOFF-COSTE
         "Lectures on finite Markov chains"

Printing: Weihert-Druck GmbH, Darmstadt
Binding: Buchbinderei Schäffer, Grünstadt